JACARANDA
SCIENCE QUEST **8**
AUSTRALIAN CURRICULUM | THIRD EDITION

T0342878

jacaranda
A Wiley Brand

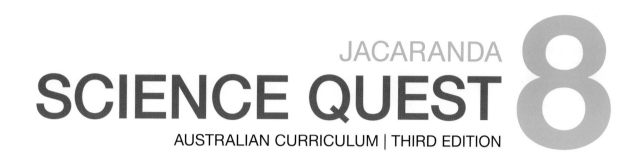

JACARANDA SCIENCE QUEST 8

AUSTRALIAN CURRICULUM | THIRD EDITION

GRAEME LOFTS | MERRIN J. EVERGREEN

CONTRIBUTING AUTHORS

Kahni Burrows | Marian Gauci | Eileen Kennedy
Robyn Kronenberg | Peter Rozanski | Paula Taylor

jacaranda
A Wiley Brand

Third edition published 2018 by
John Wiley & Sons Australia, Ltd
42 McDougall Street, Milton, Qld 4064

First edition published 2011
Second edition published 2015

Typeset in 11/14 pt Times LT Std

ISBN: 978-0-7303-4677-7

Front cover image: Michal Ninger/Shutterstock

Cartography by Spatial Vision, Melbourne and MAPgraphics Pty Ltd, Brisbane

Illustrated by various artists, diacriTech and the Wiley Art Studio.

Typeset in India by diacriTech

All activities have been written with the safety of both teacher and student in mind. Some, however, involve physical activity or the use of equipment or tools. **All due care should be taken when performing such activities**. Neither the publisher nor the authors can accept responsibility for any injury that may be sustained when completing activities described in this textbook.

NATIONAL
LIBRARY
OF AUSTRALIA

A catalogue record for this book is available from the National Library of Australia

Printed in Singapore
M WEP237382 021123

CONTENTS

OVERVIEW

Jacaranda Science Quest 8 Australian Curriculum Third Edition has been completely revised to help teachers and students navigate the Australian Curriculum Science syllabus. The suite of resources in the *Science Quest* series is designed to enrich the learning experience and improve learning outcomes for all students.

Science Quest is designed to cater for students of all abilities: no student is left behind and none is held back. *Science Quest* is written with the specific purpose of helping students deeply understand science concepts. The content is organised around a number of features, in both print and online through Jacaranda's *learnON* platform, to allow for seamless sequencing through material to scaffold every student's learning.

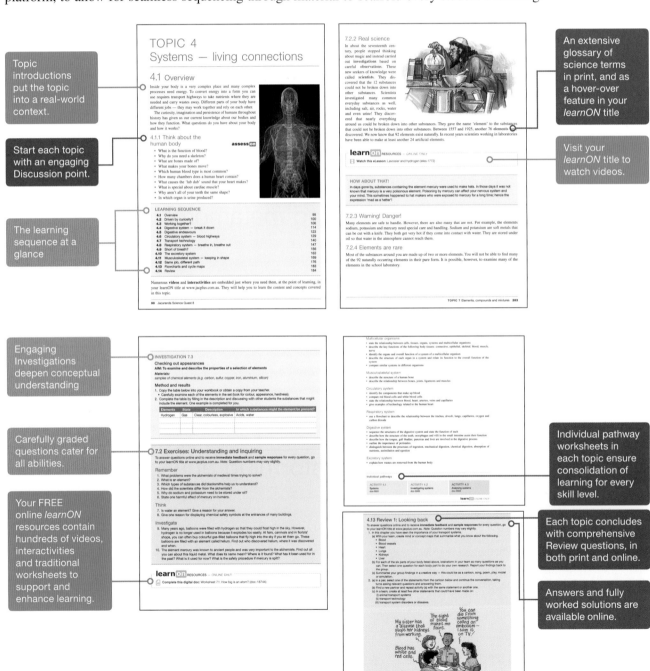

Topic introductions put the topic into a real-world context.

Start each topic with an engaging Discussion point.

The learning sequence at a glance

Engaging Investigations deepen conceptual understanding

Carefully graded questions cater for all abilities.

Your FREE online *learnON* resources contain hundreds of videos, interactivities and traditional worksheets to support and enhance learning.

An extensive glossary of science terms in print, and as a hover-over feature in your *learnON* title

Visit your *learnON* title to watch videos.

Individual pathway worksheets in each topic ensure consolidation of learning for every skill level.

Each topic concludes with comprehensive Review questions, in both print and online.

Answers and fully worked solutions are available online.

LearnON is Jacaranda's immersive and flexible digital learning platform that transforms trusted Jacaranda content to make learning more visible, personalised and social. Hundreds of engaging videos and inter-activities are embedded just where you need them — at the point of learning. At Jacaranda, our 'learning made visible' framework ensures immediate feedback for students and teachers, with customisation and collaboration to drive engagement with learning.

Science Quest contains a free activation code for *learnON* (please see instructions on the inside front cover), so students and teachers can take advantage of the benefits of both print and digital, and see how *learnON* enhances their digital learning and teaching journey.

learnON includes:

- Students and teachers connected in a class group
- Hundreds of videos and interactivities to bring concepts to life
- Fully worked solutions to every question
- Immediate feedback for students
- Immediate insight into student progress and performance for teachers
- Dashboards to track progress
- Collaboration in real time through class discussions
- Comprehensive summaries for each topic
- Dynamic interactivities help students engage with and work through challenging concepts.
- Formative and summative assessments
- And much more …

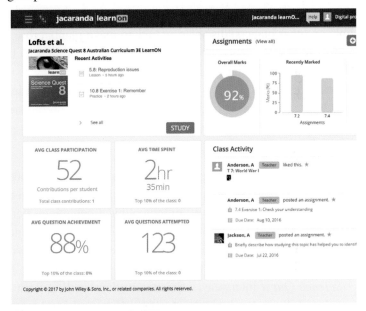

PREFACE

To the science student

Science is both a body of knowledge and a way of learning. It helps you to understand the world around you: why the sun rises and sets every day, why it rains, how you see and hear, why you need a skeleton and how to treat water to make it safe to drink. You can't escape the benefits of science. Whenever you turn on a light, eat food, watch television or flush the toilet, you are using the products of scientific knowledge and scientific inquiry.

Global warming, overpopulation, food and resource shortages, pollution and the consequences of the use of nuclear weapons are examples of issues that currently challenge our world. Possible solutions to some of these challenges may be found by applying our scientific knowledge to develop new technologies and creative ways of rethinking the problems. It's not just scientists who solve these problems; people with an understanding of science, like you, can influence the future. It can be as simple as using a recycling bin or saving energy or water in your home.

Scientific inquiry is a method of learning. It can involve, for example, investigating whether life is possible on other planets, discovering how to make plants grow faster, finding out how to swim faster and even finding a cure for cancer. You are living in a period in which knowledge is growing faster than ever before and technology is changing at an incredible rate.

Learning how to learn is becoming just as important as learning itself. *Science Quest* has been designed to help you learn how to learn, enable you to 'put on the shoes of a scientist' and take you on a quest for scientific knowledge and understanding.

To the science teacher

This edition of the *Science Quest* series has been developed in response to the Australian curriculum for Science. The Australian curriculum focuses on seven **General capabilities** (literacy, numeracy, ICT competence, critical and creative thinking, ethical behaviour, personal and social competence, and intercultural understanding). The history and culture of Aboriginal and Torres Strait Islanders, Australia's engagement with Asia, and sustainability have been embedded with the general capabilities where relevant and appropriate.

Science Quest interweaves **Science understanding** with **Science as a human endeavour** and **Science inquiry skills** under the umbrella of six **Overarching ideas** that 'represent key aspects of a scientific view of the world and bridge knowledge and understanding across the disciplines of science'.

The Australian Science curriculum provides the basis for the development of a Science curriculum in schools throughout Australia. However, it does not specify what you do in your classroom and how to engage individual classes and students.

We have attempted to make the *Science Quest* series a valuable asset for teachers, and interesting and relevant to the students who are using it. *Science Quest* comes complete with online support for students, including answers to questions, interactivities to help students investigate concepts, and video eLessons featuring real scientists and real-world science.

Exclusively for teachers, the online *Science Quest* teacher resources provides teaching advice and suggested additional resources, testmaker questions with assessment rubrics, and worksheets and answers.

Graeme Lofts and Merrin J. Evergreen

ACKNOWLEDGEMENTS

The authors and publisher would like to thank the following copyright holders, organisations and individuals for their assistance and for permission to reproduce copyright material in this book.

Images

• AAP Newswire: **144** (bottom right)/Victor Chang Cardiac Research Institute • Alamy Australia Pty Ltd: **29**/Sueddeutsche Zeitung Photo; **47** (C)/M I Spike Walker; **88** (bottom)/blickwinkel; **100** (top)/FineArt; **101** (bottom right)/Heritage Image Partnership Ltd; **108** (E)/Science History Images; **126** (top left)/PHOTOTAKE Inc.; **179** (bottom)/Stephen Barnes; **214**/Nucleus Medical Art Inc; **224** (top)/Phanie; **329**/Don B. Stevenson; **408**/© Phanie • Alamy Stock Photo: **218** (top)/67photo • Annie Cavanagh: **47** (bottom right B) • Carol Grabham: **407** (middle) • Coeliac Society of Australia: **127** (top B), **127** (top A) • Creative Commons: **47** (A)/Zatelmar; **129** • Dr. Stefan Eberhard: **48** (top right) • Eacham Historical Society: **371** (bottom), **372** (top left), **372** (top right) • Getty Images: **379** (bottom)/Corey Ford/Stocktrek Images • Getty Images Australia: **1**/Andrew Brookes; **40**/SPL/Andrew Syred; **47** (bottom left), **81** (left), **86**/Science Photo Library; **67** (C), **87** (bottom)/Andrew Syred/SPL; **82** (top)/BIOPHOTO ASSOCIATES; **94** (top), right/Dr. Gerald Van Dyke; **142** (top)/Viaframe; **144** (bottom left)/RIA NOVOSTI/SCIENCE PHOTO LIBRARY; **157**/Andrew Syred; **192**/Ed White; **213**/Science Photo Library RF; **216** (middle left)/Stockbyte; **223**/DESHAKALYAN CHOWDHURY/Stringer; **224** (middle)/David Parker; **227** (right)/Science Photo Library - STEVE GSCHMEISSNER.; **242**/Tom Branch; **243**/F Stuart Westmorland; **257**/Andrew Lambert Photography; **285**/Hulton Archive/Stringer; **314**/Per Breiehagen; **369**/Auscape; **411** (bottom)/Purestock; **432**/BanksPhotos; **404**(BOTTOM LEFT)/Dante Fenolio • Gustaaf Hallegraeff: **88** (middle)/University of Tasmania • HEARing CRC: **104** (top)/Courtesy of Dr. Jin Xu of HEARing CRC. • JAKKS Pacific: **42** • John Wiley & Sons Australia: **293**/Photo by Coo-ee Picture Library; **298**(b), **298**(c), **298** (a)/Janusz Molinski; **404** (TOP), **409**/Photo by Renee Bryon • John Wiley & Sons, Inc: **206** (top)/Figure 28.8 from Principles of Anatomy & Physiology by Tortora & Grabowski, 10th edition, © 2003 John Wiley & Sons, Inc; **209** (top), **210**/Figure 29.14 c from Principles of Anatomy & Physiology by Tortora & Grabowski, 10th edition, © 2003, John Wiley & Sons, Inc. This material is reproduced with permission of John Wiley & Sons, Inc.; **209** (bottom), **210**/Figure 29.14e from Principles of Anatomy & Physiology by Tortora & Grabowski, 10th edition © 2003, John Wiley & Sons, Inc. This material is reproduced with permission of John Wiley & Sons, Inc. • Melbourne Recital Centre: **426**/Pia Johnson 2013 • Miguel de Salas, Dr.: **89** (top right) • NASA: **32** (E), **332** (top); **330**/Tony Gray/Tom Farrar • National Archives of Australia: **328** • Newspix: **32** (A)/Shannon Morris; **32** (B)/Morne de Klerk; **32** (C)/Iain Gillespie; **32** (D)/Patrick Hamilton; **32** (F)/Richard Cisar-Wright; **80**/John Andersen; **88** (top)/Nathan Edwards; **89** (top left)/John Grainger; **127** (bottom)/David Geraghty; **229**/David Crosling • Out of Copyright: **105** • Pascale Warnant: **2, 26, 27** (top), **43** (middle left); **10**/Photograph in banner c Julie Stanton • Photodisc: **364**(top left) • Prame Chopra: **430**/Map derived from the GEOTHERM database of Chopra and Holgate 2005. © 2007 Dr Prame Chopra, Earthinsite.com Pty Ltd • Public Domain: **30** (top), **41** (bottom), **43** (left), **100** (bottom), **101** (top), **101** (bottom left), **103** (top right), **140** (left), **140** (right); **47** (bottom right A)/Janice Haney Carr; **48** (top left)/National Cancer Institute • Science Photo Library: **41**(a)/STEVE GSCHMEISSNER; **41**(b)/POWER AND SYRED; **41**(c)/SUSUMU NISHINAGA • Shutterstock: **13**/Marco Govel; **15**/shipfactory; **24**/Patricia Hofmeester; **27** (bottom)/gali estrange; **31**/Georgios Kollidas; **38**/Daniel Padavona; **45** (top), **67** (D)/Jubal Harshaw; **47** (B)/Lebendkulturen.de; **52** (bottom B)/Hack_bsh; **52** (bottom A)/Billion Photos; **61**/Stanislav Duben; **82** (middle)/Vizual Studio; **82** (bottom)/D. Kucharski K. Kucharska; **98, 108** (A)/Sebastian Kaulitzki; **102** (left)/Jakub Krechowicz; **102** (right)/EcoPrint; **108** (B), **108** (C), **108** (D), **108** (F)/Jose Luis Calvo; **124**/Vladimir Melnik; **125**/Creativa Images; **126** (middle left)/Juan Gaertner; **128**/Beloborod; **133** (right)/SGM; **136** (bottom)/khuruzero; **141** (top)/Neveshkin Nikolay; **143** (A), **158, 216** (middle right)/Image Point Fr; **143** (B)/ruigsantos; **167** (left)/Melissa Brandes; **167** (right), **232** (middle)/Ethan Daniels; **172, 313** (bottom)/wavebreakmedia; **176** (bottom)/Daryl H; **180**/Panachai Cherdchucheep; **193**/pathdoc; **193**/picturepartners; **193**/Karyna Che; **193**/Goldencow Images; **193**/rnl; **197**/Henrik Larsson; **197, 202** (right)/vetpathologist; **200**/Joel OBrien; **200** (top), **200** (top)/Suzanne Tucker; **202** (left)/Maya Kruchankova; **206**/Sashkin; **211** (bottom)/Mita Stock Images; **216**/Carolina K. Smith MD; **216** (top left)/Roman Sigaev; **216** (top right)/Hmelnitkaia Ana; **217**/JPC-PROD; **220** (bottom left)/Monkey Business Images; **224** (bottom)/bikeriderlondon; **225**/Li Wa; **230**/bluehand; **231** (top)/Peter Waters; **231** (middle)/chinahbzyg; **231** (bottom)/Lurin; **232** (top)/Brandon Alms; **232** (bottom)/Susan Flashman; **242**/© Anton Harder; **244**/JSseng; **259**/hipproductions; **260**/melis; **262** (top)/R_Szatkowski; **262** (bottom)/MarcelClemens; **266**/Mike Blanchard; **268**/You Touch Pix of EuToch; **269** (top left)/Suppakij1017; **269** (top right)/Stylus photo; **280**/molekuul.be; **281**/Andrei Shumskiy; **282** (top)/Wolfgang Zwanzger; **286**/Ivan Cholakov; **293** (bottom)/Andreas Bjerkeholt; **306** (bottom)/canismaior; **307**/FikMik; **309**/donsimon; **312**/Unknown; **313** (top), **389** (bottom)/James Steidl; **314**/Ilya Kirillov; **315**/Dominik Hladik; **316**/Dana Heinemann; **317, 403**/imagedb.com; **320**/ggw1962; **321**/Val Thoermer; **324**/Dmitri Melnik; **325** (top)/Constantine Pankin; **325** (bottom)/Gayvoronskaya_Yana; **326**/Simone van den Berg; **332** (bottom)/3Dsculptor; **350, 374** (top)/Dinoton; **351**/tab62; **353**/Aleksandras Naryshkin; **365**(a), **365**(b), **355** (middle)/Tyler Boyes; **355** (middle left)/Rob kemp; **355** (middle left)/Tamara Kulikova; **355** (bottom), **357**/www.sandatlas.org; **360** (top)/sonsam; **360** (bottom)/Sumikophoto; **361**/Artography; **361** (top)/XXLPhoto; **361** (bottom)/Matthijs Wetterauw; **363** (left)/guentermanaus; **364** (top right)/Corepics VOF; **369**/John Carnemolla; **371** (top right)/Stephanie Frey; **374** (bottom)/Baciu; **376**/APaterson; **383**/beboy; **383**/Rainer Albiez; **387**/Kat Clay; **388**/Maksym Gorpenyuk; **389**/Oleg Mikhaylov; **389**/wellphoto; **390**/Natursports; **393** (top left),

407 (top)/Olivier Le Queinec; **393** (bottom left)/buruhtan; **393** (bottom right)/ppart; **399**/Katarina Christenson; **403**/Maxim Khytra; **403**/Elzbieta Sekowska; **403**/nikkytok; **404** (BOTTOM RIGHT), **406** (BOTTOM LEFT)/Molodec; **406** (TOP)/ David Benton; **411** (top)/zimmytws; **416**/StevenRussellSmithPhotos; **420**/hurricanehank; **423**/kornilov007; **425**/Mutita Narkmuang; **365**(c)/michal812 • SuperStock: **379** (middle)/Stocktrek Images • Taronga Zoo: **35** • University of Adelaide: **221** (top)/photograph by David Ellis • University of Gothenburg: **145** • University of Western Australia: **126** (bottom)

Text

• © Australian Curriculum, Assessment and Reporting Authority (**ACARA**) 2010 to present, unless otherwise indicated. This material was downloaded from the Australian Curriculum website (www.australiancurriculum.edu.au) (**Website**) (accessed October, 2017) and was not modified. The material is licensed under CC BY 4.0 (https://creativecommons.org/ licenses/by/4.0). Version updates are tracked on the 'Curriculum version history' page (www.australiancurriculum.edu.au/ Home/CurriculumHistory) of the Australian Curriculum website.

Every effort has been made to trace the ownership of copyright material. Information that will enable the publisher to rectify any error or omission in subsequent reprints will be welcome. In such cases, please contact the Permissions Section of John Wiley & Sons Australia, Ltd.

TOPIC 1
Science is ...

1.1 Overview

Although science is a body of knowledge, it is also a way of solving problems and finding the answers to questions. Scientific knowledge is always growing and changing because scientists design and perform new investigations.

Observing, measuring, constructing tables, drawing graphs and forming conclusions are just some of the skills used in conducting scientific investigations.

1.1.1 Think about Science

assesson

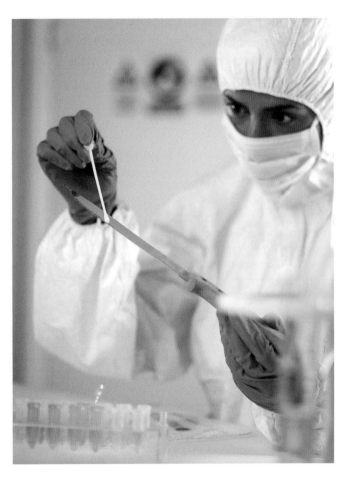

- What is the connection between a pendulum, a playground swing and a metronome?
- What can I measure with a computer?
- Which famous scientist saved the French wine industry from collapse?
- Why are graphs useful to scientists?
- How do you start your own scientific investigation?
- What do the letters CSIRO stand for?

LEARNING SEQUENCE

Numerous **videos** and **interactivities** are embedded just where you need them, at the point of learning, in your learnON title at www.jacplus.com.au. They will help you to learn the content and concepts covered in this topic.

1.1.2 Your quest
Researching the CSIRO
Investigate

A data logger can be used for Investigation 1.1.

1. What is the CSIRO?
2. The CSIRO's website describes some of the research done by CSIRO scientists. Read the information provided for one area of research that the CSIRO is involved with and summarise this research in point form.
3. Form groups of three. Explain to the other two students the area of research you have just read about. Try doing this without referring to your notes.

INVESTIGATION 1.1

Milk now or later?

You have just finished making yourself a cup of coffee when the phone rings. For your coffee to stay as warm as possible, should you add the milk now or after you have finished talking on the phone? Does your answer depend on the length of the phone call?

AIM: To compare the rate of cooling of hot coffee with and without the addition of milk

Materials:
kettle
2 identical cups
instant coffee
milk
2 thermometers or a data logger with 2 temperature probes
2 measuring cylinders

Method and results

- Your teacher will assign a particular 'phone call' time to each group of students.
- Heat some water in a kettle and use it to make two cups of instant coffee. Use the same type of cup and the same amount of hot water and coffee powder.
- Place a thermometer or temperature probe in each cup of coffee. If you are using a data logger, set it to collect results for at least 10 minutes.
- Add 40 mL of milk to one of the cups.
- If you are using thermometers, record the temperature of the coffee in both cups every 30 seconds.
- After your phone call time has passed, add 40 mL milk to the second cup.
- Continue measuring the temperature in both cups every 30 seconds until 10 minutes has passed since you added the milk to the first cup.

1. If you used thermometers, record your results in a table.

| | Temperature (°C) | |
Time (minutes)	Milk added at time 0	Milk added after 'phone call'
0.0		
0.5		
1.0		
1.5		

2. Plot line graphs of your results on the same set of axes. Put time on the horizontal axis and temperature on the vertical axis.
3. If you used a data logger, a graph is plotted automatically. If necessary, adjust the settings so that the graph shows the temperatures measured by both probes on the same set of axes. Put the graph into the results section of your experiment report.

Discuss and explain

4. Does hot coffee cool faster than warm coffee? How can you tell from your graph?
5. Did the two lines on the graph cross at any stage? What does this indicate?
6. Write a conclusion based on your results.
7. Does the length of the 'phone call' affect the results? Compare your graphs with those of other groups.
8. Why was it important to put exactly the same amount of water in both cups and to use the same type of cup?
9. What are the advantages and disadvantages of using a data logger for this experiment?
10. Explain how this experiment could be improved.

learn on RESOURCES — ONLINE ONLY

🔗 **Explore more with this weblink:** CSIRO

1.2 Safety first

1.2.1 Take care

Conducting scientific investigations in a laboratory can be exciting, but accidents can happen if experiments are not carried out carefully. There are certain rules that must be followed for your own safety and the safety of others.

ALWAYS . . .

- follow the teacher's instructions
- wear safety glasses and a laboratory coat or apron, and tie back long hair when mixing or heating substances
- point test tubes away from your eyes and away from your fellow students
- push chairs in and keep walkways clear
- inform your teacher if you break equipment, spill chemicals, or cut or burn yourself
- wait until hot equipment has cooled before putting it away
- clean your workspace — don't leave any equipment on the bench
- dispose of waste as instructed by your teacher
- wash your hands thoroughly after handling any substances in the laboratory.

NEVER . . .

- enter the laboratory without your teacher's permission
- run or push in the laboratory
- eat or drink in the laboratory
- smell or taste chemicals unless your teacher says it's ok. When you do need to smell substances, fan the odour to your nose with your hand.
- leave an experiment unattended
- conduct your own experiments without the teacher's approval
- put solid materials down the sink
- pour hazardous chemicals down the sink (check with your teacher)
- put hot objects or broken glass in the bin.

Handy hints

- Use a **filter funnel** when pouring from a bottle or container without a lip.
- Never put wooden test-tube holders near a flame.
- Always turn the tap on before putting a **beaker, test tube** or **measuring cylinder** under the stream of water.
- Remember that most objects get very hot when exposed to heat or a naked flame.
- Do not use tongs to lift or move beakers.

1.2.2 Working with dangerous chemicals

Your teacher will tell you how to handle the chemicals in each experiment. At times, you may come across warning labels on the substances you use.

Always wear gloves and **safety glasses** when using chemicals with the 'Corrosive' symbol. Corrosive substances can cause severe damage to skin and eyes. Acid is an example of a **corrosive** substance.

Flammable substances are easily set on fire, so keep them away from flames. Ethanol is **flammable**.

Chemicals with the 'Toxic' label can cause death or serious injury if swallowed or inhaled. They are also dangerous when touched without gloves because they can be absorbed by the skin. Mercury is a **toxic** substance.

1.2.3 Heating substances

Many experiments that you will conduct in the laboratory require heating. In school laboratories, heating is usually done with a Bunsen burner. A Bunsen burner provides heat when a mixture of air and gas is lit.

Bunsen burners heat objects or liquids with a naked flame. Always tie hair back, and wear safety glasses and a laboratory coat or apron when using a Bunsen burner.

A GUIDE TO USING THE BUNSEN BURNER

1. Place the Bunsen burner on a heatproof mat.
2. Check that the gas tap is in the 'off' position.
3. Connect the rubber hose to the gas tap.
4. Close the airhole of the Bunsen burner collar.
5. Light a match and hold it a few centimetres above the barrel.
6. Turn on the gas tap and a yellow flame will appear.
7. Adjust the flame by moving the collar until the airhole is open and a blue flame appears.
8. Remember to close the collar to return the flame to yellow when the Bunsen burner is not in use.

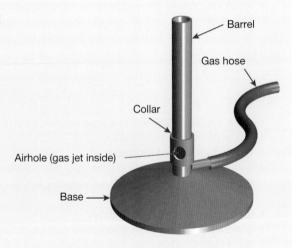

Heating containers

Beakers and evaporating dishes can be placed straight onto a gauze mat for heating. Never look directly into a container while it is being heated. Wait until the equipment has cooled properly before handling it.

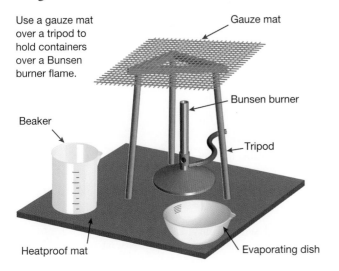

Use a gauze mat over a tripod to hold containers over a Bunsen burner flame.

Gauze mat

Bunsen burner

Beaker

Tripod

Heatproof mat

Evaporating dish

Heating a test tube

Tripods and gauze mats are not used when heating test tubes. Hold the test tube with a test-tube holder. Keep the base of the test tube above the flame. Make sure that the test tube points away from you and other students.

learn **on** RESOURCES — ONLINE ONLY

🔗 **Explore more with this weblink:** Robert Bunsen

1.2.4 Glassware
Pouring a liquid into a test tube

Pour liquids carefully into the test tube from a beaker or measuring cylinder. Use a filter funnel when pouring from bottles or containers without a lip.

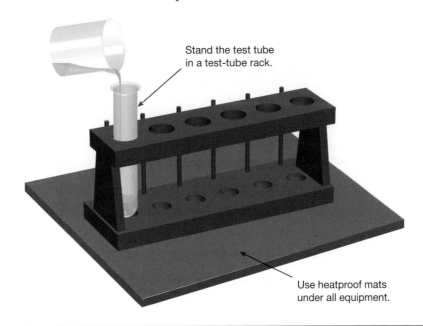

Stand the test tube in a test-tube rack.

Use heatproof mats under all equipment.

Shaking a test tube

There are two ways to shake substances in a test tube.

Method 1

Hold the top of the test tube and gently move its base in a sideways direction. This method is good to use with non-hazardous substances that do not need to be shaken vigorously. This is the method you will use most of the time.

Method 2

Use a stopper when a substance needs to be mixed by shaking vigorously. Place an appropriately sized stopper into the mouth of the test tube. With your thumb over the stopper and your hand securely around the test tube, shake the test tube with an up and down motion. Shake a test tube in this way only if instructed to do so by the teacher.

1.2.5 Using electricity safely

Electrical equipment in the science laboratory should be used with great care, just as it should be in the home or workplace. Never:

- place heavy electrical appliances near the edge of a bench or table
- allow water near electrical cords, plugs or power points
- place objects other than the correct electrical plug into a power point
- use appliances with damaged cords or exposed wires.

1.2 Exercises: Understanding and inquiring

To answer questions online and to receive **immediate feedback** and **sample responses** for every question, go to your learnON title at www.jacplus.com.au. *Note:* Question numbers may vary slightly.

Understand

1. Explain, with the aid of a diagram, how to safely heat a liquid in a test tube using a Bunsen burner. Ensure all relevant safety rules are included in your explanation.
2. How should a substance in a test tube be shaken if you are not instructed to shake it vigorously?
3. Methylated spirits is a flammable liquid. What does this mean?

Think

4. List the dangers of each of the following examples of unsafe behaviour in the science laboratory.
   (a) Not wearing safety glasses while heating a liquid in a beaker
   (b) Using an electronic balance to measure the mass of a substance before cleaning up some spilled water on the bench next to it
5. Why should you always wear gloves when working with:
   (a) corrosive substances
   (b) toxic substances?
6. Long hair should be tied back when heating or mixing substances in the laboratory. Why is this so?
7. Explain why a test tube should be standing in a test-tube rack when you are pouring a liquid into it.

Create

8. Draw a flowchart to illustrate the correct method for lighting a Bunsen burner. Use pictures or cartoons as well as words.
9. Which one safety rule do you feel is the most important when you are mixing two liquids and heating them? Create a poster to illustrate the rule.

1.3 Planning your own investigation

1.3.1 Planning your own investigation

Scientists learn new things by asking questions and then conducting investigations to find answers. You will take on the role of a scientist by planning an investigation of your own.

Discuss your ideas with others.

Before you define your question in detail, you need to find a topic that interests you. Selecting your topic is the first step and one of the most crucial parts in conducting your research project.

1. Start by searching for a general area of interest. List your hobbies and other interests.
2. Do you have a friend or relative who might be able to help you in a scientific investigation? Write down the topic areas in which you could get help.
3. Discuss the possible research topics you have already written down with a group of fellow students. Listen carefully to their ideas. They might help you to decide on your own topic. Write down your ideas.
4. Have a look through the list of ideas below. Even if none of the suggested topics appeals to you, they may help you to think of other ideas. For example, 'How strong is sticky tape?' could lead you to consider topics such as the strength of glass, wood, paper, plastics or some other material. Brainstorm possible topics with your friends and make your own list of suggested investigations.
5. Search in a library for resources about the topic areas that you have already written down. You might also find magazines or journals that include articles about these topic areas. Conduct an internet search. Use reliable websites and do not rely on just one source.

SOME IDEAS FOR TOPICS

How do fertilisers affect the growth of plants?
Can plants grow without soil?
What makes algae grow in an aquarium?
What is the best shape for a boomerang?
What type of wood gives off the most heat while burning?
What makes iron rust?
Which paint weathers best?
Which battery lasts longest?
How strong is sticky tape?
Which type of glue is best?
How much weight can a plastic bag hold?
Which food wrap keeps food freshest?
How effective are pre-wash stain removers?
Which fabrics burn faster?

How can the growth of mould on fruit be slowed down?
Which concrete mixture is strongest?
What type of fishing line is the strongest?
Does the thickness of a rubber band affect how far it stretches?
What type of paper aeroplane flies furthest?
What is the best recipe for soap bubble mixture?
Do tall people jump higher and further than short people?
What type of fabric keeps you warmest in winter?

1.3.2 The aim of the game

Your investigation should have a clear and realistic aim. Your aim should be very specific. The aim of an investigation is its purpose, or the reason for doing it. Some examples of aims are:

• to find out how the weight and shape of paper aeroplanes affects how far they fly
• to compare the effect of different fertilisers on the growth of pea plants
• to find out whether different coloured lights affect the growth of algae in an aquarium
• to find out how exposing iron to salty water affects how quickly it rusts.

'To find out if the weight of paper planes makes them fly better' is not a suitable aim because 'fly better' has not been defined. 'Fly better' could mean fly further, fly in a straighter line or stay in the air longer. A better aim would be 'To find out how the weight of paper planes affects their flight distance and time in the air'.

When you have decided what your aim is, make sure that it is realistic. You should be able to answer 'yes' to each question below.

• Is my aim simple and clear enough?
• Will I be able to get the background information that I need?
• Is the equipment I need for my experiments available or can it be made?
• Is the question a safe one to investigate?

If you answer 'no' to any of these questions you need to rethink your aim.

1.3.3 Forming a hypothesis

A **hypothesis** is a sensible guess about the outcome of an experiment. Your hypothesis should relate to your aim and should be testable with an experimental investigation. The results of your investigation will either support (agree with) or not support (disagree with) the hypothesis. It is not possible to prove conclusively that a hypothesis is correct.

When scientists make a hypothesis, they usually carry out a number of experiments to test it. Sometimes, a number of teams of scientists test the same hypothesis with slightly different experiments. Even if the results of each experiment agree with the hypothesis, the scientists could never say that the hypothesis is proven to be correct. They would say that each experiment has provided further evidence to support the hypothesis.

Your hypothesis should be based on what you know about the topic or what you have already observed. For example, if you are trying to design the best parachute for a toy, you should read about parachutes before writing your hypothesis.

Will nylon be better? Your own experience might help you form a hypothesis.

You might also recall that when you are walking in the rain, a cotton T-shirt soaks up a lot of water and becomes heavy, whereas a nylon jacket does not soak up water. As a result, your hypothesis might be: 'Closely woven nylon is a better fabric to use for a parachute than loosely woven cotton.'

A statement that cannot be tested with a scientific experiment is not a suitable hypothesis.

The table below shows how problems and observations can lead to hypotheses.

Problem	Observation	Hypothesis
The television remote control doesn't work.	If I press the 'on' button on the remote control, the television doesn't come on.	The batteries in the remote control are flat.
My hair is sometimes dry and frizzy.	My hair is driest soon after washing it with Mum's shampoo.	Mum's shampoo dries out my hair.
No parrots come to our bird feeder.	There is bread in the bird feeder, and magpies and miner birds feed there.	Parrots prefer wheat seeds.

1.3 Exercises: Understanding and inquiring

To answer questions online and to receive **immediate feedback** and **sample responses** for every question, go to your learnON title at www.jacplus.com.au. *Note:* Question numbers may vary slightly.

Understand

1. List four questions you should ask about your aim before it is final.
2. Define the term 'hypothesis'.
3. How can a hypothesis be tested?

Think

4. Why is 'to find out which glue is best' not a suitable aim? Write a more suitable aim for an investigation about glue.
5. Is each of the following statements a suitable hypothesis? If not, justify your answer.
   (a) White chocolate tastes better than dark chocolate.
   (b) Washing powder X removes tomato sauce stains faster than washing powder Y.
   (c) Plants grow faster under red light than under green light.
   (d) Sagittarians are nicer people than Leos.
   (e) Playing video games increases the muscle strength in your thumbs.
   (f) Playing video games affects the development of social skills.
   (g) Science teachers are more interesting people than English teachers.
   (h) Science teachers perform better in IQ tests than English teachers.
6. Consider the table above. Describe how you could test each of the three hypotheses.

1.4 Record keeping and research

1.4.1 Background research

Scientists do experiments to test hypotheses, which are based on observations as well as the previous discoveries of other scientists.

Before designing their experiments, scientists do background research, which usually includes reading reports written by other scientists. Scientists also need to keep records of all their observations and any changes they make to the design of their experiments. When you conduct your own research investigation, you will probably be asked to do this by keeping a logbook.

1.4.2 What is a logbook?

A logbook is a document in which you keep a record of all the work you do towards an investigation. Each entry should be dated like a diary. In your logbook, you might include the following items.

Part of a blog site used by a researcher to share the results of her investigations into acid–base indicators

- A timeline or other evidence of planning your time
- Notes about conversations you had with teachers, friends, parents or experts and how these conversations affected your project. Make sure you record each person's details so you can acknowledge their contribution in your report.
- Notes from library research you did. Include all the details you need for your bibliography.
- A plan or rough outline of the method you will use for your experiment(s)
- Notes about any problems you encountered during your project and how you dealt with these
- Information on any changes you made to your original plan
- Results of all your experiments (these may be presented roughly at this stage)
- A plan or storyboard for your presentation if you will present your research to your class

A logbook can be written by hand on paper, with a word-processing program on a computer, or it can even be written in an app or as a website. A blog is a website that has dated entries so it can be used as a logbook. It has the added advantage that you can invite other people, such as your friends, parents and teachers, to look at your work and post comments. You should check with your teacher on the format required for your logbook.

Make notes on your topic.

1.4.3 Researching your topic

Before you start your own experiments, you should find out more about your topic.

As well as increasing your general knowledge of the topic, you need to find out whether others have investigated your problem. Information already available about your topic might help you to design your experiments. It might also help you to explain your results.

Make notes on your topic as you find information. You may be able to include some relevant background information in your report.

The internet

The internet provides a wealth of information on almost every topic imaginable. Use a search engine such as Google or Yahoo! The success of your search will depend on a thoughtful choice of keywords.

Using the library

Another good place to start is the school library. There are several different types of information sources in the library, including those listed below.

Nonfiction books

Use the subject index catalogue to learn where to find books with information about your topic. Your library catalogue is most likely to be stored in a computer database. You might need to ask the librarian to help you use the catalogue at first. It is a good idea to browse through the contents list of science textbooks. Your topic may appear.

Reference books

These include encyclopaedias, atlases and yearbooks. The index of a good encyclopaedia is a great place to start looking for information.

Journals and magazines

There are quite a few scientific journals that are suitable for use by school students. They provide up-to-date information. Your library may have an index for journals, such as 'Guidelines', which you can use to find articles on your topic. You may, however, need to browse. Some journals to look for are: *New Scientist*, *Ecos*, *Australasian Science*, *Habitat*, *Popular Science*, *Choice* and *Helix*.

Information file

Many school libraries keep collections of digital files of newspaper articles on topics of interest. Ask your school librarian if you don't know how to access these resources.

Audiovisual resources

The library may have slides, videos and audio tapes that can be used or borrowed. These resources can be located using the subject index catalogue.

Beyond the library

Information on your topic may also be available from the following sources.

Look for information beyond the library.

Your science teacher

This may seem obvious, but many people don't even think to ask. Your science teacher may also be able to direct you to other sources of information.

Government departments and agencies

Federal, state and local government departments and agencies may be able to provide you with information or advice on your topic. Try searching government webpages, which usually list contact details. A polite email to the appropriate department or agency is the best way to ask for help.

Industry

Information on some topics can be obtained from certain industries. For example, if you were testing glues for strength or batteries to find which ones last longest, the manufacturers might have useful information. Use the internet to find contact details. A polite email is often the best way to ask for help.

Relatives or friends

Perhaps you or a relative know somebody who works in your area of interest. Let your friends and relatives know about your intended research.

In your logbook, complete a checklist like the one below to see if you have thoroughly searched sources of information.

The internet:	☐
School library:	
• nonfiction books	☐
• reference books	☐
• journals and magazines	☐
• information files	☐
• audiovisual resources	☐
Beyond the library:	
• your science teacher	☐
• government departments and agencies	☐
• industry	☐
• relatives or friends	☐
• other sources	☐

In your logbook, keep an accurate list of resources that you have used.

How to use information

Make notes on information that is relevant to your research topic. Think about what you really need to know. You need information that will help you to:
- plan your experiments
- understand your results later on
- show in your report how your research relates to everyday life or why your research is important.

You will need to keep an accurate list in your logbook of the steps you have taken and the resources you have used.

1.4 Exercises: Understanding and inquiring

To answer questions online and to receive **immediate feedback** and **sample responses** for every question, go to your learnON title at www.jacplus.com.au. *Note:* Question numbers may vary slightly.

Understand

1. Why is a logbook a bit like a diary?
2. Define the term 'blog'.

3. List the resources that you could use to research your investigation topic:
   (a) in your school library
   (b) outside the school library.

Think

4. Imagine you are a scientist. Assess the advantages and disadvantages of maintaining a blog rather than keeping a logbook in your office.
5. You can find information about science topics in science textbooks and on the internet.
   (a) Explain why you would not find the results of scientific research that was done last month in a science textbook.
   (b) Outline some advantages and disadvantages of using the internet as a source of information.

1.5 Controlling variables

1.5.1 In the swing of things

When was the last time you were on a swing? A playground swing is simply a large **pendulum**. Pendulums are used mainly as measuring instruments. Their most well-known use is in clocks, such as grandfather clocks.

To answer a question scientifically, you need to perform a controlled investigation. The investigation must also be reliable.

In the simple investigation of a swinging pendulum in Investigation 1.2 variables are controlled. However, to be reliable as well, measurements in the investigation need to be accurate, repeated and averaged.

A playground swing is simply a large pendulum. A pendulum is a suspended object that is free to swing to and fro. Each complete swing is called an **oscillation**. The time taken for one complete oscillation of a pendulum is called its **period**.

INVESTIGATION 1.2

The period of a pendulum

AIM: To investigate the effects of mass and length on the period of a pendulum

Materials:
length of string (at least 80 cm long)
set of slotted masses
retort stand with bosshead
pair of scissors and a one-metre ruler
stopwatch or clock with a second hand

Method and results

Part 1: The effect of mass
- Set up your pendulum so it can swing freely. Start with the largest possible length and the smallest weight.
1. Copy the table below into your logbook, and record the mass and the length of the pendulum in it. The length should be measured from the top of the pendulum to the bottom of the swinging mass, as shown in the diagram.
- Pull the mass aside so that the angle of release is about 20°. Take note of the height from which the mass is released so that this angle of release is used throughout the experiment.

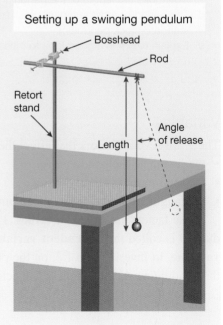

Setting up a swinging pendulum

- Release the pendulum. Measure the time taken for 10 complete swings of the pendulum. Repeat your measurement at least twice to find the average time for 10 swings. Use this average to calculate the time taken for one complete swing (the period).

The effect of mass on the period of a pendulum

Length of pendulum = _____ cm   Angle of release = 20°

	Time taken for 10 complete swings (seconds)				
Mass (grams)	Trial 1	Trial 2	Trial 3	Average	Period (seconds)

2. Record all the measurements in your table.
   - Repeat this procedure for three larger masses, completing the table as you go.

Part 2: The effect of length

3. Construct a table like the one on page 15 to identify all of the variables that need to be considered for an investigation of the effect of length on the period of a pendulum.
4. Construct a second table in which to record your measurements. Remember that this time you'll be testing four different lengths without changing the mass. Use the same procedure as you did in part 1 for measuring the period.
5. Draw a line graph to show how the period of the pendulum is affected by its length.

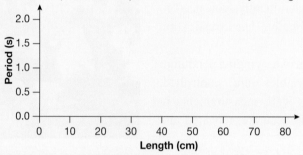

Discuss and explain

6. How does the mass of the pendulum affect its period?
7. How does the length of the pendulum affect its period?
8. The period of most standard clock pendulums is one second. Use your graph to predict the length of a standard clock pendulum.
9. Explain why it is a good idea to measure the time for 10 swings rather than just one.

Variables

There are several factors that affect the period of a pendulum. They include:
- the length of the pendulum
- the total mass that is swinging
- the height from which the pendulum is released.

These factors are called **variables**. The variable that you are measuring (in this case the period of the pendulum) is called the **dependent variable**. The variable that you are investigating is called the **independent variable**. In Investigation 1.2, on the previous page you investigated two independent variables, the mass of the pendulum and the length of the pendulum.

Fair testing

Scientific investigations must be **fair tests**. In a fair test, only one variable is changed at a time — the independent variable. In the first part of Investigation 1.2, the independent variable is the mass of the pendulum. All variables other than the independent variable must be controlled; that is, they must be kept the same. If they were not, you couldn't tell which variable was affecting the period of the pendulum. You might find it helpful when designing your own investigations to use a table like the one below to identify all the variables.

Experiment: How does mass affect the period of a pendulum?

Independent variable	• The mass of the pendulum
Dependent variable	• The period of the pendulum
Controlled variables	• The length of the pendulum • The angle of release • The method of release

1.5 Exercises: Understanding and inquiring

To answer questions online and to receive **immediate feedback** and **sample responses** for every question, go to your learnON title at www.jacplus.com.au. *Note:* Question numbers may vary slightly.

Understand

1. What is a variable?
2. Explain the difference between a dependent variable and an independent variable.
3. Why is it important to control variables in a scientific investigation?

Think

4. In Investigation 1.2 you conducted three trials for each measurement and calculated an average. List two or more reasons for the repetition.
5. A metronome is an 'upside-down' pendulum. To make the period of the metronome longer, should you move the sliding mass up or down?
6. What are (i) the independent variable and (ii) the dependent variable in:
   (a) part 1 of Investigation 1.2
   (b) part 2 of Investigation 1.2?

Investigate

7. Predict whether the angle of release affects the period of a pendulum and write down your hypothesis. Perform an investigation to test your hypothesis and write a brief report. In your conclusion, state clearly whether your results supported your hypothesis.

A metronome's period is changed by moving the sliding mass up or down.

 learn on RESOURCES — ONLINE ONLY

 **Complete this digital doc:** Worksheet 1.4: Fair testing (doc-18698)

1.6 The main game

1.6.1 Getting approval

Almost all scientists need the approval of their employer before they commence an investigation. As a student, you should not commence an investigation until your plan has been approved by your teacher.

1. Title

Choose a likely title — you may decide to change it before your work is completed.

2. The aim or problem

Briefly state what you intend to investigate or the question you intend to answer.

Aim: To study the behaviour of slaters

Problem: What makes algae grow in an aquarium?

3. Hypothesis

Make an educated guess about the answer to your problem or what you expect to find out. It is important to be creative and objective, and to use logical reasoning when devising a hypothesis and testing it.

4. Outline of experiment

Explain how you intend to test your hypothesis, and briefly outline the experiments you intend to conduct.

5. Equipment

List any equipment you need for your experiments.

6. Resources

List the sources of information that you have used or intend to use. This list should include library resources, organisations and people.

1.6.2 Performing your experiments

Once your teacher has approved your plan, you may begin your experiments. Detail how you conducted your experiments in your logbook. All observations and measurements should be recorded. Use tables where possible to record your data. Use graphs to display your data.

Some information about using tables, graphs and data loggers is provided on pages 19–25.

Where appropriate, measurements should be repeated and an average value determined. All measurements — not just the averages — should be recorded in your logbook.

Photographs should be taken if appropriate.

You might need to change your experiments if you get results you don't expect. If things go wrong, record what happened. Knowing what went wrong allows you to improve your experiment and technique. Any major changes should be checked with your teacher.

All observations and measurements should be recorded.

1.6.3 Writing your report

You can begin writing your report as soon as you have planned your investigation, but it cannot be completed until your observations are complete. Your report should be typed or neatly written on A4 paper. It should begin with a table of contents, and the pages should be numbered. Your report should include the following headings (unless they are not applicable to your investigation).

Abstract

Briefly describe your experiments and your main conclusions. Even though this appears at the beginning of your report, it is best not to write it until after you have completed the rest of your report.

Introduction

Present all relevant background information. Include a statement of the problem that you are investigating, saying why it is relevant or important. You could also explain why you became interested in the topic.

Aim

State the purpose of your investigation — that is, what you are trying to find out.

Hypothesis

Using the knowledge you already have about your topic, make a guess about what you will find out by doing your investigation.

Materials and method

Describe in detail how you carried out your experiments. Begin with a list of the equipment used and include photographs of your equipment if appropriate. The description of the method must be detailed enough to allow somebody else to repeat your experiments. It should also convince the reader that the variables in your investigation are well controlled. Labelled diagrams can be used to make your description clear. Using a step-by-step outline makes your method easier to follow.

Results

Observations and measurements (data) are presented in this section. Wherever possible, present data as a table so that they are easy to read. Graphs can be used to help you and the reader interpret data. Each table and graph should have a title. Ensure that you use the most appropriate type of graph for your data (see pages 19–25).

Discussion

Discuss your results here. Begin by stating what your results indicate about the answer to your question. Explain how your results might be useful. Outline any weaknesses in your design or difficulties in measuring here. Explain how you could improve your experiments. What further experiments are suggested by your results?

Conclusion

This is a brief statement of what you found out and may link with the final paragraph of your 'Discussion'. It is a good idea to read your aim again before you write your conclusion. Your conclusion should also state whether your hypothesis was supported. Don't be disappointed if it is not supported. Some scientists deliberately set out to reject hypotheses!

Bibliography

Make a list of books and other printed or audiovisual material to which you have referred. The list should include enough detail to allow the source of information to be easily found by the reader. Arrange the sources in alphabetical order.

For each printed resource, list the following information in the order shown:
- author(s) (if known)
- title of book or article
- publisher or name of journal/magazine (if not in title)
- place of publication (if given)

- date of publication
- chapter or pages used.

For example:

Breidahl, H., <u>Australia's Southern Shores</u>, Lothian, Melbourne, 1997, Chapter 2.

For websites, list the following:

- name of the website
- date the site was updated
- URL address
- date accessed.

For example:

'Millipede Mayhem', last updated 14 March 2008, http://www.csiro.au/csiro/channel/pchgb.html, accessed 30 November 2011.

Acknowledgements

List the people and organisations who gave you help or advice. You should state how each person or organisation assisted you.

1.6 Exercises: Understanding and inquiring

To answer questions online and to receive **immediate feedback** and **sample responses** for every question, go to your learnON title at www.jacplus.com.au. *Note:* Question numbers may vary slightly.

Understand

1. In which section of your investigation report should you write each of the following?
   (a) A list of the books and other resources you used to find information for your project
   (b) A table showing all the measurements you recorded
   (c) A diagram of the equipment you used
   (d) The purpose of the experiment
   (e) A brief summary of your investigation and findings
   (f) A statement that relates the results back to the aim and outlines what your results show

Think

2. When scientists write up their investigations for publication in a scientific journal, the abstract is one of the most important parts of the report. Explain why the abstract is usually read by many more people than the full report.
3. Explain why it is important for scientists to publish their investigations in scientific journals and to read the reports written by other scientists.

Investigate

4. There have been instances where scientists have faked their results or committed other types of scientific misconduct.
   (a) Enter the words 'scientific misconduct' in a search engine to find examples of such instances.
   (b) Why do you think that some scientists might be tempted to fake or fabricate their results?
   (c) Explain why cases of scientific misconduct are damaging to all scientists.
   (d) What do you think might happen to scientists who are found to have faked their results?

learnon RESOURCES — ONLINE ONLY

Complete this digital doc: Worksheet 1.5: Scientific reports (doc-18699)

1.7 Presenting your data

1.7.1 Presenting your data

Observations and measurements obtained from an investigation are called **data**. Having collected the data, it is important to present them clearly in a way that another person reading or studying them can understand. Tables and graphs are a great way to organise data.

1.7.2 Using tables

When data are organised in a table, they are easier to read and trends are more easily identified. An example of a simple table is shown below; it includes all the features you need to remember when constructing a table.

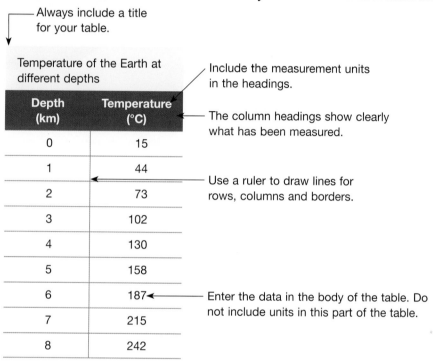

Always include a title for your table.

Include the measurement units in the headings.

The column headings show clearly what has been measured.

Temperature of the Earth at different depths

Depth (km)	Temperature (°C)
0	15
1	44
2	73
3	102
4	130
5	158
6	187
7	215
8	242

Use a ruler to draw lines for rows, columns and borders.

Enter the data in the body of the table. Do not include units in this part of the table.

You may need to construct more complex tables, like the one below, to present your research project results.

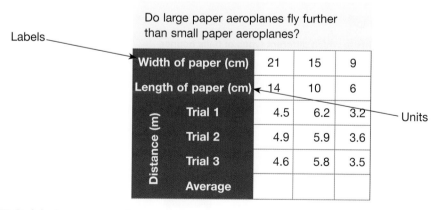

Labels

Do large paper aeroplanes fly further than small paper aeroplanes?

	Width of paper (cm)	21	15	9
	Length of paper (cm)	14	10	6
Distance (m)	Trial 1	4.5	6.2	3.2
	Trial 2	4.9	5.9	3.6
	Trial 3	4.6	5.8	3.5
	Average			

Units

1.7.3 Using graphs

Organising data as a graph is a widely recognised way to make a clear presentation. Graphs make it easier to read and interpret information, find trends and draw conclusions.

A graph, especially a line graph, can also be used to find values other than those used in the investigation. This can be done by interpolation or extrapolation (see pages 22–23).

Types of graphs

Five different types of graphs commonly used in scientific reports are pie charts, column or bar graphs, divided bar graphs, histograms and line graphs.

Pie charts (or sector graphs)

A pie chart (also known as a sector graph) is a circle divided into sections that represent parts of the whole. This type of graph may be used when the data can be added as parts of a whole. The example at right shows the food types, vitamins and minerals that make up the nutrients in a breakfast cereal.

Divided bar graphs

Divided bar graphs are also used to represent parts of a whole. However, the data are represented as a long rectangle, rather than a circle, divided into sections. The example at right shows the type of footwear worn to school today by male and female students.

Column graphs and bar graphs

A column graph (sometimes called a bar graph) has two axes and uses rectangles (columns or bars) to represent each piece of data. The height or length of the rectangles represents the values in the data. The width of the rectangles is kept constant. This type of graph can be used when the data cannot be connected and are therefore not continuous.

The example below shows data on the average height to which different balls bounced during an experiment. Each column represents a different type of ball.

A pie chart

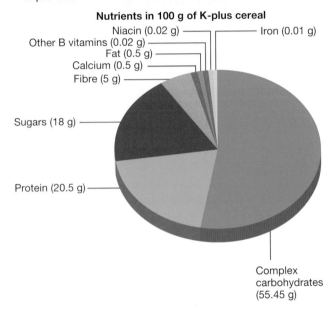

A divided bar graph

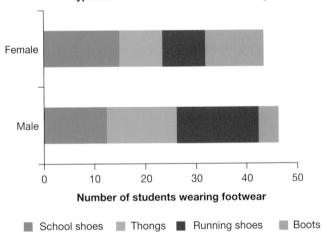

A column graph

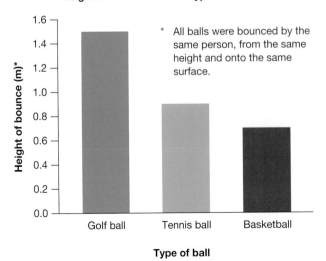

The example below shows the lengths of different metal bars when heated. Each bar represents a different metal.

A bar graph

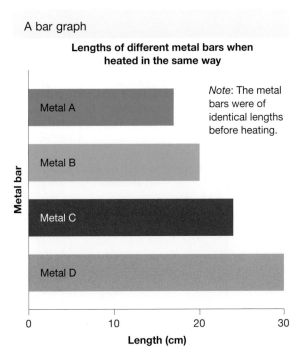

Lengths of different metal bars when heated in the same way

Note: The metal bars were of identical lengths before heating.

Histograms

Histograms are similar to column graphs except the columns touch because the data are continuous. They are often used to present the results of surveys. In the histogram below, each column represents the number of students of a particular height.

A histogram

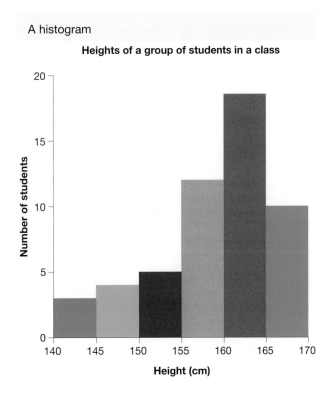

Heights of a group of students in a class

Line graphs

A line graph has two axes — a horizontal axis and a vertical axis. The horizontal axis is known as the *x*-axis, and the vertical axis is known as the *y*-axis. A line graph is formed by joining a series of points or drawing a line of best fit through the points. Each point represents a set of data for two variables, such as height and time. Two or more lines may be drawn on the same graph. Line graphs are used to show continuous data — that is, data in which the values follow on from each other. The features of line graphs are shown below.

3. Setting up and labelling the axes

Graphs represent a relationship between two variables. When choosing which variable to put on each axis, remember that there is usually an independent variable (which the investigator chooses) and a dependent variable. For example, if students wish to find out how far a runner could run in 15 seconds, they may choose to measure the distance covered every 5 seconds. The time of each measurement was chosen by the students and is the independent variable. The distance measured is therefore the dependent variable. Usually the independent variable is plotted on the *x*-axis and the dependent variable on the vertical *y*-axis.

After deciding on the variable for each axis, you must clearly label the axes with the variable and its units. The unit is written in brackets after the name of the variable.

2. Title

Tell the reader what the graph is about! The title describes the results of the investigation or the relationship between variables.

1. Grid

Graphs should always be drawn on grid paper so values are accurately placed. Drawing freehand on lined or plain paper is not accurate enough for most graphs.

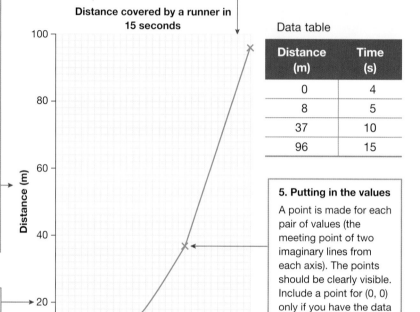

Data table

Distance (m)	Time (s)
0	4
8	5
37	10
96	15

5. Putting in the values

A point is made for each pair of values (the meeting point of two imaginary lines from each axis). The points should be clearly visible. Include a point for (0, 0) only if you have the data for this point.

4. Setting up the scales

Each axis should be marked into units that cover the entire range of the measurement. For example, if the distance ranges from 0 m to 96 m, then 0 m and 100 m could be the lowest and highest values on the vertical scale. The distance between the top and bottom values is then broken up into equal divisions and marked. The horizontal axis must also have its own range of values and uniform scale (which does not have to be the same scale as the vertical axis).

The most important points about the scales are:
• they must show the entire range of measurements
• they must be uniform, that is, show equal divisions for equal increases in value.

6. Drawing the line

A line is then drawn through the points.

A line that follows the general direction of the points is called a 'line of best fit' because it best fits the data. It should be on or as close to as many points as possible. Some points follow the shape of a curve, rather than a straight line. A curved line that touches all the points can then be used.

The type of data you graph may lead you to expect either a straight line or a curve. For example, you might expect the increase in temperature of water being boiled to be a straight line because the temperature increases at a steady rate. A graph of the growth rate of a red panda (see page 36) would be curved and irregular because pandas have growth spurts. Inspecting the data will help you decide whether your line should be straight or smooth and curved.

1.7.4 Interpolation

Line graphs can be used to estimate measurements that were not actually made in an investigation. The table at the top of the next page shows the results of an experiment in which a student measured how many spoons of sugar dissolved in a cup of tea at various temperatures.

Amount of sugar that dissolves in one cup of tea at different temperatures

Temperature (°C)	Mass of sugar dissolved (g)
0	4
20	30
40	60
60	98
80	120
100	160

Using a line graph for interpolation

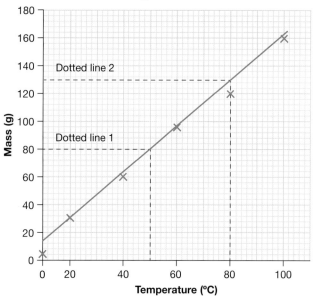

The student did not measure how much sugar dissolved at 50 °C, but we can work this out by interpolation. First we need to plot the data collected in the experiment. Then we read off the graph the amount of sugar that would dissolve at 50 °C (shown by dotted line 1 in the graph at right). The same procedure can be used to work out the water temperature that would be needed to dissolve 130 g of sugar in one cup of tea. This is shown by dotted line 2.

1.7.5 Extrapolation

In many cases it is also possible to assume that the two variables will hold the same relationship beyond the values that have been plotted. This is called extrapolation. Consider the table below, which shows the results obtained when different masses were attached to a spring and the increase in length of the spring was measured.

Amount a spring stretched when various masses were attached

Mass attached to the spring (kg)	Length by which spring stretched (cm)
0.0	0
0.5	8
1.0	16
1.6	26
?	32

Using a line graph for extrapolation

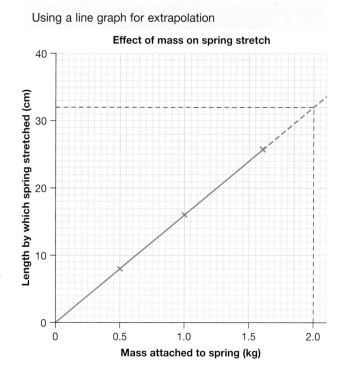

If you want to predict the mass needed to stretch the spring by 32 cm, you need to plot the data on a graph and extrapolate the value.

The data in the table above are plotted on the graph at right. Values have been plotted up to a mass of 1.6 kg and an increase in length of 26 cm. The line on the graph has been projected onwards (as the dotted lines show). This extrapolation shows that a mass of 2 kg will stretch the spring 32 cm.

INVESTIGATION 1.3

Drawing a line graph

AIM: To use a line graph to record data obtained in an experiment

A student conducted an experiment to see how temperature affected the amount of sugar that would dissolve in a cup of tea. Each cup contained the same volume of tea, and the sugar was stirred in at an equal rate for each cup. The results obtained are shown in the table below.

Graph the data in the table using the steps and diagrams below.

Amount of sugar dissolved in one cup of tea

Temperature (°C)	Mass of sugar dissolved (g)
0	4
20	30
40	60
60	98
80	120
100	160

1. Set up the grid.

2. Give the graph a title.

Effect of temperature on the amount of sugar dissolved in tea

3. Set up the axes and label them.

Effect of temperature on the amount of sugar dissolved in tea

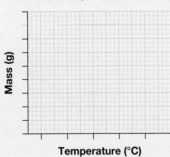

4. Place the scales on the axes.

Effect of temperature on the amount of sugar dissolved in tea

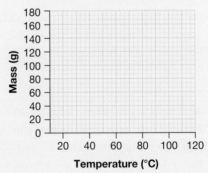

5. Plot each pair of values as a point marked with an x. Make sure each point is clearly visible. Don't forget to plot (0, 4) because you have the data for this point.

6. Draw a line of best fit; that is, a line drawn in between the points so that some points are on the line, some are below it and some are above.

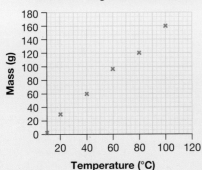

Effect of temperature on the amount of sugar dissolved in tea

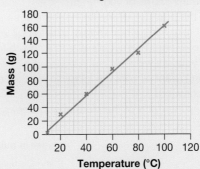

Effect of temperature on the amount of sugar dissolved in tea

1.7 Exercises: Understanding and inquiring

To answer questions online and to receive **immediate feedback** and **sample responses** for every question, go to your learnON title at www.jacplus.com.au. *Note:* Question numbers may vary slightly.

Analyse and evaluate

1. The following table shows the uses of plastics in Australia.
   (a) Select a suitable graph type and prepare a graph from this table.
   (b) Choose two uses of plastic from your graph. For each use, state a particular item that is made of plastic.
   (c) There has been controversy about the waste products that humans create.
       (i) List some uses of plastics that contribute to waste products.
       (ii) Suggest some action people can take to reduce the amount of plastic waste products.

2. The data in the following table relate the speed of a car to its stopping distance (the distance the car travels after the brakes are applied).
   (a) Graph the data.
   (b) Make a conclusion about the information in the graph.
   (c) How could this information be applied to your everyday life?

Uses of plastics in Australia

Use	Percentage (%)
Agriculture	4.0
Building	24.0
Electrical/electronic	8.0
Furniture and bedding	8.0
Housewares	4.0
Marine, toys and leisure	2.0
Packaging and materials handling	31.0
Transport	5.0
Others	14.0

Relationship between the speed of a car and its stopping distance

Speed of car (m/s)	Stopping distance (m)
10	12
20	36
30	72
40	120

3. The boiling point of water changes with air pressure. For example, water does not boil at 100 °C at the top of Mount Everest, where the air pressure is less than the pressure at sea level. The following data show the boiling point of water at various air pressure values.
   (a) Graph the data.
   (b) Describe the shape of your graph.
   (c) What is the pressure of the atmosphere at sea level?
   (d) Would it take a longer or shorter time to boil water at the top of Mount Everest compared with sea level? Explain your answer.

4. The following graph shows the increase in mass of a growing pondweed.
   (a) What was the mass of the plant after 3 weeks of growth?

Boiling point of water at different air pressures

Air pressure in kilopascals (kPa)	Boiling point of water (°C)
1	20
7	40
21	60
45	80
101	100
200	120

Increase in mass of pondweed with time

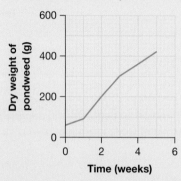

(b) How long did it take for the plant to grow to 250 grams?
(c) Predict the mass of the plant after 6 weeks of growth.
(d) Can you be sure that your extrapolation for part (c) is accurate? Suggest reasons why it may not be accurate.
(e) Would the interpolations from parts (a) and (b) be more accurate than your extrapolation? Discuss your ideas in class.

1.8 Using data loggers

1.8.1 What is a data logger?

A data logger is a type of scientific recording instrument. It collects and stores measurements that are called data because they are numbers. A data logger has to be attached to a measuring instrument called a **sensor**. The sensor does the measuring and sends the measurements to the data logger.

The real advantage of working with a data logger is that it can store thousands of individual measurements. The measurements can be taken in quick succession or over a long period of time, and the data logger can be programmed to do this automatically. This is why scientists often use data loggers in their work.

Data loggers also tend to be portable and battery-powered, and can therefore be used for applications such as remote weather monitoring and car crash testing. You may have been in a car that has driven over two closely placed rubber strips on the road — these strips are connected to a data logger used to count traffic.

Of course, to be useful, the stored measurements must be easy to access. That is why the data logger is also attached to either a computer or a graphics calculator. The computer or calculator takes the data and, using special software that comes with the data logger, shows the data as a table, a graph or both.

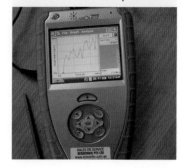
Some data loggers have their own touch screen and work like mini computers.

Other uses for data loggers

Data loggers can be used for just about any experiment where measurements are taken. All that is needed is the appropriate sensor to be plugged in. It is even possible to plug in several sensors to take different measurements at the same time.

Some of the many different sensors that are available include:
- temperature sensors capable of measuring up to several hundred degrees Celsius
- light intensity sensors
- soundwave sensors (microphones)
- motion sensors
- magnetic field sensors
- acceleration sensors
- force sensors
- electric current and voltage sensors
- humidity sensors
- blood pressure sensors
- heart rate sensors.

One type of sensor that isn't necessary is a time sensor (stopwatch) because the data logger has its own inbuilt clock that is very accurate. In fact, one of the most useful things about data loggers is their ability to collect measurements at very small and precise time intervals, even as many as a thousand measurements in one second!

More basic data loggers require the use of a computer to analyse the results.

A data logger for measuring blood pressure

1.8.2 Data loggers in temperature measurement

In Investigation 1.1 on page 2, the measuring instrument you used was a thermometer. You looked at the thermometer every 30 seconds and observed the temperature, which you wrote down in a table. You then made a line graph of temperature against time. If you had used a data logger with a temperature sensor instead of the thermometer, it could have taken the temperature every second and sent it to a computer that automatically tabulated the temperature data and graphed it as well.

1.8 Exercises: Understanding and inquiring

To answer questions online and to receive **immediate feedback** and **sample responses** for every question, go to your learnON title at www.jacplus.com.au. *Note:* Question numbers may vary slightly.

Remember

1. Match each of the words listed below with its meaning.

Word	Meaning
a. Sensor	A You may need to download the data from the data logger to one of these.
b. Data logger	B Piece of information
c. Computer	C These are plugged into the data logger and take the measurements.
d. Data logger software	D Allows you to input data into the data logger or computer by touching it with your finger or a stylus
e. Touch screen	E Allows you to process the data collected by the data logger
f. Data	F Collects and stores data from sensors connected to it

2. Sensors are devices that take the measurements that the data logger collects. Outline scientific investigations that could use data collected by sensors that measure:
   (a) electric current
   (b) heart rate
   (c) motion
   (d) sound waves
   (e) light intensity.

Analyse and evaluate

3. The graph at right shows data collected by a data logger for an experiment in which water was heated to boiling point in a beaker. A temperature sensor was used to take the measurements.

   If you were at this computer, you could scroll through every temperature measurement in the table. The computer has graphed all this data. Now let's see how much you've learned about interpreting line graphs.
   (a) How long did the whole experiment go for?
   (b) About when did the heating of the water begin?
   (c) What was the temperature of the water when heating began?
   (d) What was the temperature of the water when heating finished?
   (e) About when did the water begin to boil?
   (f) Between 100 and 400 seconds, at what rate (in degrees per second) did the water temperature rise?
   (g) The water continued to be heated even when its temperature reached boiling point, yet its temperature did not rise beyond 100 °C. What has happened to all the energy that was being put into the water if it isn't causing the water temperature to rise? (*Hint:* Think about what happens to water while it is boiling.)

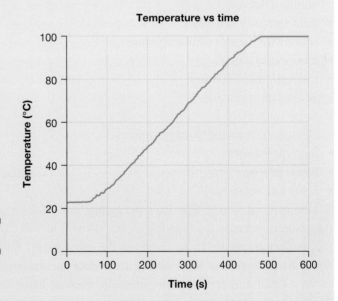

Temperature vs time

1.9 Greats from the past

Science as a human endeavour

1.9.1 Who's the greatest?

Who is the greatest scientist of all time? Is it Curie, Einstein, Newton or Pasteur? Or is it one of the people who saved millions of lives by discovering X-rays, penicillin or vaccination?

learn on RESOURCES — ONLINE ONLY

Watch this eLesson: Career spotlight: Scientist (eles-0766)

1.9.2 The slow starter

Albert Einstein (1879–1955) is most well known for his theory of relativity (there are actually two theories of relativity) and the equation $E = mc^2$, which describes how mass can be converted into energy.

Albert Einstein was certainly a slow starter. Although he was fascinated by mathematics, Einstein performed badly at school and left at the age of 15. He returned later and trained as a teacher in Switzerland. Einstein often failed to attend lectures and passed university exams by studying the notes of his classmates.

Einstein's first job was as a junior clerk in a patent office. His work was not demanding and he spent a lot of time doing 'thought' experiments.

At the age of 26, Einstein began to publish his ideas. These ideas altered our view of the nature of the universe by changing existing laws and discovering new ones.

Einstein explained the photoelectric effect, in which light energy is transformed into electrical energy, and received the Nobel Prize in Physics in 1921 for this.

Einstein's theories of relativity were so different from earlier theories that they were not believed or understood by most scientists. His theory of special relativity explains the behaviour of objects that travel at speeds close to the speed of light. His theory of general relativity explains the effect of gravity on light and predicts that time 'slows down' in the presence of large gravitational forces. These theories provide useful clues about the development and future of the universe.

Einstein's theories suggested that mass could be converted into energy. This idea led to the development of the atomic bomb and nuclear power. Einstein, who was Jewish, fled Germany in 1933 to live and work in the United States. He was an active opponent of nuclear weapons and was involved in the peace movement long before atomic bombs destroyed Hiroshima and Nagasaki at the end of World War II.

Einstein's first wife, Mileva, was a mathematician. He discussed many of his new ideas with her.

1.9.3 Did that apple really fall on his head?

Sir Isaac Newton (1642–1727) is probably most well known for his laws of gravitation, which explain the motion of the planets around the sun. According to some historians, his ideas about gravity arose after an apple fell on his head. We'll probably never know if this is true.

Isaac Newton was sent to Cambridge University at the age of 18. When the university closed down in 1665 as a result of the Great Plague, young Isaac went home for two years. There he developed his laws of gravitation and his three laws of motion. During his life, he also made discoveries about the behaviour of light and invented a whole new branch of mathematics, called calculus. Much of the scientific knowledge that has been acquired since the seventeenth century is built upon Newton's discoveries during that amazing two-year period.

1.9.4 A family affair

Marie Curie (1867–1934) became the first scientist to win two Nobel Prizes when she was awarded the Nobel Prize in Chemistry in 1911 for her discovery of two new elements: polonium and radium. Radium was used in the treatment of cancer until cheaper and safer radioactive materials were developed. Marie Curie's first Nobel Prize, for the study of radioactivity, was shared with her husband, Pierre, and fellow scientist Antoine-Henri Becquerel in 1903.

As a child, Marie Sklodowska (her birth name) wanted to study science. However, girls were forbidden to attend university in her native country of Poland. She worked as a private tutor for 3 years to earn enough money to study at the University of Paris, where she met her future husband, Pierre. They were very poor and spent most of their money on laboratory equipment, leaving very little money for food. In fact, they often couldn't afford to eat. After Pierre was knocked down and killed by a speeding wagon, Marie continued their research in radioactivity, pioneering the development of radioactive materials for use in medicine and industry. She became the first female teacher at the University of Paris and worked hard to raise money for scientific research.

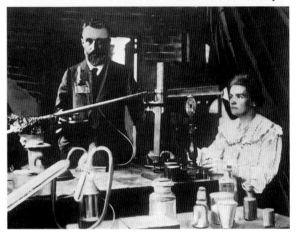
Marie Curie with husband Pierre in her laboratory

1.9.5 The germ of an idea

Louis Pasteur (1822–1895) proved that infectious diseases were caused by microbes. His ideas became known as 'germ theory'. He also developed several vaccines that made people immune to diseases such as rabies and smallpox. In doing this he has been responsible for saving the lives of millions of people and countless animals.

Pasteur began his scientific career in physics and chemistry, but became interested in microbes when he was using light to investigate the differences between chemicals in living and non-living things.

Pasteur's next challenge was to rescue the French wine industry. Wine (and beer) became sour very quickly and this was beginning to have an impact on the French economy, which relied heavily on the export of wine. Pasteur showed that the souring was caused by acids produced by the action of bacteria in the wine. Pasteur invented a process that rapidly heated some of the ingredients of the wine. The rapid heating killed most of the offending microbes without altering the flavour of the wine. The process, known as pasteurisation, was later adapted to slow down the souring of milk.

One of Pasteur's experiments

Time elapsed

Boil

No microbial growth

Boil

Stem broken, allowing air to enter flask

Microbial growth

1.9 Exercises: Understanding and inquiring

To answer questions online and to receive **immediate feedback** and **sample responses** for every question, go to your learnON title at www.jacplus.com.au. *Note:* Question numbers may vary slightly.

Think

1. Make a quick list of your 'Top 3' scientists of all time. For each one, answer the following questions.
   (a) What impact does their work have on your life?
   (b) Did they just happen to be in the 'right place at the right time'?
   (c) Did they work under adverse conditions?
   (d) Did their work save lives?
   (e) Did their work have any destructive influence?
   (f) What other special qualities make them great?
2. Is it fair to select the single 'greatest' scientist of all time? Explain your answer.
3. Louis Pasteur conducted many of his experiments on animals. Many of them would now be considered cruel. However, the experiments saved many human lives.
   (a) Present the arguments for and against the use of animals in such experiments.
   (b) Were the animal experiments justified? Write a brief statement supporting your opinion.

Imagine

4. Imagine that you are one of the three scientists you chose as the greatest scientists of all time. Write a short speech (3–5 minutes) about your life and work, and deliver it to your class. Illustrate your speech with models, diagrams or photographs.

Investigate

5. Write a biography similar to the four presented in this section about one of the following scientists: Michael Faraday (1791–1867), Charles Darwin (1809–1982), Lise Meitner (1878–1968), Barbara McClintock (1902–1992), Peter Doherty (1940–), Stephen Hawking (1942–).

1.10 Project: An inspiration for the future

Scenario

The Florey Medal was established in 1998 by the Australian Institute of Policy and Science in honour of the Australian Nobel Prize–winning scientist Sir Howard Florey, who developed penicillin. It is awarded biennially to an Australian biomedical researcher for significant achievements in biomedical science and human health advancement.

In a similar spirit, the Australian Academy of Science (AAS) hopes next year to establish an award for outstanding science students. The AAS wishes to name the medal after an Australian scientist who provides the greatest inspiration for young people considering a future career in science. After months of consultation, they have narrowed the choices down to the following:

• David Unaipon (1872–1967): Inventor

- Peter Doherty (1940–): Veterinarian and immunologist

- Fred Hollows (1929–1993): Ophthalmologist

- Andrew Thomas (1951–): Astronaut

- Fiona Wood (1958–): Plastic surgeon and burns specialist

- Ian Frazer (1953–): Immunologist

- Graeme Clark (1935–): Otolaryngeal surgeon and engineer

Your task

You will create an 8–10 minute podcast in the format of an interviewer discussing with a number of different people which of these scientists would be the best choice to name the AAS medal after. The interviewees (played by group members) should be people who would be likely to have an interest or stake in the award. Examples could include a member of the AAS medal panel, the Minister for Industry, the head of a university science or science education department, a high school science teacher, or even a high school science student. Each interviewee should have their own preference as to which scientist should be selected and at least four scientists should be discussed during the interview.

1.11 Review

1.11.1 Study checklist

The laboratory

- identify and safely use a range of equipment to perform scientific investigations
- use specialised equipment to make accurate observations and measurements
- use digital technology such as data loggers to make and record measurements

Planning and conducting investigations

- work individually and with others to identify a problem to investigate
- use information from investigations and scientific knowledge to make predictions and form hypotheses
- undertake research using a variety of sources
- develop a logical procedure for undertaking a controlled experiment
- recognise the need to control variables and distinguish between dependent and independent variables
- use repetition of measurement to increase the reliability of data

Processing and analysing data and information

- accurately record observations and measurements
- organise data clearly using tables and spreadsheets
- construct an appropriate type of graph to present your data
- use tables, spreadsheets and graphs to identify trends and patterns, and assist in forming conclusions
- identify data that support or discount a hypothesis
- form conclusions based on experimental results

Evaluating and communicating

- reflect on your methods and make suggestions for improvements to your investigations
- use information from investigations and scientific knowledge to evaluate claims
- discuss ideas and investigations with others
- use a scientific report with scientific language, clear diagrams, tables and graphs where necessary to describe your investigations and their findings

Science as a human endeavour

- identify the contributions of individual scientists, including Australians, to scientific knowledge
- describe some scientific discoveries that have had a major impact on our understanding of the world

---

Individual pathways

ACTIVITY 1.1	ACTIVITY 1.2	ACTIVITY 1.3
Investigating	Analysing investigations	Designing investigations
doc-2861	doc-2862	doc-2863

learnon ONLINE ONLY

1.11 Review 1: Looking back

To answer questions online and to receive **immediate feedback** and **sample responses** for every question, go to your learnON title at www.jacplus.com.au. *Note:* Question numbers may vary slightly.

1. The affinity diagram below organises some of the ideas used by scientists into four groups. Each category name is a single word and represents an important part of scientific investigations. However, the category names have been jumbled up. What are the correct categories for groups A, B, C and D?

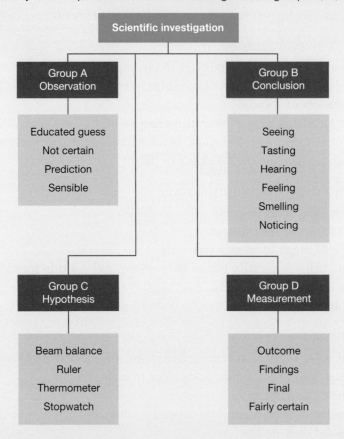

2. Bahir was sick of being bitten by mosquitoes. He counted several bites each evening when he sat outside to have dinner. He had heard that burning a citronella candle was a good way to keep mosquitoes away. Design an experiment to test Bahir's idea. List the independent and dependent variables, and the controlled variables needed to make this a fair test. Suggest a control for your experiment.

3. Four students each measured the temperature in the same classroom using a thermometer. Their results are shown in the table below.

Temperature as measured by each of four students in the same classroom

Student	Temperature (°C)
1	23.5
2	24.0
3	25.0
4	22.0

(a) Construct a bar graph of these results.
(b) Propose some possible reasons for the differences between measurements.

4. Jane and Greg decided to test how quickly water would boil when using either the yellow flame or blue flame of the Bunsen burner. They set up identical experiments, except that Jane used a blue flame and Greg used a yellow flame. Their results are graphed on the next page.
(a) How long did it take for Jane's water to boil?

(b) What was the temperature of Greg's water when Jane's water boiled?

(c) In your own words, explain how you worked out the answers for these two questions.

(d) Jane removed her beaker and Greg quickly placed his beaker over Jane's Bunsen burner. Assuming that the temperature of Greg's beaker did not drop while swapping Bunsen burners, predict the time at which his water would boil. Using your own words, explain how you predicted this.

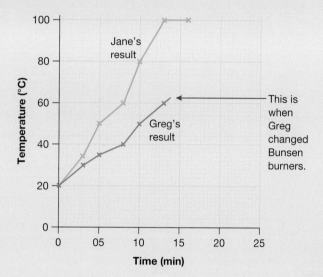

5. Singalia and Sallyana are two red panda cubs born at Sydney's Taronga Zoo. The table on the next page shows their masses during their first 22 weeks. The photograph below shows one of the cubs being weighed.

(a) Graph both sets of data onto a grid. Use different symbols for the points for each panda and label each line with the panda's name. You may have to extend the vertical axis to fit in the scale for the pandas' masses (or convert the masses to kilograms and plot in kilograms).

(b) Describe the growth of each of the panda cubs. How do they compare with each other?

(c) How long did it take the cubs to double their mass measured in week 1?

(d) Did the pandas grow at the same rate during the 22 weeks?

(e) Which were the fastest and slowest growth periods for each panda?

(f) What age was each of the cubs when it reached 1 kg?

(g) At what age would you predict each cub to reach 1.5 kg? Explain how you made your prediction. What assumption did you make to answer the question?

Red panda cubs' masses (grams)		
Week	Singalia	Sallyana
1	213	219
2	285	290
3	330	349
4	365	377
5	403	408
6	465	452
7	536	514
8	564	576
9	594	610
10	650	637
11	703	680
12	714	740
13	814	796
14	872	812
15	956	806
16	1111	786
17	1043	890
18	1130	1000
19	1163	1083
20	1182	1162
21	1225	1218
22	1335	1270

6. The table below shows the winning times for the men's 400 m freestyle swimming event. The data are from various Olympic Games from 1896 to 2012.

Year	Name, country	Time (min:s)
1896	Paul Neumann, Austria	8:12.60
1908	Henry Taylor, Great Britain	5:36.80
1920	Norman Ross, USA	5:26.80
1932	Buster Crabbe, USA	4:48.40
1948	Bill Smith, USA	4:41.00
1960	Murray Rose, Australia	4:18.30
1972	Bradford Cooper, Australia	4:00.27
1984	George DiCarlo, USA	3:51.23
1996	Danyon Loader, New Zealand	3:47.97
2000	Ian Thorpe, Australia	3:40.59
2004	Ian Thorpe, Australia	3:43.10
2008	Tae-Hwan Park, Korea	3:41.86
2012	Sun Yang, China	3:40.14

(a) Are data available for each Olympics every 4 years?

(b) Construct a line graph of the times for the men's 400 m freestyle over these years. Take into account your answer to part (a).

(c) Use your graph to estimate the winning time for this event in the 1956 Melbourne Olympic Games.

(d) Discuss how the winning times have changed over the 112-year period.

(e) Suggest some reasons for the change in winning times.

(f) Discuss how you believe the times for the men's 400 m freestyle might change over the next 40 years.

7. Create a storyboard that tells the story of the main events in the life of one of these famous scientists.
   (a) Albert Einstein
   (b) Sir Isaac Newton
   (c) Marie Curie
   (d) Louis Pasteur

8. On the right is part of a report on an experiment about dissolving sugar.
   (a) Write a 'Discussion' section for this report.
   (b) Write a conclusion for this report.
   (c) How could this investigation be improved?

Date: 29 February

Dissolving sugar

Aim:

To find out how much sugar will dissolve in hot water compared with cold water

Materials:

Beaker, heatproof mat, Bunsen burner, tripod, gauze mat, matches, spatula, stirring rod, sugar, water

Method:

1. A spatula was used to add sugar to 100 mL of cold water in a beaker. The sugar was stirred and more added until no more would dissolve. The amount of sugar dissolved was recorded.

2. The mixture of sugar and water was heated with a Bunsen burner for 4 minutes and the extra amount of sugar that could be dissolved was recorded.

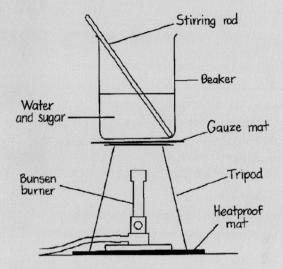

Results:

Amount of sugar dissolved in cold water = 2 spatulas
Extra amount of sugar dissolved in hot water = 4 spatulas
Total amount of sugar dissolved in hot water = 6 spatulas

TOPIC 2
Language of learning

2.1 Overview

What do you know about 'big picture' learning? Your brain may have trouble taking on a whole concept or big idea at once ... so you can 'nibble away' at a concept and build a model up over time. To do this, it often helps to be able to map out your learning journey so that you know where you are going.

2.1.1 Think about learning

assesson

- Which coloured hat should you 'wear' to think creatively?
- What's the difference between a 'fat' and a 'skinny' question?
- What can you do to make a good first impression?
- How do we communicate our feelings without using words?
- Who gave dinosaurs their names and why?
- What do emotions have to do with memory?

LEARNING SEQUENCE

Numerous **videos** and **interactivities** are embedded just where you need them, at the point of learning, in your learnON title at www.jacplus.com.au. They will help you to learn the concepts covered in this topic.

2.1.2 Your quest
The three-floor thinking model

On what floor is your thinking?

First-floor thinkers gather information. It is on this floor that the groundwork is laid.

Second-floor thinkers process the information. On this floor, thinkers decide which information is relevant and which is not, and then try to make some sense of it. This may involve brainstorming and playing with the ideas, looking for patterns or analysing data.

Third-floor thinkers apply information. They understand what needs to be done and complete it. On this floor, tasks are prioritised and further ideas are synthesised or evaluated. This may also be the floor for assembling the parts and adding the creative finishing touches.

Investigate, design and create

1. Select a project topic from one of the following and use the three-floor thinking model to gather, process and apply your information. Include information on the chemical and physical properties of your selected topic. Invent, design and construct a model of a device or method that would help to:
   (a) identify a range of common rock types
   (b) sort household wastes
   (c) recycle household wastes
   (d) test the effectiveness of detergents
   (e) test the effectiveness of toothpaste.

2. Use the three-floor thinking model to gather, process and apply information about an example of how science informs laws and guidelines about health or our environment. Present your findings as an advertisement that incorporates multimedia or animation to effectively communicate the relevant scientific understanding behind the law or guideline. You may select one of the following examples or identify your own example. Some examples of laws and guidelines influenced by our scientific knowledge include:
   - quarantine laws
   - food handling laws
   - bushfire safety guidelines
   - laws about the wearing of seatbelts
   - chemical storage guidelines
   - fire restriction laws.

Are you on your first, second or third thinking floor?

TOPIC 3
Cells

3.1 Overview

Cells are the basic units of all living things. The first cell appeared on Earth about 3.5 billion years ago. Today, there are many different kinds of cells. The differences in the cells of organisms are sometimes used to classify them into groups. Although cells may vary in their size, shape, contents and organisation, they all perform functions that are involved in keeping the organism to which they belong alive.

3.1.1 Think about cells   **assess**on

- How can you make small things look bigger?
- Which are bigger, viruses or bacteria?
- What does Schwann have to do with cells?
- Why are beaches tested for the presence of *E.coli*?
- How does a cell become a clone?
- Why don't all cells look the same?

LEARNING SEQUENCE

Numerous **videos** and **interactivities** are embedded just where you need them, at the point of learning, in your learnON title at www.jacplus.com.au. They will help you to learn the content and concepts covered in this topic.

3.1.2 Your quest

Who am I?

Microscopes are responsible for opening a whole new world to us. They have allowed us to see beyond our own vision. The more developed these microscopes become, the more detail and wonder we are able to observe — but often, rather than answering our questions, they provide us with many more.

The three photos on this page and the one on the previous page show parts of different animals. They were taken with a scanning electron microscope, which allows us to see more detail of the surface of specimens.

Observe, think and share

1. Look carefully at the photos of each animal part and think about:
   (a) what they could be
   (b) what they may do
   (c) what animal they may belong to.
2. Talk through your suggestions with your partner, adding all of the details that you have both observed onto a sheet of paper.
3. Two of these photos show parts of one type of animal, and the other two are of different animals. Does that information change the way that you look at the details? Which animal do you think two of the parts belong to? Brainstorm to decide which two animals the other parts could belong to.
4. Suggest other sorts of information that may be helpful in determining which animal these parts belong to and what they are used for.

3.2 A whole new world

Science as a human endeavour

3.2.1 The discovery of cells

A whole new world was discovered just over 400 years ago when an English inventor and scientist used magnifying lenses to observe the basic units of which all living things are made. This led to a new way of thinking about living things that required a new scientific language, new classifications and new inventions to find out more about this new world.

In the seventeenth century, Robert Hooke looked at thin slices of cork under a **microscope** that he had made himself from lenses. He observed small box-like shapes inside the cork. He called the little boxes that he saw **cells**. Microscopes opened up a whole new world that had never been seen before.

Using microscopes to carefully observe different living things showed that they were all made up of cells. Observations also showed that many of these cells shared common features, such as the presence of a structure called the nucleus.

As the magnification provided by microscopes increased, it was seen that although cells shared similar basic structures, there could also be differences between them. Groups of organisms could be made up of cells that differed from the cells of other groups. Some

An early microscope used by Robert Hooke

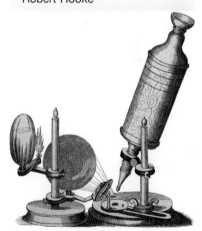

HOW ABOUT THAT!

There are many different types of scientists who study cells. Examples include bacteriologists, cell biologists, clinical microbiologists, cytologists, electron microscopists, genetic scientists, medical microbiologists and virologists.

WHAT DOES IT MEAN?

The word *microscope* comes from the Greek words *micrós*, meaning 'small', and *skopein*, meaning 'to view'.

organisms were made up of a single cell (unicellular), whereas others were made up of many cells (multicellular). Different types of cells were also observed within an individual multicellular organism.

3.2.2 Little, littler, littlest ...

With the development of instruments such as microscopes, scientists needed to find words to describe some of the tiniest lengths and time scales in nature. They wanted some simple names to describe, for example, a billionth of a billionth of a metre.

In the microscopic world, there is often a need to describe things in much smaller terms than the units of measurement that you already know, such as metre, centimetre and millimetre. In describing cells, other units of measurement, such as micrometre (μm, also called micron) and nanometre (nm), are often used.

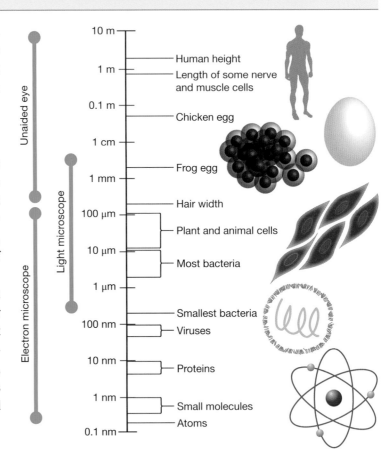

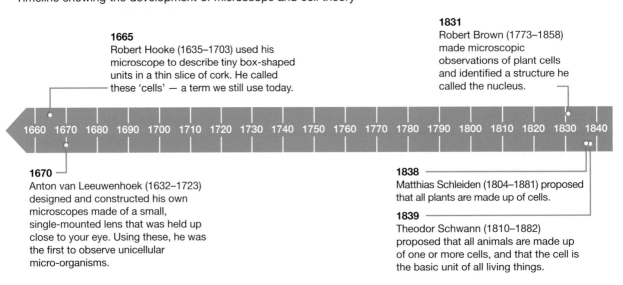

Timeline showing the development of microscope and cell theory

1665
Robert Hooke (1635–1703) used his microscope to describe tiny box-shaped units in a thin slice of cork. He called these 'cells' — a term we still use today.

1831
Robert Brown (1773–1858) made microscopic observations of plant cells and identified a structure he called the nucleus.

1660 1670 1680 1690 1700 1710 1720 1730 1740 1750 1760 1770 1780 1790 1800 1810 1820 1830 1840

1670
Anton van Leeuwenhoek (1632–1723) designed and constructed his own microscopes made of a small, single-mounted lens that was held up close to your eye. Using these, he was the first to observe unicellular micro-organisms.

1838
Matthias Schleiden (1804–1881) proposed that all plants are made up of cells.

1839
Theodor Schwann (1810–1882) proposed that all animals are made up of one or more cells, and that the cell is the basic unit of all living things.

Unit	Symbol	No. units in 1 m
Millimetres	mm	1000
Micrometres	μm	1 000 000
Nanometres	nm	1 000 000 000

3.2.3 Present day

A new generation of three-dimensional microscopes are being developed that provide even further details of objects. Superfast electron microscopes enable scientists to capture movement of atoms. Newly invented portable microscopes are becoming important field tools in research and diagnosis of diseases.

Portable microscope for spotting and tracking disease

Looking like a grotesque eyeball, this handheld microscope magnifies your specimens to two hundred times their normal size.

1858
Rudolf Virchow (1821–1902) proposed that all cells arise from cells that already exist.

1937
The first scanning electron microscope was built. These microscopes show images of cell surface features such as the blood cells shown opposite.

1981
Gerd Binnig and Heinrich Rohrer invented the scanning tunneling micrscope. This microscope can provide 3D images of objects at an atomic level.

1850 1860 1870 1880 1890 1900 1910 1920 1930 1940 1950 1960 1970 1980 1990 1990

1933
Ernst Ruska (1906–1988) built the first electron microscope, which uses a beam of electrons and provides higher magnification and resolution than light microscopes.

1957
Marvin Minsky (1929–) invented the first confocal microscope, which became commercially available in the 1980s.

3.2 Exercises: Understanding and inquiring

To answer questions online and to receive **immediate feedback** and **sample responses** for every question, go to your learnON title at www.jacplus.com.au. *Note:* Question numbers may vary slightly.

Remember

1. Match the scientist with their cell discovery contribution in the table below.

Scientist	Cell discovery contribution
(a) Anton van Leeuwenhoek	A. built the first electron microscope
(b) Robert Hooke	B. proposed that all plants are made up of cells
(c) Robert Brown	C. proposed that all animals are made up of cells
(d) Matthias Schleiden	D. designed and constructed microscopes and was the first to observe unicellular microscopic organisms
(e) Theodor Schwann	E. proposed that all cells arise from cells that already exist
(f) Rudolf Virchow	F. used the term 'cell' to describe the tiny box-like units in cork
(g) Ernst Ruska	G. used the term 'nucleus' to describe a structure found in plant cells

2. Identify:
   (a) a feature that all living things have in common
   (b) two units often used to describe cells.
3. Provide two examples of:
   (a) types of scientists who study cells
   (b) types of electron microscopes
   (c) things that can be seen with an electron microscope but not a light microscope.
4. Suggest why Hooke called the objects that he observed under the microscope 'cells'.

Think and calculate

5. Use the timeline on pages 42–43 to answer the following questions.
   (a) In which year did Hooke use the term 'cells' to describe his observation of cork slices?
   (b) In which year did Ruska build the electron microscope?
   (c) How many years were between:
       (i) Hooke first using the term 'cells' and Ruska building the first electron microscope
       (ii) Leeuwenhoek's first observation of unicellular microscopic organisms and Schwann's suggestion that all animals are made up of cells?
   (d) What did Virchow suggest in 1858?
   (e) Credit for developing the cell theory that 'all living things are made up of cells and that cells come from pre-existing cells', is usually attributed to three scientists. Who are they?
   (f) Suggest how the development of the microscope has contributed to our understanding of cell structure.
   (g) Suggest possible uses for portable microscopes.
6. Use the diagram and table on page 43 to complete the table below.

Object	Size in nanometres (nm)	Size in micrometres (µm)	Size in millimetres (mm)
Frog egg			
Hair (width)			
Plant cell		100	
Bacteria		10	
Protein	10		

Investigate and create

7. If you were living 300 years ago, how might you react to being told that you were made up of cells? What might cause these reactions? Construct a story, play or cartoon that answers these questions.
8. Find out how a light microscope works and design and construct your own model.
9. Create your own picture book or PowerPoint presentation to teach primary students about the discovery of cells.

Investigate and report

10. Robert Hooke has been described as having 'a mechanic's mind and an artist's heart'. He wove these traits into his work as a scientist and inventor. Research and report on one of the following:
    * his work with Robert Boyle
    * his debate with Isaac Newton on the nature of light and gravity
    * Hooke's Law
    * his invention of the iris diaphragm in cameras
    * his contribution to micrography.
11. Research and report on one of the following:
    * Ernst Ruska's Nobel Prize for Physics
    * reasons why Frits Zernike has been described as 'a pioneer in forensic science'
    * examples of recent microscope inventions
    * examples of Australian research using electron microscopes.
12. Research one of the following types of scientists: bacteriologist, cell biologist, clinical microbiologist, cytologist, electron microscopist, genetic scientist, medical microbiologist, virologist.
    Imagine you are such a scientist and create a journal that describes a discovery you have made.
13. In 2009, expatriate Australian Professor Elizabeth Blackburn was jointly awarded the Nobel Prize for Medicine for her discovery of telomeres and telomerase. Find out more about her discovery and its relevance to cells and microscopes. Report your findings in a PowerPoint presentation.
14. Investigate how the development of the microscope has:
    (a) contributed to our understanding of cell structure and function
    (b) affected health and medicine.

learnon RESOURCES — ONLINE ONLY

Complete this digital doc: Worksheet 3.1: History of the light microscope (doc-18708)

3.3 Focusing on a small world

Science as a human endeavour

3.3.1 Types of microscopes

The two main types of microscopes are light microscopes and electron microscopes. **Light microscopes** use light rays whereas **electron microscopes** use small particles called electrons.

Transmission electron microscopes show the internal structures of cells whereas scanning electron microscopes show images of the surface features of the specimen. New electron microscope technologies are being developed, such as superfast electron microscopy which enables scientists to capture the movement of electrons, and a variety of three-dimensional microscopes which have exciting research and medical applications.

You may have light microscopes at your school. These may be either **monocular** (using one eye) or **binocular** (using two eyes). It is important that the specimen you observe is very thin, so that the light can pass through it. However, one type of binocular microscope, a **stereo** microscope, allows you to see the detail of much larger specimens. Stereo microscopes can be used to observe various objects including living organisms or parts of them.

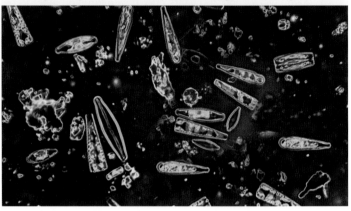

This electron micrograph shows examples of diatoms, which belong to a group of photosynthetic, single-celled algae.

Some comparisons between light microscopes and electron microscopes

Type of microscope	Magnification (how many times bigger)	Resolution (how much detail can be seen)	Advantage(s)	Disadvantage(s)	Examples of detail that can be seen
Light microscope	Up to × 2000	Up to about 500 times better than the human eye	Samples prepared quickly; coloured stains can be used; living cells can be viewed	Limited visible detail	Shapes of cells; some structures inside cells, e.g. nucleus and chloroplasts
Electron microscope	Up to × 2 000 000	Up to about 5 million times better than the human eye	High magnification and resolution	Only dead sections can be viewed; specimen preparation is difficult; very expensive	All parts of cells; viruses

Stereo light microscope

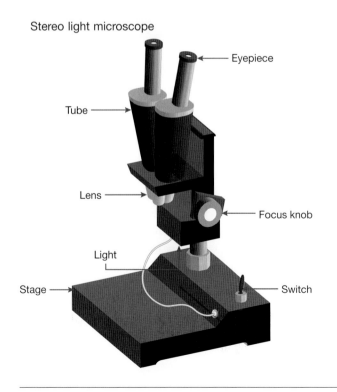

Scanning electron microscope

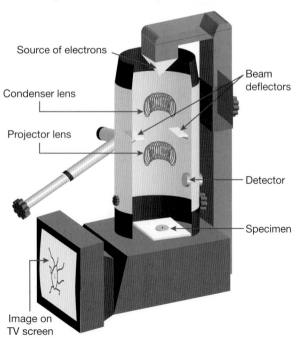

What's in your water? These images show zooplankton viewed through a scanning electron microscope.
(a) *Chaetognath* (b) *Daphnia* (c) A rotifer

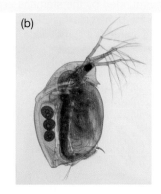
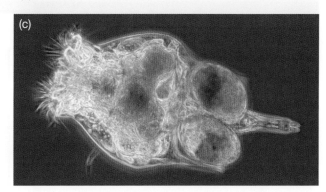

3.3.2 Award-winning images

Microscopes are used not just to observe images of organisms, but also in many other areas of science. Some microscope images win awards recognising not just expertise but also creativity. For example, the Wellcome Trust, a charity that funds health research, presents awards for images that creatively explore the fields of medicine, health care and biology. The figures below show examples of some of the 2014 winners.

A flower bud (1 mm in diameter): the petals are shown in red, the anthers in blue and purple, and the style and stigma in yellow.

Staphylococcus aureus bacteria under a scanning electron microscope

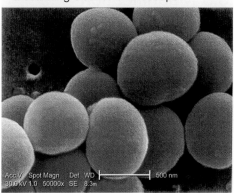

Electron micrograph of a four-day-old zebrafish embryo. The embryo was about 1 cm in length, which is too big for a single image, so three different images have been combined to create this photo. (Image: Annie Cavanagh and David McCarthy)

Human breast cancer cells under a scanning electron microscope

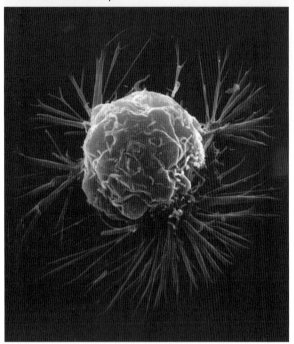

Electron micrograph of a flower. The blue feathery structure on an olive green stalk is the pistil, which is surrounded by the stamens — the light green pods on the brown stalks. The petals are coloured purple. The image is 1.2 mm wide. (Image: Stefan Eberhard)

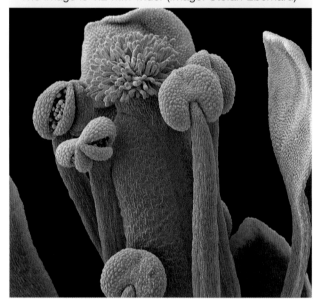

3.3.3 Magnification

The two lenses that determine the **magnification** of your microscope are the eyepiece lens and the objective lens. Each lens has a number on it that signifies its magnification. Multiplying the eyepiece number by the objective lens number will give you the magnification of the microscope. For example:

- eyepiece: ×10
- objective: ×40
- magnification = ×400.

As the field of view gets smaller, the magnification gets larger.

Field of view 4 mm (4000 µm) Magnification x40

Field of view 1.6 mm (1600 µm) Magnification x100

Field of view 0.4 mm (400 µm) Magnification x400

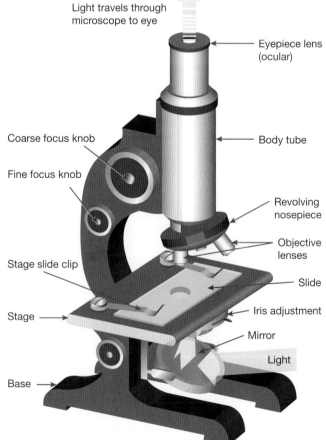

Light travels through microscope to eye

Eyepiece lens (ocular)

Coarse focus knob

Body tube

Fine focus knob

Revolving nosepiece

Objective lenses

Stage slide clip

Slide

Stage

Iris adjustment

Mirror

Light

Base

Important points to remember when using a microscope

1. When lifting the microscope, put one hand on the body of the microscope and one hand under its base.
2. The microscope should be used on a flat surface and not too close to the edge.
3. Take care that the light intensity is not too high, or it might damage your eye.
4. When you have finished using the microscope, return the shortest objective lens into position.
5. Remove the slide, and ensure that the stage is clean.
6. Make sure that when your microscope is not in use, it is always clean and carefully put away.

Using a microscope

1. Adjust your mirror so the appropriate amount of light passes through the hole in the stage.
2. Place the glass microscope slide (with a single hair specimen on top) onto the stage.
3. While watching from the side, use the coarse focus knob to lower the objective lens until it is just above the slide. Moving it down too far may shatter the slide.
4. While looking through the eyepiece lens, carefully turn the coarse focus knob until the specimen is seen clearly.
5. Carefully use the fine focus knob so that you can see the details of your specimen as clearly as possible.
6. Sketch what you see.
7. Suggest by how many times your specimen has been magnified.

How to focus your microscope — and how not to!

INVESTIGATION 3.1

Getting into focus with an 'e'

AIM: To practise focusing a monocular light microscope

Materials:
1 cm square piece of newsprint containing the letter 'e'
monocular light microscope
microscope slide
clear sticky tape
1 cm square piece of a coloured magazine or newspaper picture
hair strands (from different individuals)
spatula
selection of white powders and crystals (e.g. flour, salt, sugar, baking soda)
different brands or types of spices and leaf tea
fibres (e.g. cotton, linen, silk, wool, nylon)

Method and results

- Carefully stick the 1 cm square of newsprint onto a clean microscope slide using sticky tape.
- Using the microscope directions, get the paper into focus using the coarse focus knob and the lowest power objective lens (smallest magnification).
- Carefully move the slide until you have a letter 'e' in focus.
- Change to a higher level of magnification by rotating to a higher power objective lens.

1. In which direction did the paper under the microscope move when you moved the slide (a) towards you or (b) to the left?
2. What does the letter 'e' look like under the microscope? Draw a pencil sketch of what you see.
3. Record the magnification that you use, and estimate how much of the viewed area is covered by the letter 'e' at this magnification.
   - Using sticky tape, stick a selection of sample specimens onto microscope slides.
4. View your taped specimens using low power under the microscope and record your observations. Include detailed descriptions, the magnification used and an estimate of size next to your diagrams.

Discuss and explain

5. Suggest what the letters 'P' and 'R' would look like under the microscope. Sketch your predictions, and then view examples of these under the microscope. Were your predictions correct?
6. Summarise all your results in a table, using descriptive diagrams.
7. Summarise your microscopic observations of the sample specimens. Identify ways in which they were similar and ways in which they differed.
8. Identify details observed using the microscope that you could not see without the microscope.
9. On the basis of your experience, suggest advantages and disadvantages associated with light microscopes.
10. Propose a research question that you could explore using a light microscope, and describe how you could investigate it.

3.3 Exercises: Understanding and inquiring

To answer questions online and to receive **immediate feedback** and **sample responses** for every question, go to your learnON title at www.jacplus.com.au. *Note:* Question numbers may vary slightly.

Remember

1. Identify whether the following statements are true or false. Justify your response.
   (a) A light microscope can produce a greater magnification than an electron microscope.
   (b) Only dead sections can be viewed on an electron microscope.
   (c) Viruses can be viewed on a light microscope.
   (d) Resolution refers to how many times bigger a specimen is, whereas magnification refers to how much detail you can see.
   (e) More detail can be seen in thicker specimens when using a monocular light microscope.
2. Suggest why it is important not to have the light intensity setting too high on a light microscope.
3. Explain the importance of watching from the side of the microscope while using the coarse focus knob.
4. As the field of view of your microscope gets smaller, what happens to the magnification?
5. Use Venn diagrams to distinguish between:
   (a) a monocular microscope and a stereo microscope
   (b) a light microscope and an electron microscope
   (c) a transmission electron microscope and a scanning electron microscope
   (d) resolution and magnification
   (e) field of view and magnification.

Think

6. Use the 'field of view' diagrams on page 49 to answer the following questions. (1000 µm = 1 mm)
   (a) Estimate the length of the specimen shown in the diagram at ×40, ×100 and ×400 magnification.
   (b) Describe the differences in your observations of the three different magnifications.
7. When you are looking down the microscope, what happens when you move the microscope slide
   (a) to the left,   (b) to the right,   (c) towards you or   (d) away from you?
8. If a specimen is 1 mm in length, how big will it appear if it is magnified ×100?
9. If a specimen takes up the entire field of view at ×100, how much of it will be seen at ×400?
10. (a) Sketch a line diagram or take a photo of your microscope and label as many of its parts as you can, using the diagram on page 49.
    (b) Compare your labelled microscope figure with the diagram on page 49.
    (c) Suggest the advantages and disadvantages of the differences.
11. Copy and complete the table below.

Ocular lens (eyepiece)	Objective lens	Magnification
×5	×5	×25
×5	×10	
×10		×100
	×40	×400

12. Match the part of the microscope with its function.

Part	Function
(a) Objective lens	A. where the slide is placed
(b) Slide	B. thin piece of glass where the specimen is placed
(c) Stage clip slide	C. magnifies the image
(d) Iris adjustment	D. allows large adjustments to the distance between the stage and objective lens, which helps bring images into focus
(e) Coarse focus knob	E. adjusts the amount of light reaching the eyepiece
(f) Stage	F. allows small adjustments to the distance between the stage and the objective lens, which helps bring the image into closer focus
(g) Fine focus knob	G. holds the slide in place

Create

13. Design and make a poster which shows either how a microscope should be used or what happens when you use it the wrong way.
14. Make a model of a microscope.
15. Design a microscope of the future. Prepare an instruction booklet and develop exciting promotional material.

Investigate

16. (a) Investigate developments in our understanding of cells.
    (b) Suggest how this knowledge has influenced areas in health or medicine.
17. (a) Draw up a table that summarises the functions of the different parts of the microscope.
    (b) Construct a crossword that could be used to help students learn this information.
    (c) Share copies of your crossword with others in the class to try to find the solutions, and attempt to solve the crosswords of others in your class.
    (d) In a team, act out the parts and functions of a microscope.
18. Observe ten different specimens (for example, hair, fingernail, pencil, insect, plant) under a stereo microscope. Sketch or describe what you see. Comment on the similarities and differences observed, and on any interesting findings.
19. Find out how specimens are prepared for examination under an electron microscope. Construct a PMI chart on your findings.
20. Prepare a report on the activities of scientists involved in these studies of cells: (a) cytology (b) biochemistry (c) microbiology (d) histology.

learn**on** RESOURCES — ONLINE ONLY

📄 **Complete this digital doc:** Worksheet 3.2: In focus (doc-18709)

3.4 Form and function: Cell make-up

3.4.1 Similar, but different

Cells are the building blocks that make up all living things. Organisms may be made up of one cell (**unicellular**) or many cells (**multicellular**). These cells contain small structures called organelles that have particular jobs within the cell and function together to keep the organism alive.

Cells can be categorised on the basis of the presence and absence of particular organelles and other structural differences. Organisms can be classified by the different types of cells they are made up of.

How big is small?

The size of cells may vary between organisms and within a multicellular organism. Most cells are too small to be seen without a microscope. Cells need to be very small because they have to be able to quickly take in substances they need and remove wastes and other substances. The bigger a cell is, the longer this process would take.

Very small units of measurement are used to describe the size of cells. The most commonly used unit is the **micrometre** (μm). One micrometre equals one millionth (1/1 000 000) of a metre or one thousandth (1/1000) of a millimetre. Check out your ruler to get an idea of how small this is! Most cells are in the range of 1 μm (bacteria) to 100 μm (plant cells).

Advances in technology are creating an increased need for the use of the **nanometre**

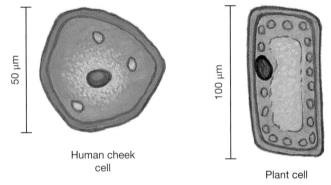

Human cheek cell

Plant cell

(nm) as a unit. One nanometre equals 1 billionth (1/1 000 000 000) of a metre. Investigating the organelles within cells and the molecules they react with requires this level of measurement.

Nanotechnology is a rapidly developing field that includes studying and investigating cells at this 'nano level'. While it requires lots of creative, exciting and futuristic 'what if' thinking, it also involves an understanding of the basics of information and ideas that are currently known.

3.4.2 Have it or not?

Prokaryotes such as bacteria were the first type of organism to appear on Earth. The key difference between prokaryotes and all other kingdoms is that members of this group do not contain a nucleus or other membrane-bound organelles. The word *prokaryote* comes from the Greek terms *pro*, meaning 'before', and *karyon*, meaning 'nut, kernel or fruit stone', referring to the cell nucleus.

Eukaryotic organisms made up of eukaryotic cells appeared on Earth billions of years later. As *eu* is the Greek term meaning 'good', **eukaryote** can be translated as 'true nucleus'. Members of the kingdoms Animalia, Plantae, Fungi and Protoctista are eukaryotes and are made up of cells containing a nucleus and other membrane-bound organelles.

Eukaryotic cells (a) contain a nucleus and membrane-bound organelles, while prokaryotic cells (b) do not.

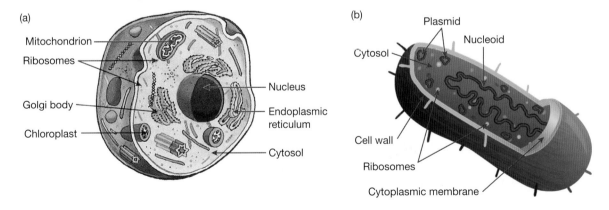

3.4.3 What do we share?

What most cells have in common is that they are made up of a **cell membrane** containing a fluid called **cytosol** and small structures called **ribosomes**. The collective term used to describe the cytosol and all the organelles suspended within it is **cytoplasm**. The hundreds of chemical reactions essential for life that occur within the cytoplasm are referred to as the cell's **metabolism**. The ribosomes are where proteins such as enzymes, which regulate the many chemical reactions important to life, are made. The cell membrane regulates the movement of substances into and out of the cell. This enables the delivery of nutrients and substances essential for reactions, and the removal of wastes.

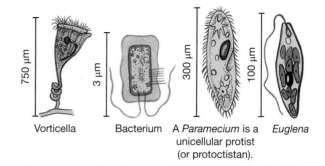

Vorticella   Bacterium   A *Paramecium* is a unicellular protist (or protoctistan).   *Euglena*

learn on RESOURCES — ONLINE ONLY

Watch this video: Inside cells (eles-0054)

3.4.4 All on my own

Unicellular organisms such as bacteria, *Amoeba*, *Euglena* and *Paramecium* need to carry out all the required processes themselves. They even reproduce themselves by dividing into two. This process is called **binary fission**.

To live long enough to reproduce, unicellular organisms need to be able to function on their own. They need to obtain their nutrients and remove their wastes. The solution to this requirement has resulted in the wonderful diversity of unicellular organisms that are alive on Earth today or have lived in our planet's history.

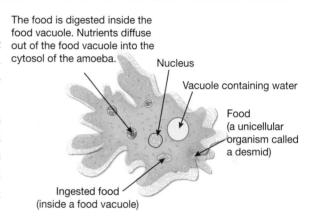

The food is digested inside the food vacuole. Nutrients diffuse out of the food vacuole into the cytosol of the amoeba.

Nucleus

Vacuole containing water

Food (a unicellular organism called a desmid)

Ingested food (inside a food vacuole)

3.4.5 Five kingdoms?

Living things can be divided into five kingdoms — **Animalia** (animals), **Plantae** (plants), **Fungi** (for example, mushrooms), **Protoctista** (also called Protista) and **Prokaryotae** (also called Monera). While this system provides an opportunity to classify organisms into these groups, information from currently developing technologies means that it will not be long until a new extended classification system evolves.

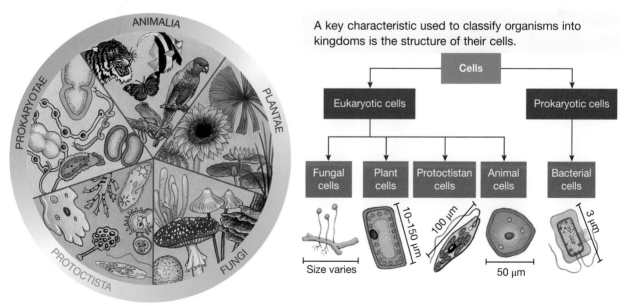

A key characteristic used to classify organisms into kingdoms is the structure of their cells.

| Cells |
| Eukaryotic cells | Prokaryotic cells |

Fungal cells — Size varies

Plant cells — 10–150 μm

Protoctistan cells — 100 μm

Animal cells — 50 μm

Bacterial cells — 3 μm

WHAT DOES IT MEAN?

The prefix *uni-* comes from the Latin term meaning 'one'. The prefix *multi-* comes from the Latin term meaning 'many'.

3.4.6 Specialist workers

Multicellular organisms are made up of many different types of cells that have different jobs to do. Each of these different types of cells has a particular structure so that it is able to do the job it is specialised for. This may include the presence and number of particular organelles or additional external structures to assist with movement (such as flagella or cilia).

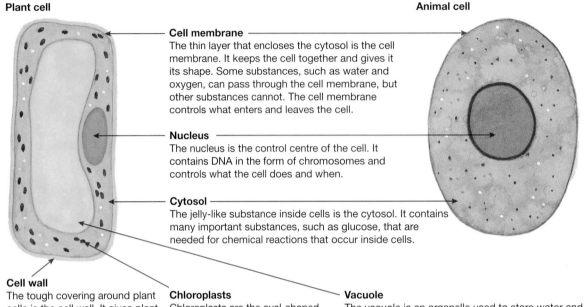

Plant cell

Animal cell

Cell membrane
The thin layer that encloses the cytosol is the cell membrane. It keeps the cell together and gives it its shape. Some substances, such as water and oxygen, can pass through the cell membrane, but other substances cannot. The cell membrane controls what enters and leaves the cell.

Nucleus
The nucleus is the control centre of the cell. It contains DNA in the form of chromosomes and controls what the cell does and when.

Cytosol
The jelly-like substance inside cells is the cytosol. It contains many important substances, such as glucose, that are needed for chemical reactions that occur inside cells.

Cell wall
The tough covering around plant cells is the cell wall. It gives plant cells strength and holds them in shape. Plant cell walls are made of a substance called **cellulose**. Water and dissolved substances can pass through the cell wall. Animal cells do not have a cell wall.

Chloroplasts
Chloroplasts are the oval-shaped organelles found only in plant cells. Chloroplasts contain a green substance called chlorophyll. Chloroplasts use energy from the sun to make food. Not all plant cells contain chloroplasts. They are found only in leaf and stem cells.

Vacuole
The vacuole is an organelle used to store water and dissolved substances. Vacuoles can look empty, like an air bubble. Plant cells usually have one large vacuole. The mixture inside a plant's vacuoles is called **cell sap**. The red, blue and violet colours that you often see in plant leaves and flowers are due to the substances stored in vacuoles. Most animal cells don't have vacuoles.

Microfactories

Mitochondria and **chloroplasts** are examples of membrane-bound organelles found in eukaryotic cells. While all eukaryotic cells contain mitochondria, because they are all involved in **cellular respiration**, only those involved in **photosynthesis** (such as those in plant leaves) contain chloroplasts. Chloroplasts contain the green pigment **chlorophyll**. This pigment is used to trap light energy so that it can be converted into chemical energy and used by the cells.

HOW ABOUT THAT!

There is a theory called the endosymbiotic theory that suggests that mitochondria and chloroplasts were once prokaryotic organisms. This theory suggests that, at some time in the past, these organisms were engulfed by another cell and over time they evolved to depend on each other.

The origin of the eukaryotic cell? Some scientists also suggest that our nucleus may have come from a giant viral ancestor.

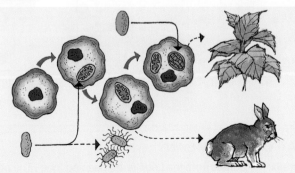

3.4.7 Some differences in the basic cell design in the five kingdoms

	Kingdom				
Characteristic	**Animalia** (animals: e.g. lizards, fish, spiders, earthworms, sponges)	**Fungi** (e.g. yeasts, moulds, mushrooms, toadstools)	**Plantae** (plants: e.g. ferns, mosses, conifers, flowering plants)	**Protoctista** (e.g. algae, protozoans)	**Prokaryotae** (bacteria and cyanobacteria)
Number of cells	Multicellular	Usually multicellular but some unicellular	Most multicellular	Unicellular or multicellular	Unicellular
Nucleus	Present	Present	Present	Present	Absent
Cell wall	Absent	Present	Present	Present in some	Present
Large vacuole	Absent	Absent	Present	Present in some	Absent
Chloroplasts	Absent	Absent	Present in leaf and stem cells	Present in some	Absent (but chlorophyll may be present in some)

3.4 Exercises: Understanding and inquiring

To answer questions online and to receive **immediate feedback** and **sample responses** for every question, go to your learnON title at www.jacplus.com.au. *Note:* Question numbers may vary slightly.

Remember

1. What do all living things have in common?
2. Why is the nucleus important to the cell?
3. State the names of the five kingdoms and use the table above to determine which kingdoms contain organisms that are eukaryotes.
4. What is the purpose of the cell membrane?
5. Identify where enzymes are made in a cell and state why they are important.
6. Construct a triple Venn diagram (three overlapping circles) to compare plants, animals and fungi.
7. Suggest why organisms are divided into five kingdoms rather than just the plant and animal kingdoms.

Think and reason

Use the table above to answer the following questions.
8. In which kingdom(s) do the cells of an organism:
    (a) not have a cell wall, large vacuole or chloroplasts
    (b) have a cell wall, large vacuole and chloroplasts
    (c) have a cell wall, but no large vacuole or chloroplasts
    (d) have a cell wall and a nucleus without a membrane around it?
9. List two examples of each of the five kingdoms.
Use the diagrams of cells on page 54 to complete activities 10 and 11.
10. Construct a table with the following headings: 'Name of organism' or 'Type of cell', and 'Cell size (μm)'.
11. Show the sizes of the cells on a graph, with the horizontal axis representing the type of cell and the vertical axis representing the size of cell. Sketch an outline of each cell as accurately as you can in the correct position on the graph. Which cell is the biggest and which is the smallest?
12. Fungi were once classified as plants. On the basis of their cells, suggest why they are no longer classified in this group.

Create

13. Make a labelled model of a cell from one of the kingdoms. Use materials available at home, such as drink bottles, egg cartons, cottonwool, wool, cotton or dry foods.

14. What does the endosymbiotic theory suggest? Formulate questions to ask about it. Research and report on your questions.
15. Research and report on:
    (a) examples of prokaryotic cells and interesting survival strategies
    (b) mitochondrial DNA and haplogroups.
16. Research two of the organelles or cells listed below. Create a play and construct puppet models for your characters. Present your play to the class.
    - Nucleus
    - Mitochondrion
    - Chloroplast
    - Prokaryotic cell
    - Protoctistan cell
    - Animal cell
    - Plant cell
17. Investigate the different types of cells and create your own picture book about them using the following steps.
    (a) Construct a matrix table (see section 10.9 Matrixes and Venn diagrams) to show the differences between the cells of the different kingdoms.
    (b) Construct a storyboard for a picture book about them.
    (c) Create the picture book.

3.5 Zooming in on life

3.5.1 Sketching what you see under the microscope

Some points to remember

1. Use a sharp pencil.
2. Draw only the lines that you see (no shading or colouring).
3. Your diagrams should take up about a third to half a page each.
4. Record the magnification next to each diagram.
5. State the name of the specimen and the date of observation.
6. A written description is also often of considerable value.
7. When you are viewing many cells at one time, it is often useful to select and draw only two or three representative cells for each observation.

An example of a sketch of a microscope specimen

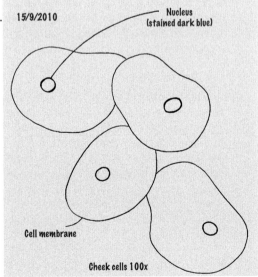

3.5.2 Preparing a specimen

Light microscopes function by allowing light to pass through the specimen to reach your eye. If the specimen is too thick, the object cannot be seen as clearly or may not be seen at all.

Careful peeling, scraping, slicing or squashing techniques can be used to obtain thin specimens of the object to be studied.

Staining a specimen

Many objects are colourless when viewed under the microscope, so specimens are often stained to make them easier to see. Methylene blue, iodine and eosin are some examples of common stains.

Each stain reacts with different chemicals in the specimen. For example, iodine stains starch a blue-black colour.

Take care when using these stains, because they can stain you as well!

INVESTIGATION 3.2

Preparing a wet mount

AIM: To prepare a wet mount and observe micro-organisms on a microscope slide

Materials:

light microscope   coverslips
pipette   toothpick
pond water
microscope slides (well slides work best for this)
culture of living microscopic organisms: Paramecium, Amoeba, rotifers, Euglena

Method and results

- Use the pipette to put a drop of pond water or microbe culture on a clean microscope slide.

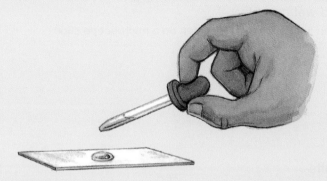

- Gently place a coverslip over the drop of water by putting one edge down first. Use a toothpick as shown below.
- Incorrect placement of the coverslip can result in air bubbles.

- Use a microscope to observe the slide.

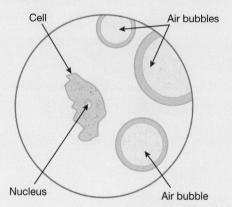

1. Draw detailed sketches of what you see. Remember to include a title, the magnifications used and as many comments as you can.
   - Remove the coverslip, rinse and dry the slide, and then prepare a new slide specimen and repeat the steps above.

Discuss and explain

2. Construct a matrix to show the similarities and differences between the specimens.
3. Suggest reasons for these differences.
4. Which kingdoms do you think each specimen may belong to? Provide reasons for your classification.
5. Identify two structures you observed in the investigation and find out more about their function (that is, what their 'job' is).
6. You have been observing living specimens. Identify advantages and disadvantages of using living rather than dead specimens or prepared slides.

INVESTIGATION 3.3

Preparing stained wet mounts

AIM: To prepare, stain and observe a specimen on a microscope slide

Materials:

light microscope	*pipette*
blotting paper	*toothpick*
forceps or tweezers	*scalpel*
microscope slides and coverslips	
water, methylene blue, iodine	
onion, ripe and unripe banana, celery stick	

Method and results

• Use the pipette to put a drop of water on a clean microscope slide.

• Use a scalpel and forceps to peel a small piece of the very thin, almost transparent onion skin from the inside surface of an onion.

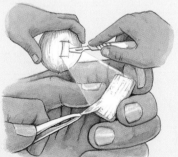

• Use the forceps to put the thin piece of the onion skin into the drop of water on the microscope slide.

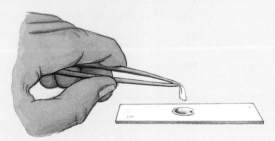

- Gently place a coverslip over the drop of water containing the onion skin by putting one edge down first. Use a toothpick as in Investigation 3.2 to avoid air bubbles. Use blotting paper to soak up any excess water outside the coverslip.

- Use a microscope to observe the slide; first use low power and then increase the magnification.
- Prepare another slide of onion skin, except this time add a drop of methylene blue instead of water to the slide. Make sure that you carefully blot excess stain from the slide after you add the coverslip.
- Observe this stained onion specimen; first use low power, then view at a higher magnification.
1. Draw detailed sketches of what you see. Remember to include a title, the magnifications used and as many comments as you can. Label any parts that you can identify.
   - Remove the coverslip, and rinse and dry the slide.
   - Use the steps outlined on the previous page to prepare the following slides:
     - celery epidermis (outer layer of the celery stem) with and without methylene blue stain
     - squashed ripe and unripe banana with and without iodine.
2. Draw detailed sketches of what you see. Remember to include a title, the magnifications used and as many comments as you can. Label any parts that you can identify.

Discuss and explain

3. Compare the cells of the stained onion epidermis and the celery epidermis. Identify their similarities and differences. Suggest reasons for the differences.
4. Compare the cells of the stained ripe and unripe banana. Identify how they are similar and how they are different. Suggest reasons for the differences.

CAUTION

The scalpel has a very sharp blade. Handle it with care.

Background information

Methylene blue is used to stain the nucleus so that it is easier to see. Iodine changes from yellow-brown to a dark blue when it combines with starch.
5. why stains are used. Include reasons for using methylene blue and iodine that relate to your observations in this investigation.
6. Investigate the functions of the structures observed in your stained specimens. Suggest how features of these structures assist their function.

Evaluation

7. Identify strengths, limitations and improvements related to this investigation.

3.5 Exercises: Understanding and inquiring

To answer questions online and to receive **immediate feedback** and **sample responses** for every question, go to your learnON title at www.jacplus.com.au. *Note:* Question numbers may vary slightly.

Think, discuss and investigate

1. Carefully observe the student sketches shown below. For each diagram, list what is wrong with it and suggest how it could be improved.

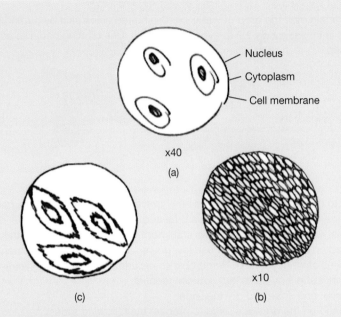

x40

(a)

x10

(c)   (b)

2. (a) Carefully observe the figure of plant cells below and construct a sketch of one of the cells.
   (b) Use references to suggest labels for the structures shown in your sketch.

Create

3. Design a poster that shows others how to prepare a variety of specimens to be viewed under a microscope.

3.6 Focus on animal cells

3.6.1 In all shapes and sizes

Cells within an organism may differ in their shape and size. This difference may be due to the particular jobs or functions that the cells carry out within the organism. The human body is made up of more than 20 different types of cells, with each type suited to a particular function.

Nerve cells develop long, thin fibres that quickly carry messages from one cell to another. Cells lining the trachea have hair-like cilia that move fluid and dust particles out of the lungs. Muscle cells contain fibres that contract and relax, and the human sperm cell has a tail or flagellum that helps it swim to the egg cell.

Cells can also differ in the organelles that they contain within them. Muscle cells, for example, contain many more mitochondria than other types of cells due to their high energy requirements. Red bloods cells also differ from many other types of cells because, as they mature, they lose their nucleus. This makes more room available for them to carry more oxygen throughout your body.

HOW ABOUT THAT!

Did you know these facts about human cells?
- Hair and nails are made of dead cells, and because they are not fed by blood or nerves you can cut them without it hurting.
- A human baby grows from one cell to 2000 million cells in just nine months.
- Red blood cells live for one to four months and each cell travels around your body up to 172 000 times.
- Some of the nerve cells in the human body can be one metre long. But that's small compared with the nerve cells in a giraffe's neck. They are two to three metres long!

INVESTIGATION 3.4

Animal cells — what's the difference?
AIM: To observe the features of different types of animal cells

Materials:
light microscope
prepared animal slides: blood cells, muscle cells, cheek cells, nerve cells

Method and results
- Use a microscope to observe the prepared slides.
1. Record detailed diagrams of your observations. Next to your diagrams, include details of the (a) source of the specimen, (b) type of specimen, (c) magnification used and (d) a detailed description of the specimen.

Discuss and explain
2. Were all of the animal cells you observed the same size? Explain.
3. Did all of the cells observed contain a nucleus? Explain.
4. Identify features that all of the observed animal cells shared.
5. Identify differences between the features of the cells observed.
6. Suggest reasons for the differences between the cells.
7. Compare your cells with those on page 63.
    (a) Do your sketched diagrams match the structures shown in the figure? Explain.
    (b) Read through the text related to the functions of the different types of cells. Do these match those you suggested in question 6? Explain.
    (c) Suggest how the shape or size of a cell may assist it in doing its job.

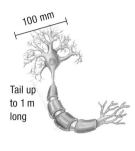

100 mm

Tail up to 1 m long

Nerve cells

Nerve cells are very long and have a star shape at one end. The long shape of nerve cells helps them detect and send electrical messages through the body at the speed of a Formula 1 racing car. There are nerve cells all over your body. They allow you to detect touch, smell, taste, sound, light and pain.

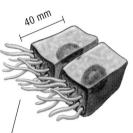

40 mm

Lung epithelial cells

The cells that line your nose, windpipe and lungs are a type of lining cell. They have hair-like tips called cilia. These cells help protect you by stopping dust and fluid from getting down your windpipe. The cilia can also move these substances away from your lungs. You remove some of these unwanted substances whenever you sneeze, cough or blow your nose.

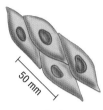

50 mm

Muscle cells

Muscle cells are long and elastic. Long thin cells can slide further over each other to allow you to move. There are different types of muscle cells. The walls of your blood vessels and parts of your digestive system have 'smooth muscle' cells. The muscles that are joined to your bones are called 'skeletal muscles'. Skeletal muscles work in pairs — one muscle contracts (shortens) and pulls the bone in one direction while the other muscle relaxes.

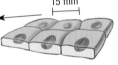

30 mm

Adipose tissue cells

Some cells store fat. Fat stores a lot of energy for cells to use later. Round shapes are good for holding a lot of material in a small space. Fat cells are mostly found underneath your skin, especially in the chest, waist and buttocks.

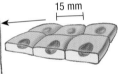

15 mm

Skin cells

Special cells line the outside surfaces of your body. These are the cells that form your skin. These cells have a flattened shape so they can better cover and protect your body.

10 mm

Red blood cells

Red blood cells carry oxygen around the body. Their small size allows them to move easily through blood vessels. The nucleus in a red blood cell dies soon after the cell is made. Without a nucleus, red blood cells live for only a few weeks. The body keeps making new blood cells to replace those that have died. Red blood cells are made in bone marrow at the rate of 17 million cells per minute! This is why most people can donate some of their blood to the Red Cross without harm. White blood cells, which are larger than red blood cells, are also made in the bone marrow. Their job is to rid the body of disease-causing organisms and foreign material.

45 mm

Sperm cells

Sperm cells have long tails that help them swim towards egg cells. Only males have sperm cells.

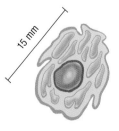

15 mm

Bone cells

Minerals such as calcium surround your bone cells. The minerals help make bone cells hard and strong. Bone cells need to be hard so that they can keep you upright.

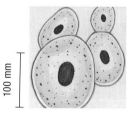

100 mm

Egg cells

Egg cells are some of the largest cells in a human body. Their large round shape helps them store plenty of food. Only females have egg cells. When a sperm cell moves into an egg cell, the egg cell is fertilised.

3.6 Exercises: understanding and inquiring

To answer questions online and to receive **immediate feedback** and **sample responses** for every question, go to your learnON title at www.jacplus.com.au. *Note:* Question numbers may vary slightly.

Remember

1. Match the types of cells with their descriptions.
2. Identify which features most animal cells have in common. Suggest reasons why.
3. Describe some ways in which cells may differ.
4. Suggest why the cells in a multicellular organism are not all the same. Give examples in your answer.
5. Distinguish between:
   (a) skin cells and sperm cells
   (b) red blood cells and nerve cells
   (c) adipose tissue cells and muscle cells.

Type of cell	Description
(a) Muscle cell	A. Has a long tail that helps it to swim towards the egg cell
(b) Skin cell	B. Long, thin elastic cells that contract and relax
(c) Red blood cell	C. A flat cell that lines the outside surface of your body
(d) Nerve cell	D. Very tiny cell that lacks a nucleus when mature and carries oxygen
(e) Sperm cel	E. Very long cell, star-shaped at one end, detects and sends messages

Think and calculate

6. (a) Summarise the information from page 63 into a table with the headings: 'Type of cell', 'Function', 'Shape' and 'Size'.
   (b) Using these data, determine the average size of an animal cell.
   (c) Use a bar graph to plot the sizes of the different types of animal cells.
   (d) Identify which animal cells are 'above average' in size and which are 'below average'. Suggest reasons for the differences.
   (e) Comment on the differences in other features between the cells.

Investigate, imagine and create

7. Find out more about a particular type of cell and use this information to write a play, poem or story about a day in the life of this type of cell.
8. Construct a model of a nerve cell using food as your construction material.
9. Using your own research and the information on page 63, construct a 'peep through' learning wheel that shows the structure and function of the different types of animal cells. Instructions for making a 'peep through' learning wheel are given at right and below.

How to make a 'peep through' learning wheel

Wheel 1

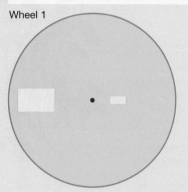

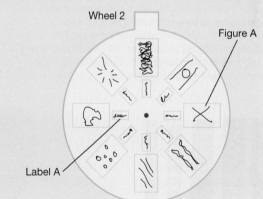

Wheel 2

Figure A

Label A

a. On an A4 piece of white paper or card draw two circles, one with a 'tab' (wheel 2) and one without (wheel 1).
b. Cut out the two rectangular box areas as shown on wheel 1.
c. Draw in the large and the small rectangles as shown on wheel 2.
d. Write the animal cell types in the small boxes on wheel 2. Sketch matching diagrams of examples of these cell types in the corresponding large box opposite.
e. Attach the two wheels, with wheel 1 on top, using a paper fastener.
f. Rotate your wheel to view examples of types of animal cells.

3.7 Focus on plant cells

3.7.1 Have or have not

Like animal cells, plant cells have cytoplasm, a membrane and a nucleus. Unlike animal cells, plant cells have a cellulose cell wall and a large central vacuole filled with cell sap. Often plant cells also contain chloroplasts, which enable them to make their own food.

Some of the types of cells found in plants

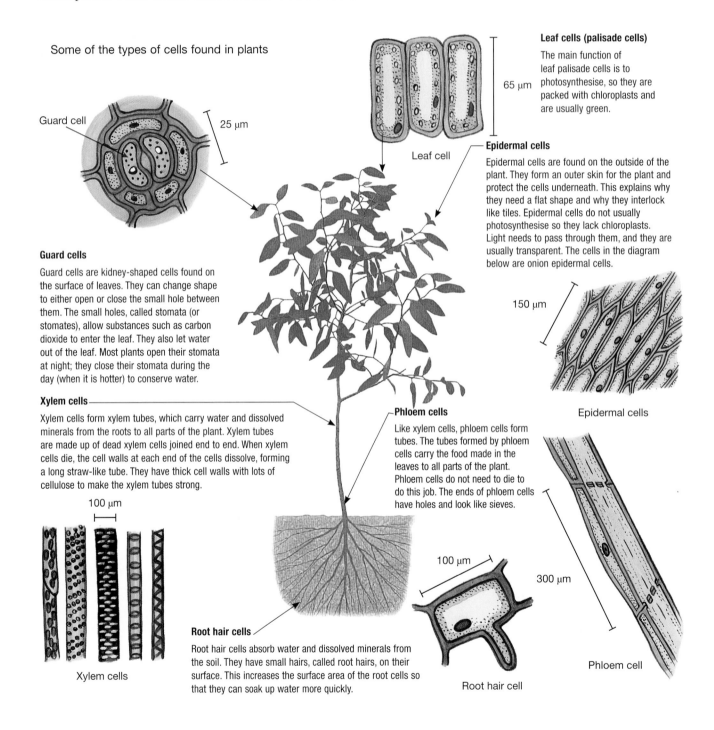

Leaf cells (palisade cells)

The main function of leaf palisade cells is to photosynthesise, so they are packed with chloroplasts and are usually green.

65 μm

25 μm

Guard cell

Leaf cell

Epidermal cells

Epidermal cells are found on the outside of the plant. They form an outer skin for the plant and protect the cells underneath. This explains why they need a flat shape and why they interlock like tiles. Epidermal cells do not usually photosynthesise so they lack chloroplasts. Light needs to pass through them, and they are usually transparent. The cells in the diagram below are onion epidermal cells.

150 μm

Guard cells

Guard cells are kidney-shaped cells found on the surface of leaves. They can change shape to either open or close the small hole between them. The small holes, called stomata (or stomates), allow substances such as carbon dioxide to enter the leaf. They also let water out of the leaf. Most plants open their stomata at night; they close their stomata during the day (when it is hotter) to conserve water.

Epidermal cells

Xylem cells

Xylem cells form xylem tubes, which carry water and dissolved minerals from the roots to all parts of the plant. Xylem tubes are made up of dead xylem cells joined end to end. When xylem cells die, the cell walls at each end of the cells dissolve, forming a long straw-like tube. They have thick cell walls with lots of cellulose to make the xylem tubes strong.

Phloem cells

Like xylem cells, phloem cells form tubes. The tubes formed by phloem cells carry the food made in the leaves to all parts of the plant. Phloem cells do not need to die to do this job. The ends of phloem cells have holes and look like sieves.

100 μm

100 μm

300 μm

Root hair cells

Root hair cells absorb water and dissolved minerals from the soil. They have small hairs, called root hairs, on their surface. This increases the surface area of the root cells so that they can soak up water more quickly.

Xylem cells

Root hair cell

Phloem cell

On the surfaces of leaves, there are pairs of special cells called guard cells, which surround tiny pores called stomata. The guard cells can change shape, opening or closing the stomata. Special cells on the roots extend into microscopic hairs that penetrate between soil particles. The hairs provide a large surface area through which water may be absorbed from the soil.

INVESTIGATION 3.5

Plant cells in view

AIM: To observe the features of different types of plant cells

Materials:
light microscope
prepared plant slides: leaf epidermal cells, root hair cells, stomata/guard cells

Method and results
- Use a microscope to observe the prepared slides.
1. Record detailed diagrams of your observations. Next to your diagrams, include details of the (a) source of the specimen, (b) type of specimen, (c) magnification used and (d) a detailed description of the specimen.

Discuss and explain
2. Were all of the plant cells the same size? Explain.
3. Did all of the cells observed contain a nucleus? Explain.
4. Identify features that all of the observed plant cells shared.
5. Identify differences between the features of the cells.
6. Suggest reasons for the differences between the cells.
7. Compare your cells with those on page 65.
   (a) Do your sketched diagrams match the structures shown in the figure? Explain.
   (b) Read through the text related to the functions of the different types of cells. Do these match your answer to question 6? Explain.
   (c) Suggest how the shape or size of a cell may help it to do its job.

WHAT DOES IT MEAN?

The word *xylem* comes from the Greek word *xulan*, meaning 'wood'. The word phloem comes from the Greek word *phloos*, meaning 'bark'.

3.7 Exercises: Understanding and inquiring

To answer questions online and to receive **immediate feedback** and **sample responses** for every question, go to your learnON title at www.jacplus.com.au. *Note:* Question numbers may vary slightly.

Remember

1. Match the types of cells with their descriptions.

Type of cell	Description
(a) Guard cell	A. Sieve-like cells that form tubes which carry food made in the leaves to other parts of the plant
(b) Phloem cell	B. Cells with small hairs that increase their surface area so that they can absorb more water
(c) Xylem cell	C. Thick-walled cells that carry water up the plant
(d) Root hair cells	D. Kidney-shaped cells that can change shape to either open or close the small hole between them, which allows gas exchange between the plant and its environment

2. Identify which features most plant cells have in common. Suggest reasons why.
3. Describe some ways in which plant cells may differ.
4. Distinguish between:
   (a) palisade cells and guard cells
   (b) xylem cells and phloem cells
   (c) epidermal cells and root hair cells.

Think and calculate

5. (a) Summarise the information from page 65 into a table with the headings: 'Type of cell', 'Function', 'Shape' and 'Size'.
   (b) Using these data, determine the average size of a plant cell.
   (c) Use a bar graph to plot the sizes of the different types of plant cells.
   (d) Identify which plant cells are 'above average' in size and which are 'below average'. Suggest reasons for the differences.
   (e) Comment on the differences in other features between plant cells.

Investigate, imagine and create

6. Find out more about guard cells or leaf epidermal cells. Write a story about what happens to them over 24 hours.
7. Construct a model of a pair of guard cells, using balloons.
8. Using your own research and the information on page 65, construct a 'peep through' learning wheel that shows the structure and function of the different types of plant cells. Instructions for making a 'peep through' learning wheel are given on page 64.

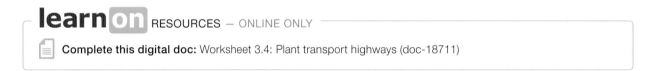

learn on RESOURCES — ONLINE ONLY

Complete this digital doc: Worksheet 3.4: Plant transport highways (doc-18711)

3.8 Plant cells — holding, carrying and guarding

3.8.1 Sweet transport: phloem

As in animals, plant cells can work together for a variety of functions to meet their survival needs. Plants have their own transport systems, which consist of many thin tubes made up of different types of cells. Other types of plant cells are involved in water regulation and exchange of important gases, such as oxygen and carbon dioxide, with their environment.

Using the process of photosynthesis, plants make sugar in their leaves. The system of thin-walled tubes that carries this sugar (in the form of glucose or sucrose) from the leaves to other parts of the plant is called **phloem**. Phloem consists of living cells called sieve tubes and companion cells. The transport of the sugar solution up and down the plant is called **translocation**.

3.8.2 Water pipes: xylem

Flowering plants also have tubes with strong, thick walls that carry water and minerals up from the roots through the stem to the leaves. These are called **xylem vessels**. These tubes are formed from the empty remains of dead cells, the walls of which are strengthened with a woody substance called **lignin**. The xylem is therefore a 'dead' one-way street, rather than a 'living' two-way highway like the other transport tubes you have studied.

Water moves up from the roots of the plant, through its stem and to its leaves, where some may pass out of the plant as water vapour through pores called **stomata**. This movement of water is called the **transpiration stream**.

Diagrams of typical cross-sections of the stem of (a) a young dicot and (b) a monocot. The photographs show how the cells of (c) a dicot (buttercup) appear when viewed under an electron microscope and (d) a monocot (sugarcane) appear under a light microscope.

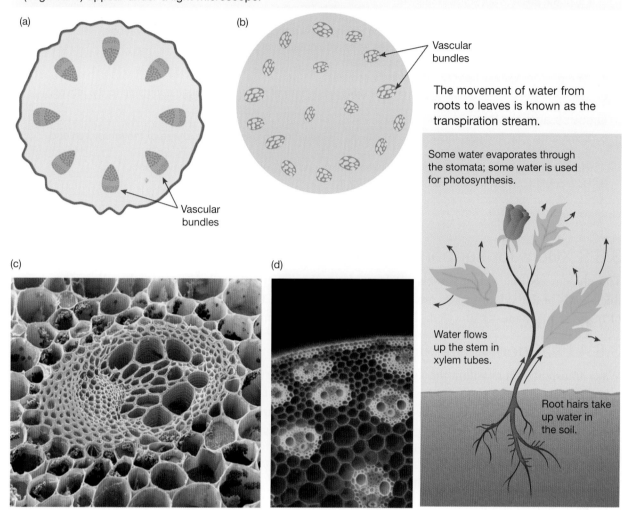

(a)

(b)

Vascular bundles

The movement of water from roots to leaves is known as the transpiration stream.

Some water evaporates through the stomata; some water is used for photosynthesis.

Vascular bundles

(c)

(d)

Water flows up the stem in xylem tubes.

Root hairs take up water in the soil.

INVESTIGATION 3.6

Stem transport systems

AIM: To identify xylem cells in celery

Materials:
celery stick (stem and leaves)
knife
two 250 mL beakers
water
blue food colouring
red food colouring
hand lens

Method and results

- Slice the celery along the middle to about halfway up the stem.
- Fill two beakers with 250 mL of water. Colour one blue and the other red with the food colouring.

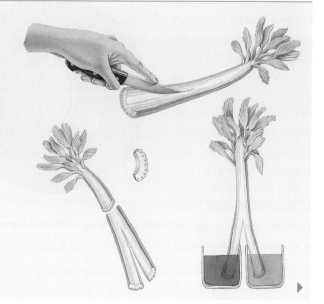

- Place the celery so that each 'side' of the celery is in a separate beaker.
- Leave for 24 hours and then observe the celery.
- Cut the celery stick across the stem.
- Use the hand lens to look at the inside of the stem.

1. Look at where the water has travelled in the celery. Draw a diagram to show your observations.
2. Draw a diagram to show what you can see when you cut across the stem.
3. Where are the different colours found in the stem?
4. Where are the different colours found in the leaves?
5. Draw a diagram of the whole celery stick and trace the path of the water through each side to the leaves.

Discuss and explain
6. Summarise your findings. Include comments on the relationship between the shape of the structures and their function.
7. Identify strengths and limitations of this investigation and suggest possible improvements.
8. How could you turn a white carnation blue? Try it.

3.8.3 Xylem for support

The phloem and the xylem vessels are located together in groups called **vascular bundles**. The strong, thick walls of the xylem vessels are also important in helping to hold up and support the plant. The trunks of trees are made mostly of xylem. Did you know that the stringiness of celery is due to its xylem tissues?

3.8.4 Leaf doorways: stomata

Water transport occurs within the xylem vessels. Some of the water that is transported through the xylem to the leaves is used in photosynthesis. Some water is also lost as water vapour through tiny holes or pores in the leaves. These tiny pores, called stomata (or stomates), are most frequently found on the underside of the leaves. Evaporation of water from the stomata in the leaves helps pull water up the plant. Loss of water vapour through the stomata is called **transpiration**.

3.8.5 Guard cells in control

Oxygen and carbon dioxide gases also move in and out of the plant through the stomata. **Guard cells**, which surround each stoma, enable the hole to open and close, depending on the plant's needs. When the plant has plenty of water, the guard cells fill up with water and stretch lengthways. This opens the pore. If water is in short supply, how-

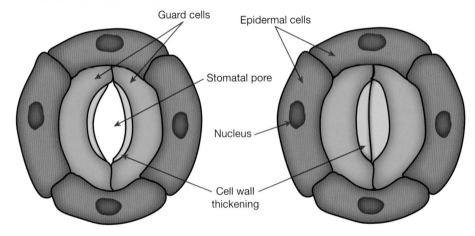

Stomata can close to conserve water.

Guard cells
Epidermal cells
Stomatal pore
Nucleus
Cell wall thickening

ever, the guard cells lose water and they collapse towards each other. The pore is then closed. This is one way in which the plant can control its water loss.

3.8.6 Dusty doors

Air pollution can result in particles of dust settling on the leaves of plants. This may limit the amount of light reaching the leaf and so reduce photosynthesis. If these dust particles block up stomata, they can also affect transpiration and gaseous exchange.

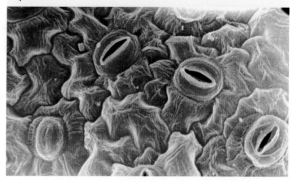

'Stomata art': the arrangement of stomata in a plant

HOW ABOUT THAT!

Scientists have used genetic engineering to produce plants that glow particular colours when they have mineral deficiencies. This provides farmers with information about which soils need extra minerals added.

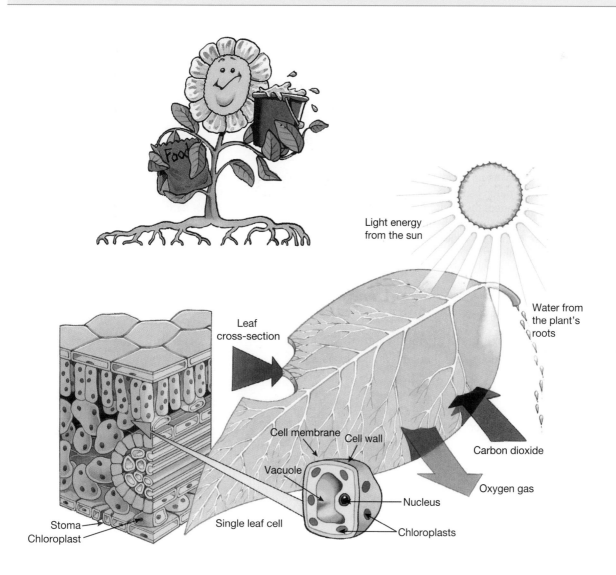

Light energy from the sun

Water from the plant's roots

Leaf cross-section

Carbon dioxide

Oxygen gas

Cell membrane

Cell wall

Vacuole

Nucleus

Single leaf cell

Chloroplasts

Stoma

Chloroplast

INVESTIGATION 3.7

Observing leaf epidermal cells

AIM: To observe leaf epidermal cells and identify stomata

Materials:

leaf   clear sticky tape
microscope slide   microscope

Method and results

- Put some sticky tape over a section of the underside of the leaf.
- Press the sticky tape firmly onto the leaf.
- Tear the tape off. Some of the lining cells should come off with the sticky tape.
- Press the tape, sticky side down, onto a microscope slide.
- View the sticky tape under the microscope.
- Try to find a pair of guard cells and one of the stomata.
1. Is the stoma (the opening) open or closed?
2. Make a drawing of a group of cells, including the guard cells. Include as much detail in your drawing as possible.
3. Label the guard cells and stomata.
4. Title and date your drawing. Write down the magnification used.

Discuss and explain

5. Summarise your findings. Include comments on the relationship between the shape of the structures and their function.
6. Identify strengths and limitations of this investigation and suggest possible improvements.

HOW ABOUT THAT!

Although water makes up about 90–95 per cent of the living tissues of plants, water is often being lost to their surroundings. As much as 98 per cent of the water absorbed by a plant can be lost through transpiration. A variety of factors affect the amount of water that plants lose. Weather is a major factor, as high temperatures, wind and low humidity can increase the evaporation of water from the stomata. It has been recorded that large trees may lose more than 400 litres of water in a day.

INVESTIGATION 3.8

Looking at chloroplasts under a microscope

AIM: To observe chloroplasts under a light microscope

Materials:

tweezers   water
moss, spirogyra or elodea   dilute iodine solution
light microscope, slides, coverslips

Method and results

- Using tweezers, carefully remove a leaf from a moss or elodea plant or take a small piece of spirogyra.
- Place the plant material in a drop of water on a microscope slide and cover it with a coverslip.
- Use a light microscope to observe the leaf.
- Put a drop of dilute iodine solution under the coverslip. (Iodine stains starch a blue-black colour.)
- Using the microscope, examine the leaf again.
1. Draw what you see before staining.
2. Label any chloroplasts that are present.

3.8.7 Moving in or out?

The movement of substances into and out of cells is controlled by the cell membrane. This enables useful substances to be delivered into the cells and waste products to be moved out. Some types of movements require energy and others do not.

Oxygen and carbon dioxide enter the cell by a process called diffusion. Diffusion moves substances from where they are in a high concentration to where they are in a low concentration and so does not require energy. Water moves across the membrane via a special type of diffusion called osmosis. This movement of water into and out of the guard cells is responsible for opening and closing the stomata.

Flaccid or firm?

If too much water is lost or not enough water is available, the plant may **wilt**. When this occurs, water has moved out of the cell **vacuoles** and the cells have become soft or **flaccid**. The firmness in the petals and leaves is due to their cells being firm or **turgid**.

If the cells of a plant do not contain enough water, they become flaccid and the plant wilts.

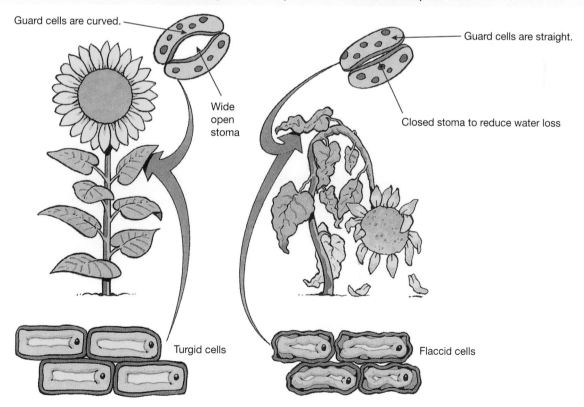

Guard cells are curved.

Wide open stoma

Turgid cells

Guard cells are straight.

Closed stoma to reduce water loss

Flaccid cells

INVESTIGATION 3.9

Moving in or out?

AIM: To make a model of a cell membrane to simulate the effect of water moving in and out of a cell

Materials:
two 20 cm lengths of dialysis tubing
scales   iodine solution
starch solution   2 beakers

Method and results

- Soak the dialysis tubing in water so it becomes soft.
- Tie a knot at one end of each piece of dialysis tubing. This will form two small bags.
- Pour water into bag A until it is one-third full. Pour the same amount of starch solution into bag B and add 10 drops of iodine solution.
- Tie a knot at the top of each bag to seal them.

1. Draw up a table to record the weights of the bags before and after being left in the beakers.
   - Weigh both bags.
   - Put bag A in a beaker of starch solution. Add enough iodine to the starch solution to produce a dark blue colour.
   - Put bag B in a beaker of water.
   - Leave the two bags undisturbed for at least two hours (or overnight).
   - Weigh the bags again.
2. What happens to iodine when it is added to starch solution?
3. Draw bags A and B in the beakers they were left in. On your diagram, label where blue and yellow colour can be seen.

Discuss and explain

4. In this experiment, we made a model of a cell. Which part represented the cell membrane?
5. Dialysis tubing allows some substances, but not others, to pass through. Which of the following substances could pass through the dialysis tubing and which could not? What evidence supports this?
   (a) Starch
   (b) Water
   (c) Iodine
6. Did the masses of the two bags change? What caused the change or lack of change?
7. When water moves in or out of cells by osmosis, it moves in the direction that balances the concentrations of substances inside and outside the cell. Use this information to explain why the masses of the bags changed.
8. Identify the strengths and limitations of this investigation and suggest possible improvements.

3.8 Exercises: Understanding and inquiring

To answer questions online and to receive **immediate feedback** and **sample responses** for every question, go to your learnON title at www.jacplus.com.au. *Note:* Question numbers may vary slightly.

Remember

1. What is the name for the tubes that carry sugar solution around the plant?
2. Describe the difference between:
   (a) phloem and xylem vessels
   (b) sugar and water transport in plants
   (c) the arrangement of vascular bundles in dicots and monocots.
3. In what ways are the vascular bundles important to plants?
4. State two things that may happen to water in a plant.
5. On which part of the plant are stomata usually found? Can you suggest why?
6. What helps 'pull' water up a plant?
7. Describe how the guard cells assist the plant in controlling water loss.

8. Describe the difference between flaccid cells and turgid cells.
9. Copy and complete the table below.

Tissue	What it carries	Direction of movement	Name of cells that form tubes	Are cells that form tubes living?
Xylem				
Phloem				

Think and discuss

10. Carefully examine the reaction for photosynthesis as shown below.

$$\text{Carbon dioxide + water} \xrightarrow[\text{Chlorophyll}]{\text{Light energy}} \text{Glucose + oxygen (+ water)}$$

(a) Suggest why water and carbon dioxide are so important to plants.
(b) Suggest why guard cells are important to plants.
(c) Predict consequences for a plant if the guard cells close the stomata for long periods of time.

Investigate

11. Describe the patterns in which the vascular tissue is arranged in the stems of different plants. Obtain your information by:
    (a) examining stained cross-sections
    (b) finding and examining diagrams of the stems of different plants in cross-section.
12. Find out the relationship between 'wood' and xylem tissue.
13. How long do you think it would take for a plant to take up 50 mL of water? What conditions might speed it up? Put forward a hypothesis, and then design an experiment to test your hypothesis.
14. Design an experiment to test the time taken for different volumes of water to be taken up by a plant.
15. Some plants have special features that help them reduce water loss. Some leaves have a thick, waxy layer (cuticle). Others have a hairy surface or sunken stomata. Plants that are able to tolerate extremely dry environments are called xerophytes. Find out some ways in which plants in dry environments, such as deserts, reduce their water loss. Present your information on a poster or as a model.
16. Using sticky tape, remove a layer of cells from the underside of a leaf and place it on a glass microscope slide. Examine the leaf cells for stomata under a light microscope. Repeat the procedure with as many different types of plants as you can. Summarise your findings in a poster.
17. Design an experiment to measure the amount of water lost through the leaves of a plant.
18. Place a plastic bag over the leaves of plants growing in the school grounds. Seal the bag and record the amount of water collected over 24 hours. What conclusions can you draw from your results?
19. Use a mind map to suggest ways in which plants can trap as much light as possible.
20. Research and report on one of the following types of science.
    • Phycology (study of seaweeds)
    • Plant marine biology
    • Plant physiology
    • Plant pathology
    • Ecophysiology

Create

21. In a group, write and then act out a play or simulation of the way water moves through a plant.
22. Write a story about a group of water molecules that travels from the soil, through a plant and then into the atmosphere as water vapour.
23. Design experiments to determine the answer to one of the following questions.
    (a) Is carbon dioxide needed for photosynthesis?
    (b) Do plants need chlorophyll for photosynthesis?
    (c) If part of a leaf is covered, will the leaf still photosynthesise?
    (d) If a plant is covered with a plastic bag, will it still photosynthesise?
    (e) How could you determine the best light (colour or intensity) for photosynthesis?

3.9 Cell division

3.9.1 Cell division in eukaryotes

Ouch! Did you burn or cut yourself? What about those skin cells you left on the towel when you dried yourself and those hairs you left behind in your brush? Have you replaced these cells? Throughout the life of multicellular organisms, cell division takes place to enable growth, development, repair and replacement of cells. Cell division also plays an important role in reproduction.

Nucleus, chromosomes and DNA

All eukaryotic cells have a **nucleus**, which contains genetic information with instructions that are necessary to keep the cell (and organism) alive. This information is contained in structures called **chromosomes**, which are made up of a chemical called **deoxyribonucleic acid (DNA)**.

Mitosis

Mitosis is the name of a process involved in cell division in eukaryotic cells. Some organisms use this type of cell division to asexually reproduce. Multicellular organisms also use mitosis to produce cells for growth, development, repair and replacement.

The cells produced by mitosis are genetically identical to each other and to the original cell. They have the same number and types of chromosomes and DNA instructions. As they have identical genetic information, they are described as being **clones**.

Cytokinesis

Mitosis is a process that involves division of the nucleus. Once a cell has undergone this process, the cell membrane pinches inwards so that a new membrane forms, dividing the cell in two. This process of dividing the cytoplasm is called **cytokinesis**.

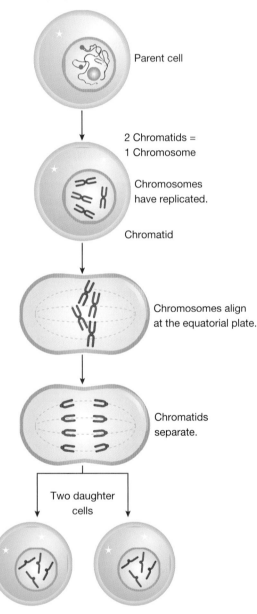

Mitosis is a type of cell division that produces identical cells.

Parent cell

2 Chromatids =
1 Chromosome

Chromosomes have replicated.

Chromatid

Chromosomes align at the equatorial plate.

Chromatids separate.

Two daughter cells

Each daughter cell contains the diploid ($2n = 4$) number of chromosomes.

DNA makes up chromosomes, which are located in the nucleus of the cell.

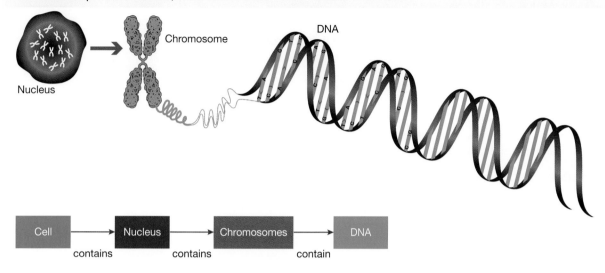

Eukaryotic unicellular organisms like (a) *Amoeba* and (b) *Euglena* divide by binary fission involving mitosis.

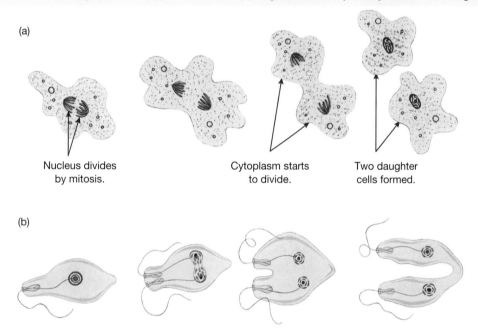

(a)

Nucleus divides by mitosis.

Cytoplasm starts to divide.

Two daughter cells formed.

(b)

3.9.2 Cell division in prokaryotes

Prokaryotes (such as bacteria) reproduce themselves by dividing into two using a process called binary fission. Although binary fission also occurs in some eukaryotes, this is much less complex in prokaryotes as they do not have a nucleus. The cells produced are clones; they are identical to each other and to the cell from which they originate.

3.9.3 Using bacteria to make human proteins

Knowledge of how bacteria reproduce can be used to get them to make human proteins. Scientists can insert genetic instructions from other organisms (including humans) into bacterial cells. When these bacterial cells divide, they produce cells that also contain the inserted foreign DNA and are able to make the protein that it codes for.

INVESTIGATION 3.10

Mitosis: Patterns of order

AIM: To observe slides showing mitosis under a light microscope

Materials:
light microscope
prepared onion root tip cells showing various stages of mitosis

Method and results

- Use a light microscope to observe the prepared slides.
1. Construct labelled diagrams to record your observations.

Discuss, explain and investigate

2. Carefully observe your diagrams, noting any similarities or differences between your observations for each slide.
3. Discuss your observations with at least two other students.
4. Construct a table that summarises similarities and differences between the slides showing various stages of mitosis.
5. Comment on any patterns that you have observed.
6. Use the mitosis diagram on page 75 to suggest a sequence or order for the mitosis slides observed.
7. Construct a PMI for this investigation that summarises the pluses (strengths), minuses (limitations) and suggested improvements (if you were to do it again).
8. Find out more about where and when mitosis occurs in plants. Create a poster to communicate your findings.
9. The cells that result from mitosis are identical to each other and the original cell. Suggest advantages and disadvantages of this feature.

This technology can be used to produce insulin, a protein used in the treatment of a type of diabetes. The rapid rate of bacterial reproduction results in many cells with the human DNA and the production of useful quantities of this important human protein.

Cell division and disease

Diseases can be divided according to whether they are infectious or non-infectious. **Infectious diseases** can be transferred from one organism to another. Tetanus and tuberculosis are examples of infectious diseases in which cells are damaged by a bacterial infection. **Non-infectious diseases** are not transferred between organisms. Cancer is an example of a non-infectious disease that can be considered as a form of uncontrolled cell division or a disease of mitosis.

Scientists use their knowledge of cell division of disease-causing organisms to control or kill them. **Antibiotics** can be used to kill bacteria inside your body. **Disinfectants** can be used to kill bacteria on surfaces of non-living objects. Disinfectants should not be used on your skin as they can damage your cells. **Antiseptics** can be used on your skin. Antiseptics that kill bacteria are referred to as **bactericidal**, and those that stop bacteria from growing or dividing (but do not kill them) are called **bacteriostatic**.

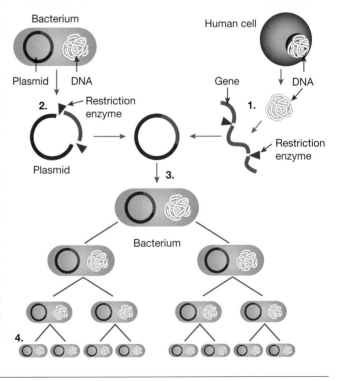

Bacteria can produce human insulin if the insulin gene is inserted into the bacterial cells.

INVESTIGATION 3.11

Where are those germs?

AIM: To observe a variety of micro-organisms from your local environment

Materials:

sterile cotton buds	*sticky tape*
nutrient agar plates in	*sterile Pasteur pipette*
Petri dishes (3 per group)	*marker pen*

CAUTION

Agar plates should not be opened after incubation.

Method

- Swipe a sterile cotton bud across a surface of your choice (such as canteen counter, computer keyboard, phone mouthpiece or bin lid).
- Swipe the cotton bud across the surface of the agar. Be careful not to push down too hard. The cotton bud should not leave a mark on the agar.
- Use sticky tape to seal the plate around the edge.
- Use a marker pen to write your group's name and where you collected the sample from.
- Use a different cotton bud to swipe a part of your body (such as the inside of your nose, your teeth, inside your ear or your scalp).
- Swipe the cotton bud on the surface of the second agar plate, then seal and label it as before.
- Use the sterile Pasteur pipette to collect about 1 mL of water from a location of your choice (such as a fish tank, puddle, local creek, school swimming pool or drain pipe).
- Pour the sample of water over the surface of the agar and swish it around. Seal and label the agar as before.
- Incubate the three plates upside down at 30 °C for 48 hours. Remove the plates from the incubator and observe the colonies of bacteria through the lid of the Petri dishes. (Do not open the Petri dishes.)

Results

1. Draw a diagram of each Petri dish showing the location and size of the colonies.

Discuss and explain

2. Colonies of bacteria tend to be smooth whereas colonies of fungus appear furry and are often larger. Do you have colonies of bacteria or fungi or both on your plates?
3. Look at the other groups' plates.
   (a) Which of the surfaces tested by your class had the most microbes? How can you tell?
   (b) Which body part tested had the most microbes?
   (c) Which of the water samples tested contained the most microbes?
4. Explain why it would be dangerous to unseal the agar plates and lift the lid to look at the colonies of microbes.
5. Find out from your teacher how the plates are disposed of safely at your school.
6. Design an experiment to test whether antibacterial surface spray really does kill bacteria.

3.9 Exercises: Understanding and inquiring

To answer questions online and to receive **immediate feedback** and **sample responses** for every question, go to your learnON title at www.jacplus.com.au. *Note:* Question numbers may vary slightly.

Remember

1. Provide three reasons for cell division.
2. Suggest the relationship and sequence of the following: nucleus, DNA, cell, chromosome.

3. State a feature that the daughter cells produced by mitosis share with the parent cell.
4. If a cell is described as being a clone, what does this mean?
5. Identify a type of unicellular organism that:
    (a) uses mitosis to reproduce
    (b) does not use mitosis to reproduce.
6. Identify the difference between mitosis and cytokinesis.
7. Suggest a way that scientists can apply their knowledge of cell reproduction to benefit humans.
8. Outline the differences between disinfectants, antiseptics and antibiotics.

Think and discuss
9. Charlotte wanted to find out if antibacterial soap really works. She prepared two agar plates. She swiped her fingers over the surface of plate A. She then washed her hands with antiseptic soap and swiped her fingers over the surface of plate B. She incubated both plates. Her results are shown below.

A   **B**

    (a) Write a conclusion for Charlotte's experiment.
    (b) Which plate was the control?
    (c) What were the independent and dependent variables in this experiment?
    (d) Which variables need to be controlled in this experiment so that it is a fair test?
10. Before mitosis begins, the DNA in the cell is replicated. Suggest why this replication step needs to occur.
11. Suggest the advantages and disadvantages of being a clone.

Investigate and create
12. Find some examples of disinfectants and antiseptics. Select one and research how it works, reporting your findings as a 'scientific journal article'.
13. Find out more details about mitosis and then use wool, plasticine or pipe cleaners to create your own model of the process of mitosis.
14. Find out where and when mitosis occurs in plants. Create a poster to communicate your findings.
15. *Entamoeba histolytica* is a unicellular organism that is a cause of diarrhoea among travellers to developing countries.
    (a) Find out more about the disease that it causes, its life cycle and what you can do to avoid being infected by it.
    (b) Prepare a brochure, poster or PowerPoint presentation that could be used to inform travellers.
16. Investigate cell division in *Amoeba*, *Euglena* or *Paramecium* and create an animation to show how they reproduce.
17. Find out how DNA is replicated in a cell and create an animation or PowerPoint presentation to communicate your findings to others.
18. Investigate the differences between the structure of chromosomes in prokaryotes and eukaryotes, creating a model of each type of chromosome.
19. Investigate the development of the microscope and its impact on our understanding of cell function and division. Present your findings as a short documentary or play.
20. Research examples of genetic engineering in which bacteria have foreign DNA inserted into them so that they produce human proteins. Communicate your findings as a newspaper article.
21. *Clostridium perfringens* is one of the fastest growing bacteria, having an optimum generation time of about 10 minutes.
    (a) If you started with one bacteria, plot on a graph how many bacteria there would be each hour over a 24-hour period.
    (b) Find out more about the structure and reproduction of this bacterium.
    (c) Find out why it is sometimes referred to as a 'flesh-eating' bacterium.
    (d) Write a story that includes features of *Clostridium perfringens* as a key part of the storyline.
22. (a) Find out why *Escherichia coli* (*E. coli*) counts at beaches are often stated in newspapers.
    (b) How is the concentration of *E. coli* measured?

(c) Find out more about the structure and reproduction of *E. coli*.

(d) Create a model of this organism.

23. (a) Find out the differences between disinfectants and antiseptics, providing examples of each.

(b) Select one of your examples and prepare an advertisement that could be used to market it.

24. Search the internet for animations, songs and games that involve mitosis or binary fission. Use these to guide you in the development of your own creative lesson on these processes.

3.10 Skin 'n' stuff

3.10.1 Skin deep

Your skin is made up of lots of cells that work together to keep you alive. A type of cell division called mitosis enables you to make skin cells for growth, repair and replacement. But what happens when something goes wrong?

Your skin is the largest organ of your body. As well as holding the insides of your body in, it also:

- protects your body from microbes that could cause disease
- is almost completely waterproof
- protects the inside of your body from chemicals and harmful radiation from the sun
- detects heat, cold, pain, pressure and movement
- helps control your body temperature
- forms vitamin D in sunlight
- releases water and other waste products.

The skin is divided into three layers.

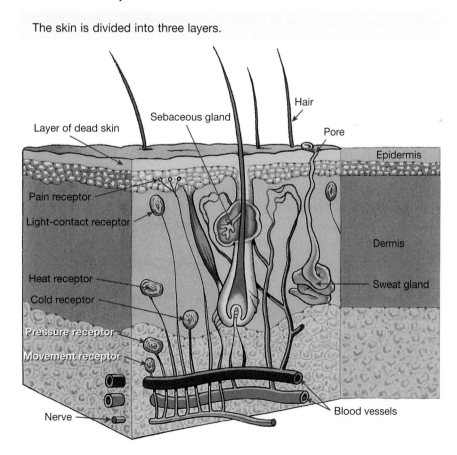

 RESOURCES — ONLINE ONLY

📼 **Watch this eLesson:** A cure? (eles-0070)

Stay in the shade whenever possible between 11 am and 3 pm.

Protect your face and neck with a hat.

Use zinc cream to provide total blockout for your nose and other sensitive areas.

Wear a long-sleeved shirt whenever possible when you are out in the sun.

Apply a broad-spectrum sunscreen labelled 30+ (or more) to uncovered skin.

Your skin varies in thickness between about 0.5 millimetres and 5 millimetres. The thickest part is on the soles of your feet. Skin consists of three layers.

The **epidermis** is the top layer. It contains several layers of cells. At the very top is a layer of dead skin cells, which flake off continually. At the bottom of the epidermis, new cells are always being produced. They push upwards on the older cells, moving them towards the surface. Below the epidermis is the **dermis**, which contains **receptors** for the sense of touch. It also contains **sweat glands** and many small blood vessels. Beneath the dermis is a thicker layer of fatty tissue, which acts as an insulator to help keep the body temperature constant. This fat has been stored by the body and can be used when needed to provide extra energy.

When you get hot, it is important that your body cools itself down so the blood remains at its constant temperature of about 37 °C. Your sweat glands produce a liquid that is released through the **pores** at the surface of your skin. When the water in your sweat **evaporates**, it takes some of the heat out of your body.

3.10.2 Are you ticklish?

Are you more ticklish on some parts of your skin than others?

Below the surface of your skin there are many receptors that are attached to nerves. The nerves send messages to the brain. There are different receptors for heat, cold, light contact, pain, pressure and movement. They are all receptors to the sense of touch.

The light-contact receptors are nearer to the surface and closer together in some parts of your skin than others. It is those parts that are most sensitive to tickling. Some parts of the skin are also more sensitive to pain, heat, cold, pressure and movement than others. Your sensitivity depends a lot on how close together the receptors are and how deep they are.

Dr Fiona Wood, pioneer of 'spray-on skin'

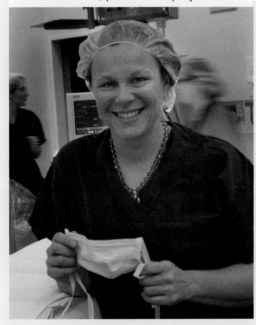

'Spray-on skin' in action

3.10.3 Sunsense

Skin cancer is the most common form of cancer in Australia. In fact, two out of three Australians are likely to get skin cancer at some time during their lives. The most serious forms of skin cancer are responsible for about 2000 deaths each year in Australia.

What is cancer?

As the body's cells die, new cells are made to replace them. In a healthy person, just the right number of new cells are formed using mitosis. Cancer can be considered as a disease of mitosis. Damage to the DNA in a cell can cause the normal regulatory processes in cell division to be ignored or overridden. This can result in uncontrolled cell division, a condition we call **cancer**.

This uncontrolled cell division can form a mass of cells called a **tumour**. The cells of a tumour are not specialised and cannot do the jobs of the cells that they are replacing. Some tumours still respond to the body's control mechanisms and do not spread to other parts of the body. These tumours are called **benign**. Others have uncontrolled cell growth and do spread, damaging vital organs. These are called **malignant** tumours or cancer. If cancer is detected early, the diseased cells can be removed or destroyed by chemotherapy or radiation. However, once cancer spreads, it is very difficult to control.

What causes skin cancer?

The main cause of skin cancer is exposure to the sun. The ultraviolet radiation reaching Earth from the sun is not visible. Ultraviolet radiation, which is also the cause of sunburn, is at its peak in the middle of the day when the sun is directly overhead. Ultraviolet radiation causes cancer in the cells of the epidermis, the top layer of the skin, because it damages the cells' genetic material.

3.10.4 Early detection

The key to curing skin cancer is early detection. Even melanomas can be cured in more than 95 per cent of patients if they are detected quickly. If you see a new lump or spot, or a changing freckle or mole, see a doctor promptly.

The three main types of skin cancer include the following.

Squamous cell carcinoma

- Less common and more dangerous than basal cell carcinoma
- Appears as a red, scaly sore
- Usually found on the hands, forearms, face and neck, but can spread to other parts of the body
- Mostly affects people over the age of 40 who have been exposed to the sun for many years
- Kills about 500 Australians each year

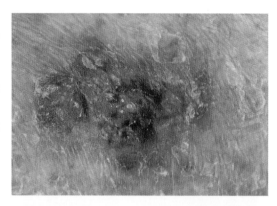

Basal cell carcinoma

- Most common form of skin cancer and also the least dangerous
- Appears as a red, flaky lump on the skin
- Rarely spreads to other parts of the body but needs to be treated before it grows large or forms a deep sore

Melanoma

- Least common but most dangerous form of skin cancer
- First sign is a change in size, shape or colour of a freckle or mole, or the appearance of a new spot on normal skin
- Can spread quickly to other parts of the body
- Most common in adults aged between 30 and 50 years, usually caused by long periods of exposure to the sun during childhood and adolescence
- Cause of the most deaths from skin cancer — about 1500 each year in Australia

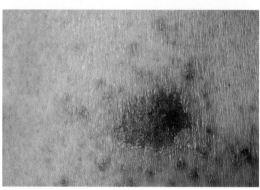

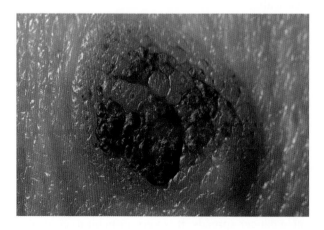

3.10.5 Some questions about fun in the sun

Q: *Is a suntan healthy?*

A: No. A suntan is evidence that you have been exposed to the sun for too long. A suntan will not protect you from skin cancer. Fake suntan lotions do not offer protection from skin cancer either.

Q: *Do I need to worry about sunburn or skin cancer when it's cloudy?*

A: Yes. Although clouds block out a lot of the sun's visible light, they do not block out enough ultraviolet radiation to protect your skin completely, especially during summer. The graphs below show that light cloud cover has little effect on the harmful ultraviolet radiation reaching the ground on a summer's day in Sydney. Heavy cloud, however, decreases the amount of ultraviolet radiation reaching the ground by over 90 per cent.

Q: *Do I need to use sunscreen when I wear a hat?*

A: Yes. The sun's radiation is reflected from the ground and from water. Snow and sand reflect a lot of radiation, even on cloudy days. In addition, many hats, including baseball caps, do not protect you from direct radiation. Wide-brimmed hats or 'legionnaire' hats provide the best protection because they shade the neck and ears.

Q: *What does SPF 30+ mean?*

A: SPF stands for 'sun protection factor'. It allows you to estimate how long you can stay in the sun before your skin starts to go red. This period can be estimated by multiplying the amount of time that it takes your skin to redden by the SPF factor. For example, if your unprotected skin starts to burn after 10 minutes in the hot sun, proper use of SPF 4 sunscreen would allow you to remain in the sun for $10 \times 4 = 40$ minutes before burning starts. After that 40 minutes, you would burn, even with more sunscreen applied. An SPF water-resistant 30+ sunscreen reapplied every 2 hours would allow you to remain in the sun for at least $10 \times 30 = 300$ minutes before burning starts. SPF 30+ sunscreen blocks out about 97 per cent of the sun's radiation.

Q: *What does 'broad spectrum' mean?*

A: The Cancer Council Victoria recommends a broad spectrum SPF 30+ (or higher) sunscreen. Broad spectrum sunscreens offer protection from the two different types of ultraviolet radiation that reach Earth's surface: UVA and UVB.

These graphs show how the ultraviolet radiation reaching the ground changes on a typical summer day in Melbourne.

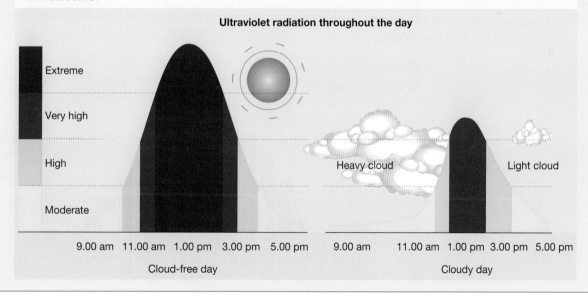

Ultraviolet radiation throughout the day

Cloud-free day — 9.00 am, 11.00 am, 1.00 pm, 3.00 pm, 5.00 pm

Cloudy day — 9.00 am, 11.00 am, 1.00 pm, 3.00 pm, 5.00 pm

Extreme / Very high / High / Moderate

Heavy cloud / Light cloud

HOW ABOUT THAT!

An Australian drug company called Peplin has developed a gel made from a common weed called *Euphorbia peplus*, which has been used successfully to treat some skin cancers. The plant contains a cancer-fighting chemical. Peplin has found a way of extracting this chemical from the plant and making it into a gel that can be applied to skin cancers. When the gel was tested, it cleared most skin lesions after just two days.

3.10 Exercises: Understanding and inquiring

To answer questions online and to receive **immediate feedback** and **sample responses** for every question, go to your learnON title at www.jacplus.com.au. *Note:* Question numbers may vary slightly.

Remember

1. Identify the most serious form of skin cancer and how it is caused.
2. Outline the difference between a benign tumour and a malignant tumour.
3. State which part of the sun's radiation is the major cause of skin cancer and sunburn.
4. Identify the most dangerous time of day to be out in the sun.
5. Describe what you should look for on the skin when checking for signs of skin cancer.
6. List five ways that you can help protect yourself from skin cancer.

Think and discuss

7. Melanomas occur mostly in adults over the age of 30. Why is it so important that young children and adolescents are aware of the dangers of the sun's radiation?
8. Daniel has very pale, sensitive skin that begins to burn after only eight minutes in the summer sun. He goes to the beach and takes a tube of SPF 6 sunscreen with him.
   (a) If he doesn't go swimming, how long would he be able to sit in the sun before getting burnt, assuming that he applies his sunscreen correctly?
   (b) If he used SPF 30+ sunscreen instead, would he be safe sitting in the sun all day? Give a reason for your answer.
9. Discuss the validity of the following statements.
   (a) Cancer can be considered as a disease of mitosis.
   (b) Melanomas are less dangerous than carcinomas.
10. Tissues are grouped together to form organs to help you function effectively. Describe how the following are important to your survival.
    (a) Muscles
    (b) Bones
    (c) Skin

Think and reason

11. Use the 2006 melanoma incidence in Victoria graphs below and over the page to complete the following.
    (a) Describe the pattern of melanoma incidence in Victoria in 2006.
    (b) Suggest reasons for this pattern.

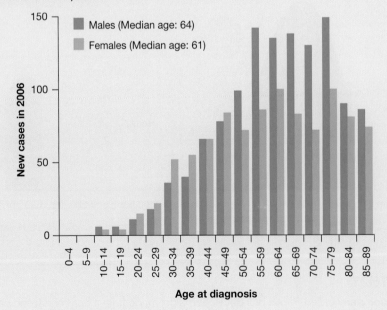

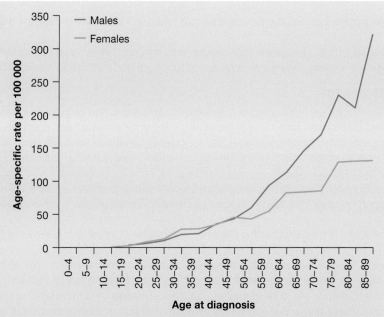

(c) Identify the age group that has the highest number of new cases of melanomas:
  (i) for males
  (ii) for females.
(d) Do the data suggest that there is a higher incidence of melanomas in males or females?
(e) Suggest possible reasons for these data.

12. Use the 1986 and 2006 melanoma incidence in Victoria graph below to complete the following.
  (a) State the age group and year with the highest incidence of melanoma cancer.
  (b) State the age group and year with the lowest incidence of melanoma cancer.
  (c) Discuss your responses to parts (a) and (b) and suggest possible reasons for your findings.
  (d) Describe differences in the patterns of melanoma incidence in Victoria between 1986 and 2006.
  (e) Discuss your observations and suggest reasons for this difference.

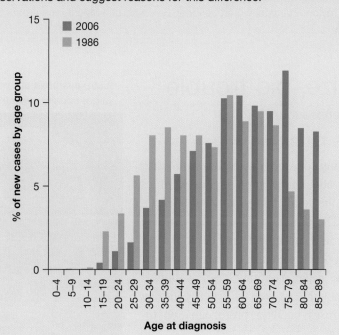

Create and design

13. Design and construct a model that shows details of the three layers of skin.
14. Construct a mind map that shows links between key scientific terms that are relevant to skin cancer. Use this map as a framework to create a song or poem about skin cancer that will increase awareness of types, causes, detection and prevention.

15. Design a colourful poster that would encourage people to protect themselves from the sun's harmful radiation.
16. Design and construct a multipurpose hat that shades the head and has at least one other purpose. Give your multipurpose hat a name. Prepare an advertising brochure and instruction manual for your hat.

Investigate and design

17. Design and carry out a survey (consisting of a series of questions) to find out whether people of different age groups protect themselves from the danger of skin cancer by wearing hats, shirts and sunscreens. By sharing your data with other members of your class, you may be able to form a sound conclusion.
18. A number of schools describe themselves as being 'sunsmart'. Find out what the criteria are and who decides whether they are 'sunsmart'.
19. Design your own 'Sunsmart' advertising campaign. Present your advertisement as a video clip with music.
20. Research and report on one of the following science careers: dermatologist, plastic surgeon, skin cancer research scientist.
21. Professor Graham Giles is a scientist at the Cancer Epidemiology Centre and is involved in Health 2020. Investigate and report on his research and that of other scientists at the centre.
22. Associate Professor Greg Woods is investigating relationships between immune cells and cancer cells. Investigate and report on his research into skin cancer at the Menzies Research Institute.
23. Professor Jonathan Cebon at the Ludwig Institute for Cancer Research is investigating responses to a melanoma vaccine. Investigate and report on his research and that of other scientists at the institute.
24. Research and report on the difference between an oncologist and a pathologist.
25. Obtain a selection of at least five different sunscreens. Identify criteria that could be used to determine their similarities and differences. Construct a matrix table with these criteria and record details for each sunscreen. Determine which sunscreen you would use. Which criteria did you use? Give reasons for your criteria. If you were to invent the 'ideal' sunscreen, what features would it have and why? Design an advertising brochure or advertisement for your 'ideal' sunscreen.

Investigate

26. With a partner, play a guessing game to see how well you can use your sense of touch alone to identify 10 unknown objects.

     You will need a blindfold and 10 objects of about the same size. Sandpaper, plastic, coins and pieces of carpet, polystyrene, nylon and wool would be ideal. See who can identify the most objects correctly.

3.11 Tiny size, big trouble

Science as a human endeavour

3.11.1 'Down the hatch'

Just because you are tiny, doesn't mean that you can't cause trouble — think about how annoying mosquitoes can be! Sometimes it's the small things in life that make the biggest difference. The presence, absence or extreme levels of particular microbes can reflect the health of their habitat, whether it be an organism or an entire ecosystem.

What's in that gulp of water? You may have swallowed a little more than you thought! As well as water, you may have also 'invited in' viruses, bacteria, protozoans, phytoplankton and zooplankton.

Giardia lamblia are pear-shaped and quite large, usually more than 6 μm in size.

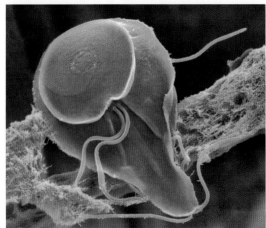

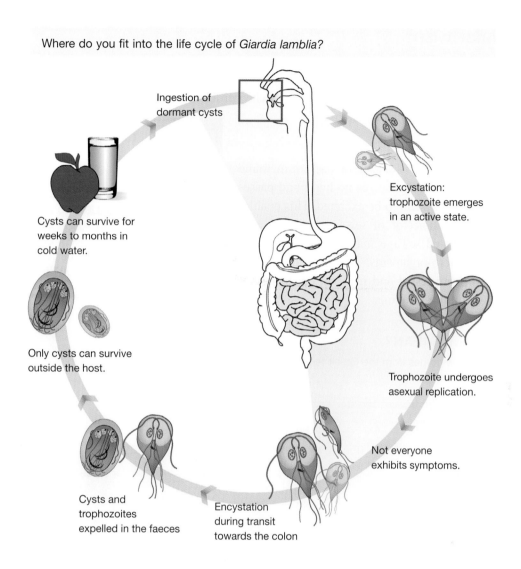

Where do you fit into the life cycle of *Giardia lamblia*?

Ingestion of dormant cysts

Cysts can survive for weeks to months in cold water.

Only cysts can survive outside the host.

Cysts and trophozoites expelled in the faeces

Encystation during transit towards the colon

Not everyone exhibits symptoms.

Trophozoite undergoes asexual replication.

Excystation: trophozoite emerges in an active state.

A protoctistan (protozoan) by the name of *Giardia lamblia* has become quite well known for 'hitching a ride' in drinking water. These parasites are the cause of one of the most common parasitic gastrointestinal infections in humans worldwide. *Giardia lamblia* made newspaper headlines in 1998, along with another protozoan, *Cryptosporidium parvum*, when high levels of both were reported to be contaminating Sydney's drinking water supply.

Giardia lamblia use their flagella to move around, and they have a complex life cycle. The parasite can survive for a long period outside the body in an inactive form called a cyst. Once swallowed, the cyst is activated by your stomach acid and develops into the disease-causing stage. Using a huge sucker, they then attach themselves to the lining of your intestine, sucking your blood as their food source. After about ten days of infection you could have a million of them living off your blood supply and causing symptoms associated with gastrointestinal complaints. Some of their reproductive cysts pass through your digestive system and are excreted, so that another host can become infected.

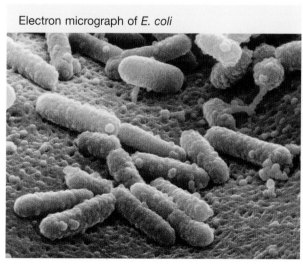

Electron micrograph of *E. coli*

3.11.2 Water wise

Living things need water to survive. Some living things also need to live in water to survive. In these ecosystems, there are links between the inhabitants to keep them balanced and healthy. Sometimes, however, these links can be broken or disrupted. This is when problems can occur that may result not only in an unbalanced ecosystem, but also in death.

Take a swim with me

Escherichia coli (*E. coli*) is a type of bacterium found in our intestine. It usually causes us no harm and passes through our digestive system to be excreted. This enables it to be used as an indicator of sewage contamination in water. Contaminated sewage may contain dangerous or even deadly micro-organisms. It is for this reason that *E. coli* levels are tested and reported on at various beaches and swimming locations.

City of Botany Bay

WARNING

THIS AREA IS SUBJECT TO LEVELS OF CONTAMINATION. THE PUBLIC ARE ADVISED NOT TO SWIM IN THIS AREA OR EAT ANY FORM OF MARINE LIFE DUE TO POSSIBLE DANGER TO HEALTH.

BY ORDER
GENERAL MANAGER

WHAT DOES IT MEAN?

The word *pathogen* comes from the Greek terms *pathos*, meaning 'disease', and *gen-*, meaning 'birth'.

HOW ABOUT THAT!

Professor Gustaaf Hallegraeff is a scientist at the University of Tasmania. In his research he uses light, scanning and transmission electron microscopes to help him classify tiny marine organisms called phytoplankton. On the basis of his findings, he has produced publications to communicate information about harmful Australian microalgae.

Professor Gustaaf Hallegraeff

Bloom'n algae

Algal blooms occur naturally and provide food for many aquatic organisms. Sometimes, however, they can cause harm. Algal blooms can cause large fluctuations in the levels of oxygen and pH (acidity) of the water and block sunlight penetrating through it. The species that cause these blooms may also be toxic to some of the aquatic organisms or even toxic or a skin irritant to humans. The presence of algal blooms can have economic as well as biological consequences.

Microcystis is an example of blue-green algae.

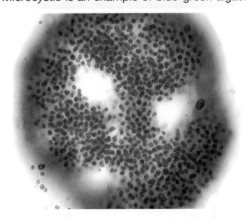

Algal bloom at Warragamba Dam, Sydney, NSW

Fluorescence microscopy can be used to identify the species that a cell belongs to, and can therefore help determine whether the cell may be of a dangerous type. In this image, the position of the chloroplasts shows as red and the nucleus as blue.

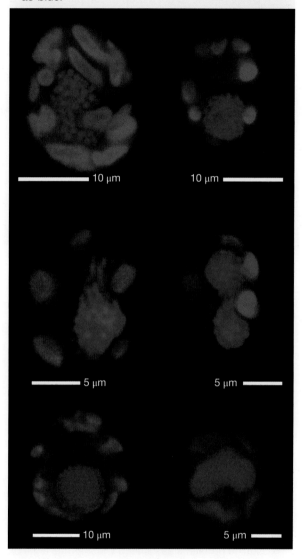

HOW ABOUT THAT!

Snap frozen

Australian Antarctic Division scientists are using a new field emission scanning electron microscope that magnifies specimens by up to 650 000 times. This allows them to observe features two million times smaller than the head of a pin. Snap-freezing technology is used, with the specimen being snap-frozen in super-cooled liquid nitrogen in temperatures of −210 °C! The snap-freeze process allows scientists to study more delicate specimens than were previously able to be observed.

Red tide of death

Scientists at the University of Tasmania have discovered that two new types of algae are killing fish in the Southern Ocean. These algae have been found in abundance and, when in full bloom, produce a distinctive red-coloured tide. The university's scientists are working with the Australian Antarctic Division to try to establish the number of fish that are suffocating from the algae. This will enable sustainable fishing levels to be maintained.

INVESTIGATION 3.12

Teeming with tiny …

AIM: To observe micro-organisms found in water on a microscope slide

Materials:
light microscope
microscope slides (well slides work best for this observation)
coverslips
pipette

toothpick
water from a variety of sources, e.g. sea water, pond water,
 stagnant water, fish tank water

Method and results

- Use a pipette to put a drop of sample water on a clean microscope slide.
- Gently place a coverslip over the drop of water by putting one edge down first. Use a toothpick.
- Observe the slide under a microscope.
1. Draw detailed sketches of what you see. Remember to include a title, the magnifications used and as many comments as you can.
   - Remove the coverslip, rinse and dry the slide, and then prepare a new slide specimen and repeat the previous steps.

Discuss and explain

2. Construct a matrix to show the similarities and differences between the specimens.
3. Suggest reasons for these differences.
4. Research common micro-organisms found in water to identify your specimens.
5. Which kingdoms do you think each specimen may belong to? Provide reasons for your classification.
6. Identify two structures observed in the investigation and find out more about their function (that is, what their 'job' is).
7. Discuss an interesting observation with your partner and comment on why it interests you.
8. Formulate three research questions that could be used for further investigation.

3.11 Exercises: Understanding and inquiring

To answer questions online and to receive **immediate feedback** and **sample responses** for every question, go to your learnON title at www.jacplus.com.au. *Note:* Question numbers may vary slightly.

Remember

1. (a) Describe the shape and size of *Giardia lamblia*.
   (b) Describe how *Giardia lamblia* obtains its food.
   (c) In which form can *Giardia lamblia* survive for weeks outside its host?
2. Identify the genus to which each of the following organisms belongs.
   (a) *Escherichia coli*
   (b) *Giardia lamblia*
   (c) *Microcystis* spp.
3. Explain why levels of *E. coli* are measured in water.
4. Are algal blooms always harmful? Explain.
5. Outline problems associated with harmful algal blooms.
6. Identify the type of organism that is responsible for causing red tides and killing fish in the Southern Ocean.
7. Explain why scientists are trying to determine the actual numbers of fish being killed by the effects of red tides.
8. State a function of fluorescence microscopy.

Investigate, think and discuss

9. Use the graph on the next page to answer the following questions.
   (a) Describe the pattern of indicator bacteria in each location between 2001 and 2007.
   (b) Identify which location has the highest bacterial counts. Suggest a reason for this observation.
   (c) Identify which location has the lowest bacterial counts. Suggest a reason for this observation.
   (d) Find examples of *E. coli* counts in your local community waters.

(e) Graph your findings and discuss patterns in your class data.

(f) Suggest reasons for the patterns that you have observed.

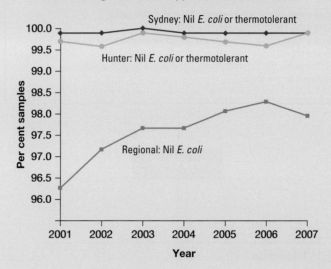

Indicator bacteria counts in drinking water, Sydney, Hunter and regional water supplies, NSW 2001 to 2007

10. Pathogens are disease-causing organisms. Research and report on microbial pathogens in Australian waters.

11. Research and report on one of the following.
    - *Escherichia coli*
    - *Microcystis* spp.
    - *Giardia lamblia*
    - *Cryptosporidium parvum*

12. Research and report on an example of Australian research on one of the following topics.
    - Algal blooms
    - Microalgal pigments
    - Harmful Australian microalgae
    - Micropalaeontological studies
    - Coastal eutrophication
    - Taxonomy using light and electron microscopy

13. Find out how *E. coli* levels are measured and examples of other micro-organisms they may indicate the presence of.

14. (a) Research and report on the diagnosis, symptoms and treatment of:
    (i) cryptosporidiosis
    (ii) giardiasis.
    (b) Construct a Venn diagram to show how the conditions are similar and how they are different.

15. (a) Identify and research an issue related to micro-organisms in Australian waters.
    (b) Construct a PMI chart to summarise your findings.
    (c) Discuss your PMI chart with others in your team and add their comments.
    (d) Organise a class or team debate on the issue.

16. Do microbes reflect marine health? Investigate this question by researching online and justify your response.

3.12 Target maps and single bubble maps

3.12.1 Target maps

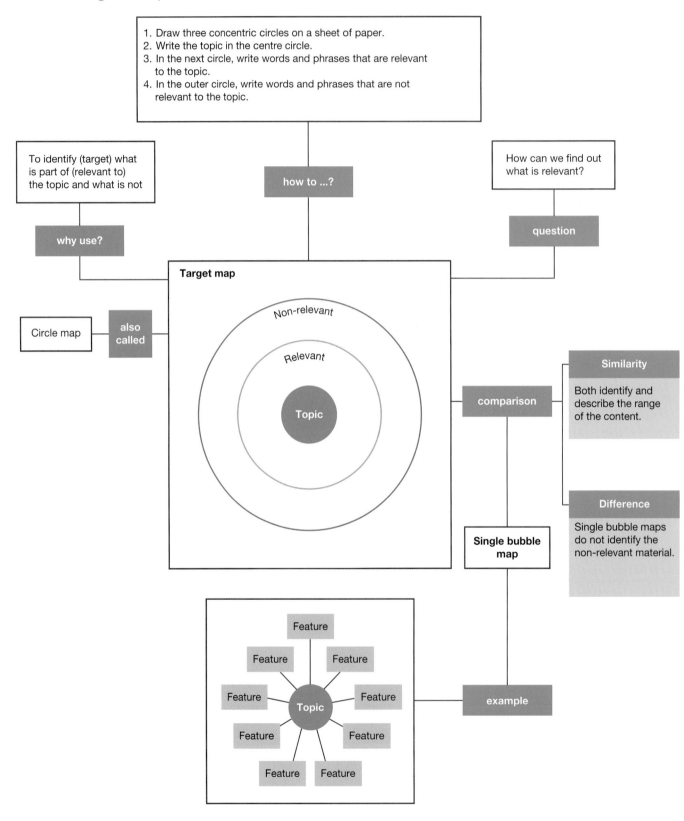

1. Draw three concentric circles on a sheet of paper.
2. Write the topic in the centre circle.
3. In the next circle, write words and phrases that are relevant to the topic.
4. In the outer circle, write words and phrases that are not relevant to the topic.

To identify (target) what is part of (relevant to) the topic and what is not

how to ...?

How can we find out what is relevant?

why use?

question

Circle map

also called

Target map

Non-relevant

Relevant

Topic

comparison

Similarity

Both identify and describe the range of the content.

Single bubble map

Difference

Single bubble maps do not identify the non-relevant material.

Feature

Feature

Feature

Feature

Feature

Topic

Feature

Feature

Feature

Feature

Feature

example

3.13 Review

3.13.1 Study checklist

Cells

- name and state the function of the parts of a light microscope
- describe how to prepare a specimen for observation under a light microscope
- observe and sketch labelled diagrams of cells and other specimens as viewed under a light microscope
- suggest why stains are used in the preparation of microscope slides
- explain the significance of the invention of the microscope to biology
- outline the contributions of three scientists to our understanding of cells
- state examples of different types of cells and relate their structure to their function
- describe the differences between unicellular and multicellular organisms
- name a type of cell division involved in growth and repair
- explain why not all cells have the same structure
- use differences in cell structure to classify organisms into groups
- explain how cell structure can provide us with evolutionary information

Ecosystems

- give examples of how unicellular organisms can have big impacts on ecosystems

Science as a human endeavour

- outline developments in the understanding of cells and how this knowledge has affected research areas such as health and medicine
- describe the development of the microscope
- outline the effect microscopes have had on our understanding of cell functions and cell division
- describe how people use understanding and skills from across the disciplines of science in their occupations
- provide an example of the role of knowledge of cells and cell divisions in the area of disease treatment and control

---

Individual pathways

ACTIVITY 2.1	ACTIVITY 2.2	ACTIVITY 3.3
Investigating cells	Analysing cells	Researching cells
doc-6048	doc-6049	doc-6050

learnon ONLINE ONLY

3.13 Review 1: Looking back

To answer questions online and to receive **immediate feedback** and **sample responses** for every question, go to your learnON title at www.jacplus.com.au. *Note:* Question numbers may vary slightly.

1. (a) Brainstorm as many 'cell'-related words as you can, writing them on a piece of paper.
   (b) Pair up with another class member and add any of their words that you missed. Ask your partner what these words mean if you are unsure.
   (c) On a new piece of paper, work with your partner to group or link words to make a concept, cluster or mind map.
   (d) Compare your map with that of another pair in the class, adding as many more bits and pieces as you can.

▶

2. Which of the following types of microscopes were used to take the photos shown?
   • Scanning electron microscope
   • Light microscope
   Give reasons for your answers.

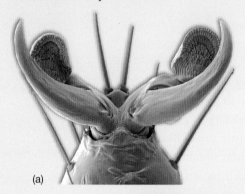
(a)

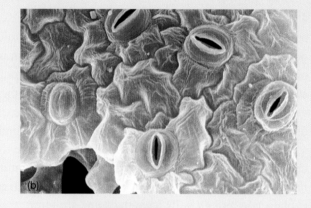
(b)

3. Make a sketch of these human cheek cells

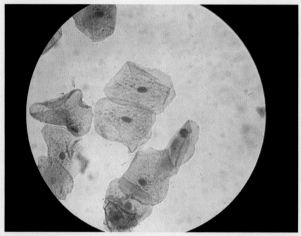

4. (a) Why do you think that cells have been described as 'living factories'?
   (b) Think of a typical plant or animal cell. Make a list of all of the different parts and organelles. If the cell was a living factory, what might be the job of each listed part?
   (c) Write a play to act out what happens in cells and perform it with others in your class. What sorts of things were easy to show? What sorts of things were hard to show? If you were to rewrite the play, what might you change and why?
   (d) Convert the classroom into a giant cell! Take photos and then add information to them on a poster.
5. Construct a Venn diagram to show the similarities and differences between light microscopes and electron microscopes.
6. Explain the significance of the invention of the microscope in terms of how we see the world.
7. Suggest why the invention of microscopes led to the development of new scientific language and classifications.
8. Carefully observe the diagram of the skin on the right. Match the letters on the diagram with the following labels: epidermis, dermis, fatty layer, light receptor, sweat gland, blood vessels, sebaceous gland, pain receptor, nerve, pore.

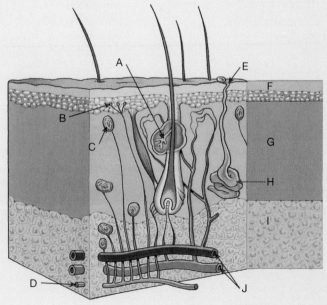

9. Copy and complete the table below:

Cell feature	Plant cells	Animal cells	Fungal cells
Cell wall	✓	×	
Cytoplasm			
Cell membrane			
Chloroplast			
Nucleus			
Large vacuole			

10. Draw and label a typical plant cell and a typical animal cell.

11. What's green and eats porridge? Identify the parts of the microscope on the right and use the code below to find out the answer to this riddle.
Code:
O = revolving nose piece; U = objective lenses; S = coarse focus knob; K = fine focus knob; D = microscope slide; L = stage slide clip; C = base; O = mirror; L = iris adjustment; I = stage; M = eyepiece lens

12. Unscramble the words using the clues provided.
(a) Control centre of the cell   SEUNCLU
(b) Surrounds the cell   ERAMMBNE
(c) Contains cell sap   OCVAUEL
(d) Part of the cell between the cell membrane and the nucleus   CATOPLMYS
(e) Building blocks of all living things   LELSC
(f) Living things   ASMOGNIRS

13. (a) Match the following cell names to the diagrams below.
    Euglena
    Paramecium
    onion epidermal cells
    nerve cell
    sperm cell
    guard cells
    root hair cell
    bacterium
(b) To which kingdom does each of these cells belong?

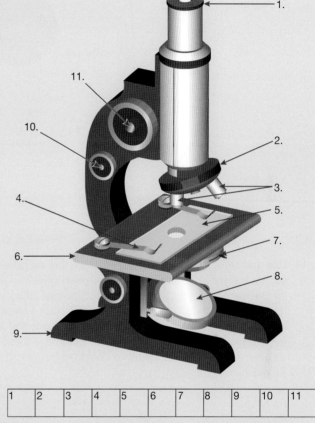

1	2	3	4	5	6	7	8	9	10	11

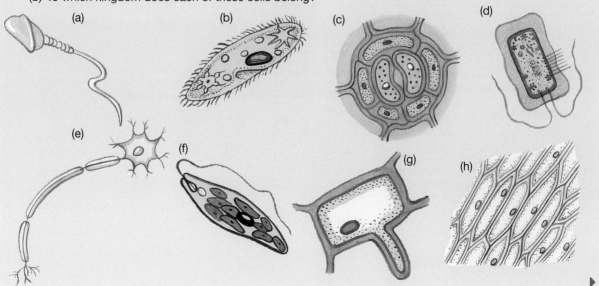

(a)
(b)
(c)
(d)
(e)
(f)
(g)
(h)

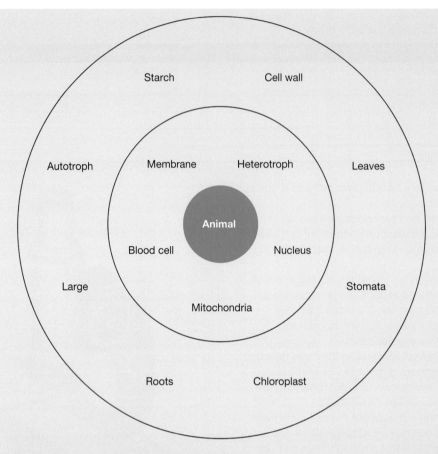

14. Use the target map above to answer the following questions.
    (a) List the content that is relevant to animal cells.
    (b) List the content that is not relevant to animal cells.
    (c) Using the words in the target map, construct a target map that is relevant to plant cells.
    (d) Identify which words are relevant to both plant and animal cells.
    (e) Suggest why plant and animal cells both have these features in common.
15. Use the terms in the box below to construct target maps that are relevant to:
    (a) plants
    (b) animals
    (c) fungi
    (d) protoctistans
    (e) prokaryotes (monera).

multicellular	chloroplast	*Euglena*
prokaryote	bacteria	mushroom
eukaryote	fern	yeast
nucleus	alga	lizard
cell wall	*Paramecium*	sponge
large vacuole	unicellular	moss
xylem	cell membrane	blood cells
possum	stomata	phloem

16. Construct a single bubble map to identify:
    (a) types of plant cells
    (b) types of animal cells
    (c) scientists who have contributed to our knowledge of cells
    (d) examples of body systems
    (e) functions of skin
    (f) issues related to stem cells.

17. Use the information in the single bubble map below to construct a target map of the parts of a light microscope.
18. (a) Use the internet to find images of at least five different types of zooplankton.
    (b) Carefully observe your zooplankton images, recording key features of each in single bubble maps.
    (c) Construct target maps for each of your zooplankton to show how they are different from the other zooplankton.

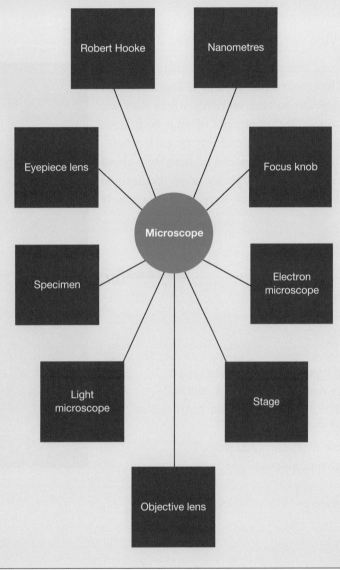

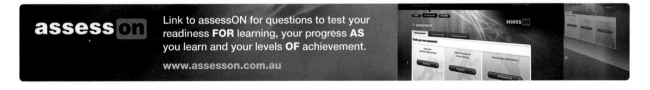

TOPIC 4
Systems — living connections

4.1 Overview

Inside your body is a very complex place and many complex processes need energy. To convert energy into a form you can use requires transport highways to take nutrients where they are needed and carry wastes away. Different parts of your body have different jobs — they may work together and rely on each other.

The curiosity, imagination and persistence of humans throughout history has given us our current knowledge about our bodies and how they function. What questions do you have about your body and how it works?

4.1.1 Think about the human body

assesson

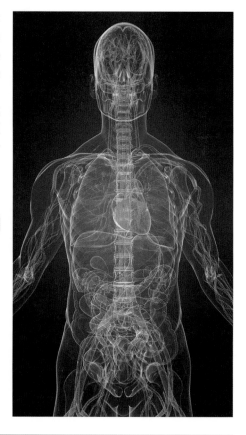

- What is the function of blood?
- Why do you need a skeleton?
- What are bones made of?
- What makes your bones move?
- Which human blood type is most common?
- How many chambers does a human heart contain?
- What causes the 'lub dub' sound that your heart makes?
- What is special about cardiac muscle?
- Why aren't all of your teeth the same shape?
- In which organ is urine produced?

Numerous **videos** and **interactivities** are embedded just where you need them, at the point of learning, in your learnON title at www.jacplus.com.au. They will help you to learn the content and concepts covered in this topic.

4.1.2 Your quest

Getting below the surface

Have a look at the other students in your classroom. How different from each other are you? Which features do you all have in common? Perhaps there are differences on the outside, but inside you are made up of all of the same bits and pieces organised in the same way.

Think and create

Some of the things that you have in common with your classmates are your body systems.

1. Use a mind map to summarise all that you know about human body systems.
2. Compare your mind map with those of at least three team members.
3. Create a new team mind map that combines all your ideas and compare that with the mind map of another team. Add any comments that you think help you learn more about human body systems.

Think and investigate

4. In your team, make a list of ten questions about human body systems. Select four questions and place these on the class noticeboard with those of other teams to make a 'class question gallery'. Arrange these questions into groups or themes.
5. Browse the class question gallery and select one question that interests you most. Research your selected question and report back to the class on your findings in an interesting and creative way.

4.2 Driven by curiosity?

Science as a human endeavour

4.2.1 Intensely curious …

> … in the medical faculty he learned to dissect the cadavers of criminals under inhuman, disgusting conditions … because he wanted [to examine and] to draw the different deflections and reflections of limbs and their dependence upon the nerves and the joints. This is why he paid attention to the forms of even very small organs, capillaries and hidden parts of the skeleton.
>
> Paolo (the first biographer of Leonardo da Vinci), 1520

Leonardo da Vinci spent hours amid rotting corpses to draw amazingly detailed observations of body structures.

Leonardo da Vinci was one of the best scientific minds of his time. He was intensely curious and painstaking in his observations. He used close observation, repeated testing and precise illustrations with explanatory notes. Using pen, chalk and brush, his scientific illustrations offered visual answers to mysteries that had escaped others for centuries. His volumes of amazing notes of scientific and technical observations in his handwritten scripts led to the birth of a new systematic and descriptive method of scientific study.

Leonardo da Vinci questioned everything. He may have been the most relentlessly curious man in history. He asked questions such as: Why do birds fly? Why can seashells be found in mountains? What is the origin of the wind and clouds? Why do people die? Where is the human soul found?

4.2.2 Dissecting, details and drawing

Leonardo's anatomical studies of human muscles and bones began around 1490. His exploration of embryology and cardiology came later, with his astonishingly detailed image of a fetus within the womb (around 1505) providing details for obstetric surgery hundreds of years later. His observations were not just of bodies — later generations have been in awe of his sketches of inventions that were centuries ahead of their time.

4.2.3 Challenging 'knowledge'

Knowledge of the human body was very different in Leonardo's day from what we

Is this a self-portrait of a young Leonardo da Vinci (1452–1519)? 'Hidden' under handwriting on a page of his *Codex on the flight of birds* for about 500 years, a combination of scientific techniques were used to 'unveil' it.

accept today. The heart was thought to be made up of two chambers and its function to warm the blood, which was thought to be made in the liver. It was also thought that sperm were produced in the marrow of the spinal column and that the human soul may be located in the spine. Leonardo had questions he wanted to answer. He wanted to find out more. His investigations challenged the accepted knowledge of his day.

4.2.4 Visions and models

Leonardo also emphasised the significance of visual observations and model making — he believed that reality needed to be reconstructed before it could be represented. His models of hands or legs were used to reveal the structural relationships between different layers of arteries, muscles and bones. Leonardo also made a glass model of the heart and used water with different coloured dyes to trace its flow through the heart. His investigations linked anatomy (structure) and physiology (function).

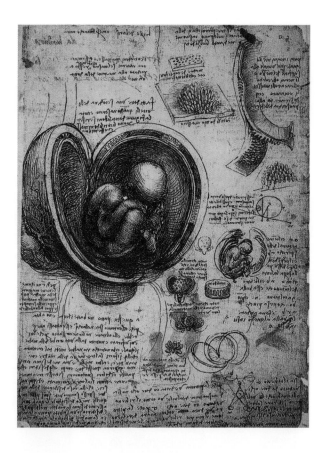

Leonardo drew this diagram around 1510 — can you see his secretive, reversed form of handwriting?

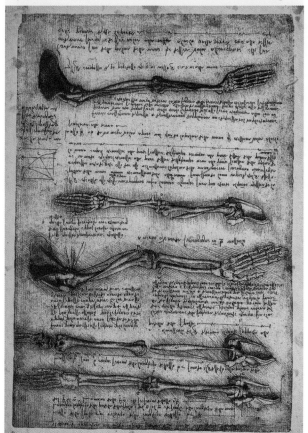

Leonardo da Vinci was a master of detail with his sketches of body parts.

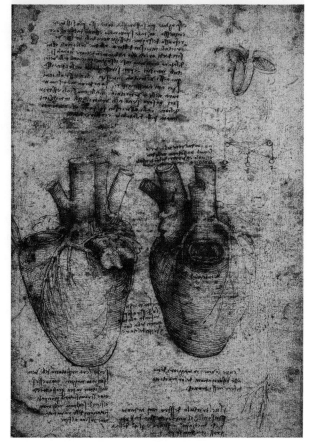

Analogies are sometimes used to help people to connect new learning to previous knowledge. Leonardo used analogies to compare arteries in human bodies to 'underground rivers in the earth' and described the bursting of blood from a vein like 'water rushing out a burst vein of the earth'.

Leonardo's dissections led to changes in the knowledge and understanding of the structure and function of the heart, including that:

- the heart was a muscle
- the heart did not warm the blood
- the heart had four chambers
- left ventricle contractions were connected to the pulse in the wrist.

To locate cavities around the brain and cranium, Leonardo used innovative techniques, such as injecting molten wax into them. Although Leonardo did not find the location of the human soul, his studies led him to the discovery that the brain and spine were connected.

Leonardo's curiosity, determination, creativity and persistence did more than make an amazing contribution to our current scientific knowledge of our bodies. These features also helped mould the way in which scientific frameworks were developed to structure our investigations to explore our questions.

Leonardo sketched this skull in 1489.

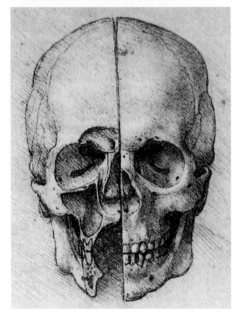

Aboriginal 'X-ray style' rock painting figure from Kakadu National Park, Northern Territory, Australia.

4.2.5 Curiosity throughout time and space

Curiosity is one of the features of humans that has contributed to our survival. Some of this curiosity has been about the structure and function of our own bodies. Evidence of this curiosity is woven throughout history and is often found in art. While Leonardo da Vinci provides one example of curiosity driving a search to find out more, he is not the only example. Nor is human curiosity limited to the place or time in which you live.

Knowledge of the internal biology and physiological process in art appears in rock paintings in caves in Australia that are thousands of years old. Examples of Aboriginal X-ray art provide evidence that this type of knowledge dates back more than 6000 years.

The culture and scientific knowledge of the times often determines the types of treatment given for various diseases of the human body. In medieval times, astrology played a key role in medicine and medical prognosis. It was believed that the 'movement of the heavens' could influence human physiology, with each part of the body being associated with a different astrological sign. An image of the 'Zodiac Man' in the medical texts of the time was used to assist practitioners in their medical treatments.

Chinese traditional medicine is an ancient medical system that has been practised for over 5000 years and applies understanding of the laws and patterns of nature to the human body. It views health as the changing flow throughout the body of vital energy (*qi*) that, if hindered, can lead to illness. Acupressure is an application of this theory that aims to release blocked energy by stimulating specific points along the body's energy channels.

Scientists are curious

Scientists are also often driven by the thirst to find answers to their questions. With increased technology and knowledge, the answers to these questions often result in even more questions.

Compared with the situation in Leonardo's day, there are now an amazing number of different types of careers that involve investigations, explorations and applications of science to the human body. Australian scientists are involved in medical research and intervention. They are also involved in the invention and development of medical equipment that assists our understanding of our body systems.

4.2.6 Australian scientists: creative inventors and explorers

Australian scientists have made significant contributions to medical discoveries and inventions. Howard Florey and his team discovered how penicillin could be extracted, purified and produced to be used as an antibiotic to help fight bacterial infections. Barry Marshall and Robin Warren showed that a certain type of bacteria caused stomach ulcers that could be treated with antibiotics. Professor Graeme Clark and his team were involved in the invention of an effective 'bionic ear'. Dr Fiona Wood pioneered a new treatment for burns in her development of spray-on skin that used the patient's own skin cells. Professor Ian Frazer developed the world's first vaccine against cervical cancer.

The 'Zodiac Man' chart was based on astrology and provided advice on when, for example, to 'bloodlet' (a medical treatment involving bleeding the patient), regulated by the position of the moon.

An example of an acupressure reflexology chart

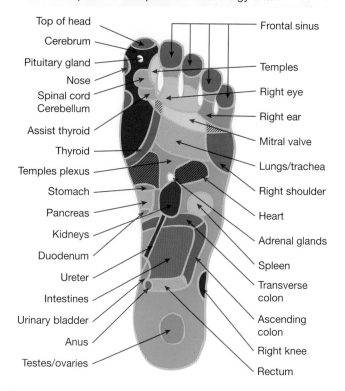

4.2.7 The bionic ear

The cochlear implant, also known as the bionic ear, has allowed some people with inner-ear problems to hear sound for the first time. When deafness results from serious inner-ear damage, no sounds are heard at all. Normal hearing aids, which make sound louder, do not help in these cases because the cochlea cannot detect the vibrations. However, the cochlear implant can often help by changing sound energy from outside the ear into electrical signals that can be sent to the brain.

An enlarged X-ray of the cochlea showing the experimental electrode array inside

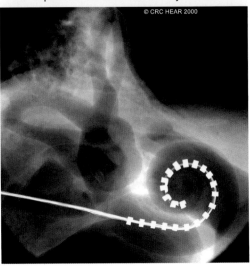

© CRC HEAR 2000

How a cochlear implant works

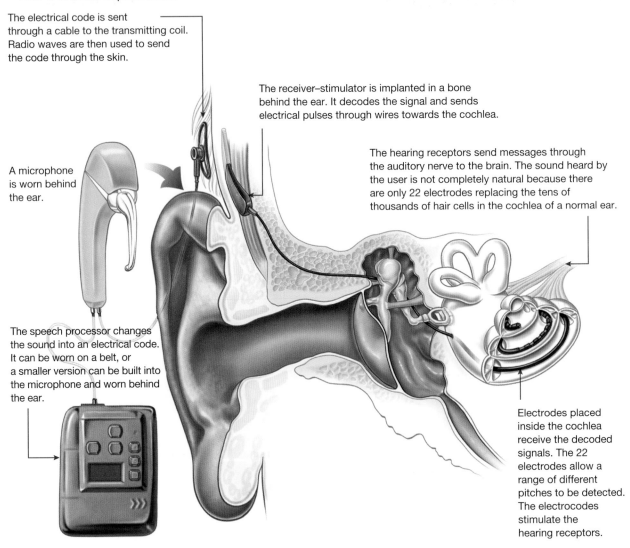

The electrical code is sent through a cable to the transmitting coil. Radio waves are then used to send the code through the skin.

A microphone is worn behind the ear.

The speech processor changes the sound into an electrical code. It can be worn on a belt, or a smaller version can be built into the microphone and worn behind the ear.

The receiver–stimulator is implanted in a bone behind the ear. It decodes the signal and sends electrical pulses through wires towards the cochlea.

The hearing receptors send messages through the auditory nerve to the brain. The sound heard by the user is not completely natural because there are only 22 electrodes replacing the tens of thousands of hair cells in the cochlea of a normal ear.

Electrodes placed inside the cochlea receive the decoded signals. The 22 electrodes allow a range of different pitches to be detected. The electrocodes stimulate the hearing receptors.

4.2.8 David Unaipon

David Unaipon (1872–1967) has been described as 'Australia's Leonardo'. He was born in South Australia, the fourth of nine children of James Ngunaitponi and his wife Nymbulda. Both of David's parents were Yaraldi speakers from the lower Murray River region.

Interested in Aboriginal mythology, philosophy and science, David was a preacher, author and inventor. He compiled his own versions of Aboriginal legends such as *Hungarrda* (1927), *Kinie Ger — The Native Cat* (1928) and *Native Legends* (1929). David's published poetry and legends pre-dated the work of other Aboriginal writers by over thirty years.

Obsessed with discovering the secret of perpetual motion, David made ten patent applications between 1909 and 1944 for inventions including a modified handpiece for shearing, a centrifugal motor, a multiradial wheel and a mechanical propulsion device.

David Unaipon

4.2 Exercises: Understanding and inquiring

To answer questions online and to receive **immediate feedback** and **sample responses** for every question, go to your learnON title at www.jacplus.com.au. *Note:* Question numbers may vary slightly.

Investigate

1. Research and construct a model of an invention from one of Leonardo da Vinci's sketches.
2. Research examples of Leonardo da Vinci's inventions and make your own variation of one of them, presenting it as a series of annotated sketches.
3. Research an invention sketch by Leonardo da Vinci that is related to something that we use today. Use a tree diagram to show how it may be linked.
4. Find out what the *da Vinci*® Mitral Valve Repair is and suggest why it is named after Leonardo da Vinci.
5. Find out about the history behind *Treatise on painting* and how it relates to science.
6. Find out more about *Codex on the flight of birds* and summarise your findings in a newspaper article.
7. Research three Australian scientists involved in medical research and intervention, and present your findings as curricula vitae.
8. Find an example of how Australian scientists have been involved in the development of medical equipment. Produce a brochure to advertise this equipment to prospective buyers.
9. Suggest how scientific understanding of human body systems can determine how we respond to public health issues such as the 2009 swine flu pandemic.
10. Research traditional Chinese medicine and find out about the knowledge of the structure and function of human body systems.
11. Research two of the following applications of traditional Chinese medicine:
    * qigong
    * herbal therapy
    * acupressure
    * healing foods
    * Chinese psychology.
12. Research Aboriginal X-ray art to investigate examples of Aboriginal knowledge of the internal biology and physiological processes of animals.
13. The Warlpiri are one of the largest Aboriginal groups in the Northern Territory. Research and report on their traditional health system and the involvement of ngangkayikirili (or ngangkari or ngangkayi) and Yawulyu ceremonies of the Warlpiri.
14. Research and report on the processes involved in the preparation of mummies in ancient Egypt. Include what happened to specific body organs and why.

4.3 Working together?

Science as a human endeavour

4.3.1 The building blocks of life

Like all matter in the universe, you are made up of atoms. Collections of atoms make up **molecules**, molecules make up organelles, which make up cells, which make up tissues, which make up organs, which make up systems, which make up you. This progression is shown in the diagram below.

The building blocks of life

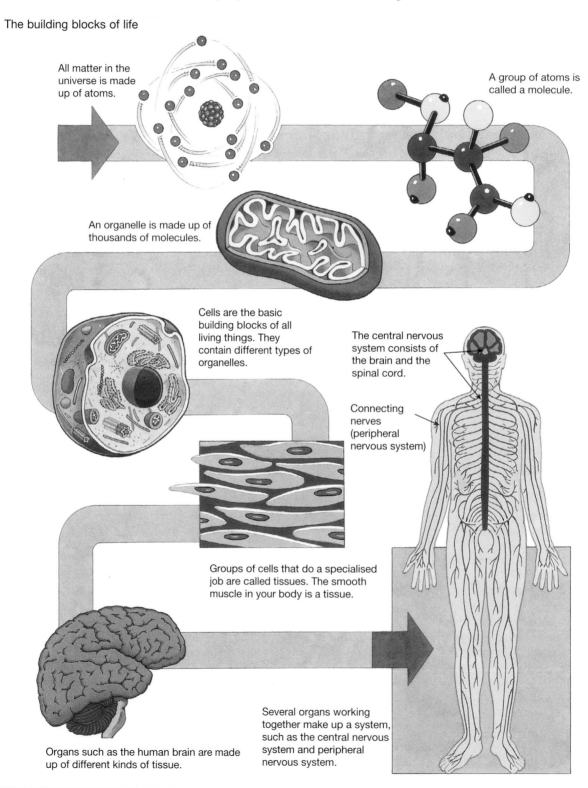

All matter in the universe is made up of atoms.

A group of atoms is called a molecule.

An organelle is made up of thousands of molecules.

Cells are the basic building blocks of all living things. They contain different types of organelles.

The central nervous system consists of the brain and the spinal cord.

Connecting nerves (peripheral nervous system)

Groups of cells that do a specialised job are called tissues. The smooth muscle in your body is a tissue.

Organs such as the human brain are made up of different kinds of tissue.

Several organs working together make up a system, such as the central nervous system and peripheral nervous system.

4.3.2 All alone? Independent!

Unicellular organisms are made up of only one cell that must do all of the jobs that are required to keep the organism alive. These single-celled organisms are small enough that essential substances (e.g. oxygen) and wastes (e.g. carbon dioxide) can be exchanged with their environment through simple diffusion.

4.3.3 One of many? Better get organised!

Like other multicellular organisms, you are made up of many cells. These cells cannot survive independently of each other. They depend on each other and work together. Working together requires organisation.

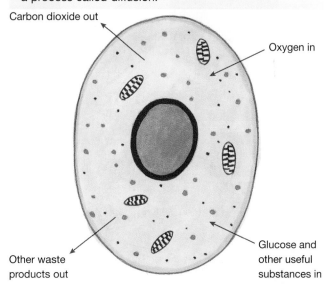

Useful substances (e.g. oxygen) can move into cells and wastes (e.g. carbon dioxide) can move out through a process called diffusion.

Carbon dioxide out

Oxygen in

Other waste products out

Glucose and other useful substances in

Pattern, order and organisation

Multicellular organisms are made up of a number of body systems that work together to keep them alive. Body systems are made up of organs, which are made up of tissues, which are made up of particular types of cells.

Organelles

Within each cell there are structures called **organelles**. Each organelle has a particular job to do. Mitochondria, for example, are organelles in which the chemical energy in glucose is transformed into energy that our cells can use.

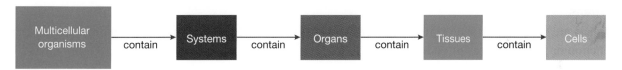

Multicellular organisms → contain → Systems → contain → Organs → contain → Tissues → contain → Cells

Cells

Multicellular organisms are made up of many different types of **cells**, each with a different job to do. Although these cells may have similar basic structures, they differ in size, shape, and in the number and types of organelles they contain. The different make-up of different types of cells and structures within them makes them well suited to their function.

Tissues

Groups of similar cells that perform a specialised job are called **tissues**. Muscle tissue contains cells with many mitochondria so that the energy requirements of the tissue can be met. Nerve tissue consists of a network of nerve cells with extensions to help carry messages throughout your body. The table below shows some examples of tissues that make up your body, what they look like and what their main functions are.

Name of tissue	Description	Main functions
Epithelial tissue	Sheets of cells	To line tubes and spaces, and form the skin
Connective tissue	Tough flexible fibres	To bind and connect tissues together
Skeletal tissue	Hard material	To support and protect the body and permit movement
Blood tissue	Runny fluid containing loose cells	To carry oxygen and food substances around the body
Nerve tissue	Network of threads with long extensions	To conduct and coordinate messages
Muscle tissue	Bundles of elongated cells	To bring about movement

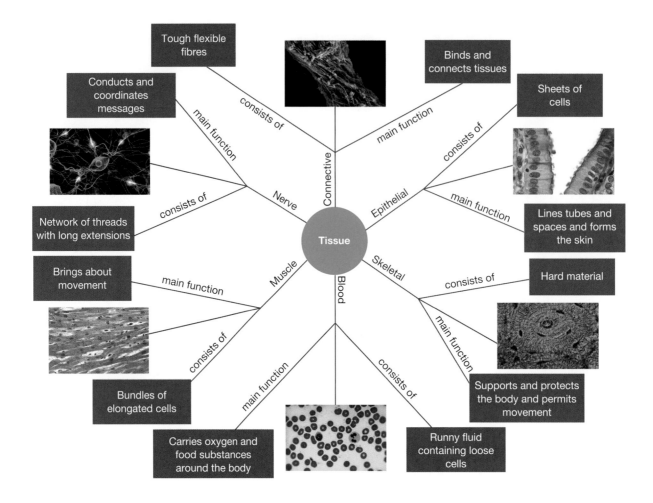

The concept map shows Tissue at the centre connecting to:

- **Nerve** — main function: Conducts and coordinates messages; consists of: Network of threads with long extensions
- **Connective** — consists of: Tough flexible fibres; main function: Binds and connects tissues
- **Epithelial** — consists of: Sheets of cells; main function: Lines tubes and spaces and forms the skin
- **Skeletal** — consists of: Hard material; main function: Supports and protects the body and permits movement
- **Muscle** — main function: Brings about movement; consists of: Bundles of elongated cells
- **Blood** — main function: Carries oxygen and food substances around the body; consists of: Runny fluid containing loose cells

Organs

Organs are made up of one or more different kinds of tissue and perform one (or sometimes more) main function or job. Examples of your organs include:

- brain
- stomach
- lungs
- heart
- skin
- kidneys.

Systems

Multicellular organisms contain organised **systems** of organs that work together to perform specialised functions. The table below provides examples of some of your systems, some organs within them and their main functions.

Name of system	Organs in system	Main functions
Digestive system	Stomach, intestine, liver, pancreas, gall bladder	To digest and absorb food
Respiratory system	Trachea and lungs	To take in oxygen and get rid of carbon dioxide
Circulatory system	Heart and blood vessels	To carry oxygen and food around the body
Excretory system	Kidneys, bladder, liver	To get rid of poisonous waste substances
Sensory system	Eyes, ears, nose	To detect stimuli
Nervous system	Brain and spinal cord	To conduct messages between body parts
Musculoskeletal system	Muscles and skeleton	To support and move the body
Reproductive system	Testes and ovaries	To produce offspring

You are made up of many different body systems that contain organs that work together to keep you alive.

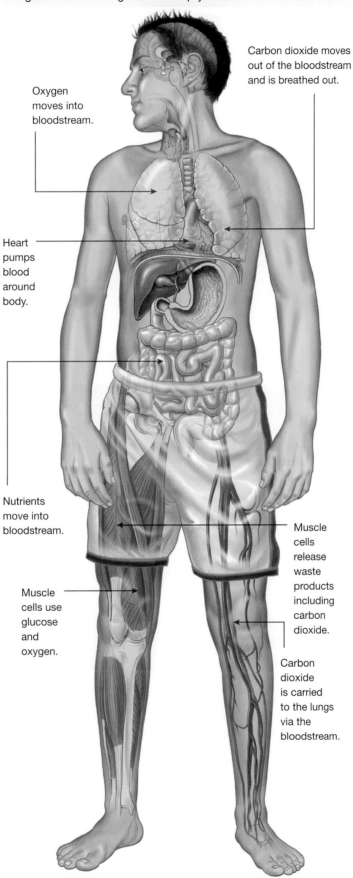

Oxygen moves into bloodstream.

Carbon dioxide moves out of the bloodstream and is breathed out.

Heart pumps blood around body.

Nutrients move into bloodstream.

Muscle cells use glucose and oxygen.

Muscle cells release waste products including carbon dioxide.

Carbon dioxide is carried to the lungs via the bloodstream.

4.3.4 Systems need to work together

Body systems within multicellular organisms work together to keep them alive. For example, cells need energy to survive. A process called **cellular respiration** breaks down glucose to release energy in a form that your cells can then use. This process also requires oxygen and produces carbon dioxide, a waste product. Your digestive, circulatory, respiratory and excretory systems work together to provide your cells with nutrients and oxygen, and to remove wastes such as carbon dioxide.

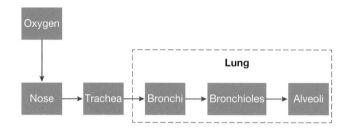

Your **respiratory system** is responsible for getting oxygen into your body and carbon dioxide out. This occurs when you inhale (breathe in) and exhale (breathe out).

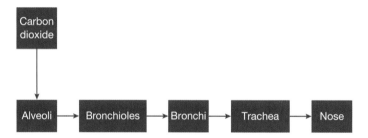

Your **circulatory system** is responsible for transporting oxygen and nutrients to your body's cells, and wastes such as carbon dioxide away from them. This involves **blood cells** that are transported in your **blood vessels** and **heart**. The major types of blood vessels are **arteries**, which transport blood from your heart; **capillaries**, through which materials are exchanged with cells; and **veins**, which transport blood back to the heart.

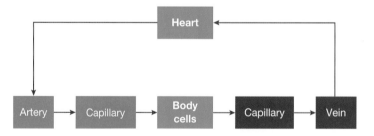

Your **digestive system** plays a key role in supplying your body with the nutrients it requires to function effectively. You ingest food, digest it, then egest it. Your digestive system is involved in breaking food down, so nutrients are small enough to be transported to, and used by, your cells. Some of the organs of the digestive system are shown in the flowchart on the opposite page.

Your **excretory system** removes wastes such as undigested food or waste products from a variety of chemical reactions that your body needs to stay alive. The main organs of your excretory system are your skin, lungs, liver and kidneys. Your skin excretes salts and water as sweat and your lungs excrete carbon dioxide when you breathe out. Your liver is involved in breaking down toxins for excretion and your kidneys are involved in excreting the unused waste products of chemical reactions (e.g. urea) and any other chemicals that may be in excess (including water), so that a balance within your blood is maintained.

Your **musculoskeletal system** consists of both your bones and various types of muscles throughout your body. Bones and muscles provide both support and protection for your organs. While the **reproductive systems** of males and females contain different organs in each gender, they both play a key role in continuation of our species. Other systems such as your **nervous system** and **endocrine system** are also involved in coordinating and regulating processes in your body. You will find out more about these later in your studies.

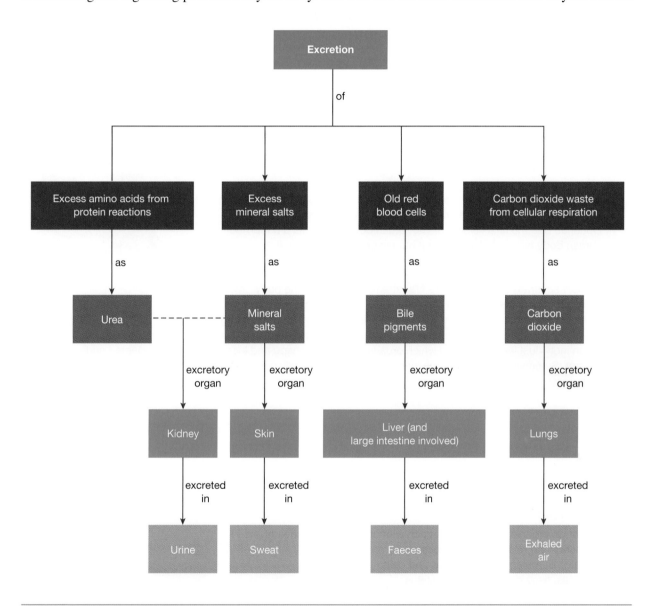

INVESTIGATION 4.1

Mapping your organs

AIM: To draw a diagram to map out the positions and shapes of some human body organs

Materials:
large sheets of paper (e.g. butcher paper)
pencils and marker pens
sticky tape
scissors
optional: light coloured material, sewing thread and needle (or stapler or craft glue), 'stuffing', various other bright coloured materials

Method and results

- Use the sticky tape to join the paper together so that it is the size of a student body outline.
- One member of the team lies down on the paper with their arms away from their body.

1. Another team member carefully draws (about 5 cm away from their body) an outline of their partner's body.

- Once the outline is drawn, the person on the paper can join the rest of the team for the remainder of the activity.
- As a team, decide where in the body outline the following organs are located: heart, lungs, small intestine, nose, oesophagus, liver, stomach, ears, kidney, large intestine, pancreas, eyes, bladder, brain, trachea, mouth.
- Once the location of each organ has been agreed upon, discuss their shape and size.

2. Once consensus is reached within the group, draw each of these organs onto the paper body outline.

- Compare your diagram to reference materials to judge its accuracy.

3. Using these references as a guide, use different coloured pens to draw in more accurate organ shapes, sizes or locations onto your paper body outline.
Optional: Use the final version of your organ body outline as a pattern to make human body organ stuffed toys or a human body organ blanket.

Discuss and explain

4. (a) How accurate was your team's first attempt at drawing the body organ outline?
   (b) Which organs were located correctly and which were not?
   (c) How closely did your team's estimate of shape and size compare to that referenced for each organ?
5. Identify the system to which each of the organs on your outline belong.
6. As an individual learner, identify which organs had a size, shape and location that you expected and which did not. Summarise what you have learned about human body organs in this investigation.

4.3 Exercises: Understanding and inquiring

To answer questions online and to receive **immediate feedback** and **sample responses** for every question, go to your learnON title at www.jacplus.com.au. *Note:* Question numbers may vary slightly.

Remember

1. What do organisms have in common with other matter in the universe?
2. Describe the relationship between:
   (a) atoms, molecules, organelles and cells
   (b) cells, tissues, organs and systems.
3. Identify two ways in which unicellular organisms differ from multicellular organisms.
4. Name an example of an organelle and state its function.
5. Suggest why different types of cells within a multicellular organism may differ in their size and shape.
6. Explain why cells in muscle tissue contain many mitochondria.
7. Identify six:
   (a) types of tissues
   (b) examples of organs
   (c) types of systems.
8. Identify two organs in the:
   (a) respiratory system
   (b) circulatory system
   (c) digestive system.

---

9. Match the type of tissue with its function in the table below.

Tissue		Function	
(a)	Blood	A	Conducts and coordinates messages
(b)	Connective tissue	B	Brings about movement
(c)	Muscle tissue	C	Binds and connects tissues
(d)	Nervous tissue	D	Lines tubes and spaces and forms skin
(e)	Skeletal tissue	E	Carries oxygen and food substances around the body

10. Match the system with its organs in the table below.

System		Organs	
(a)	Circulatory system	A	Liver, kidney, skin, lungs
(b)	Digestive system	B	Heart, blood vessels
(c)	Excretory system	C	Stomach, liver, gall bladder, intestines, pancreas
(d)	Reproductive system	D	Brain, spinal cord
(e)	Respiratory system	E	Lungs, trachea

11. Outline the overall function of the:
    (a) digestive system
    (b) respiratory system
    (c) circulatory system.

Think and discuss

12. Identify whether the following statements are true or false. Justify your response.
    (a) Cellular respiration involves production of glucose.
    (b) The respiratory system takes oxygen into your body and removes carbon dioxide from your body.
    (c) The circulatory system transports carbon dioxide and nutrients to your body cells, and transports wastes such as oxygen away from them.
    (d) Arteries transport blood to your heart and veins transport blood away from your heart.
    (e) Your kidneys, skin, liver and lungs all play a role in removing wastes from your body.

Investigate and create

13. Find out about the different tissues and systems that exist in plants. Present your information using diagrams and lots of colour in a poster or in an electronic format (such as PowerPoint or Animoto presentation). Be as creative as you can.
14. Design and construct a model of one of the following human body systems: respiratory, excretory, digestive or circulatory system.
15. Select one of the following research questions to investigate and present your findings as a labelled model(s), informative animation, picture story book or interesting class presentation. In each question, select animal (i), (ii) or (iii) to compare it with a human.
    (a) In which ways are the respiratory systems of (i) a fish, (ii) an earthworm OR (iii) an insect and a human similar, and how are they different?
    (b) In which ways are the digestive systems of (i) a starfish, (ii) a snake OR (iii) a bird and a human similar, and how are they different?
    (c) In which ways are the circulatory systems of (i) an insect, (ii) a frog OR (iii) a snake and a human similar, and how are they different?

4.4 Digestive system — break it down

4.4.1 Digestion

Digestion involves the breaking down of food so that the nutrients it contains can be absorbed into your blood and carried to each cell in your body.

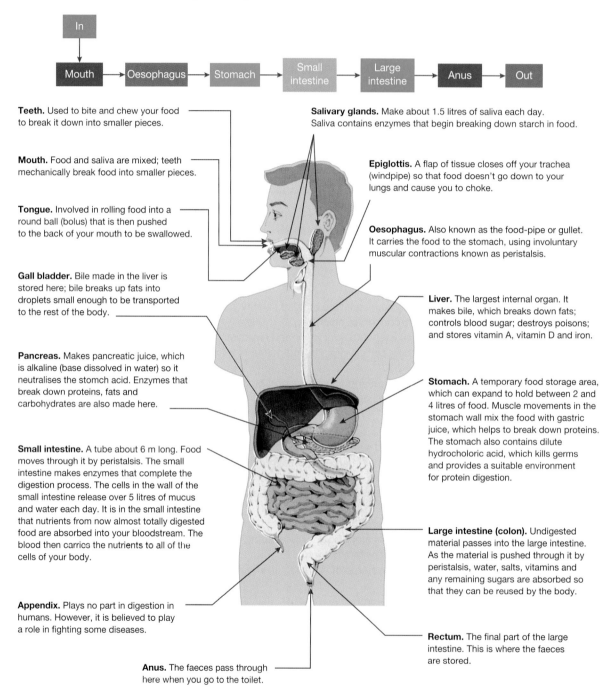

Teeth. Used to bite and chew your food to break it down into smaller pieces.

Mouth. Food and saliva are mixed; teeth mechanically break food into smaller pieces.

Tongue. Involved in rolling food into a round ball (bolus) that is then pushed to the back of your mouth to be swallowed.

Gall bladder. Bile made in the liver is stored here; bile breaks up fats into droplets small enough to be transported to the rest of the body.

Pancreas. Makes pancreatic juice, which is alkaline (base dissolved in water) so it neutralises the stomch acid. Enzymes that break down proteins, fats and carbohydrates are also made here.

Small intestine. A tube about 6 m long. Food moves through it by peristalsis. The small intestine makes enzymes that complete the digestion process. The cells in the wall of the small intestine release over 5 litres of mucus and water each day. It is in the small intestine that nutrients from now almost totally digested food are absorbed into your bloodstream. The blood then carries the nutrients to all of the cells of your body.

Appendix. Plays no part in digestion in humans. However, it is believed to play a role in fighting some diseases.

Anus. The faeces pass through here when you go to the toilet.

Salivary glands. Make about 1.5 litres of saliva each day. Saliva contains enzymes that begin breaking down starch in food.

Epiglottis. A flap of tissue closes off your trachea (windpipe) so that food doesn't go down to your lungs and cause you to choke.

Oesophagus. Also known as the food-pipe or gullet. It carries the food to the stomach, using involuntary muscular contractions known as peristalsis.

Liver. The largest internal organ. It makes bile, which breaks down fats; controls blood sugar; destroys poisons; and stores vitamin A, vitamin D and iron.

Stomach. A temporary food storage area, which can expand to hold between 2 and 4 litres of food. Muscle movements in the stomach wall mix the food with gastric juice, which helps to break down proteins. The stomach also contains dilute hydrocholoric acid, which kills germs and provides a suitable environment for protein digestion.

Large intestine (colon). Undigested material passes into the large intestine. As the material is pushed through it by peristalsis, water, salts, vitamins and any remaining sugars are absorbed so that they can be reused by the body.

Rectum. The final part of the large intestine. This is where the faeces are stored.

Five key processes are important in supplying nutrients to your cells. These are:
- ingestion — taking food into your body
- mechanical digestion
- chemical digestion
- absorption of the broken-down food into your cells
- assimilation — converting the broken-down food into chemicals in your cells

4.4.2 Mechanical and chemical digestion

Mechanical digestion (also known as physical digestion) involves physically breaking down the food into smaller pieces. Most of this process takes place in your mouth when your teeth bite, tear, crush and grind food. **Chemical digestion** involves the use of chemicals called enzymes to break down food into small molecules. These molecules can then pass through the walls of the small intestine and into the bloodstream.

The human digestive system

The key role of your digestive system is to supply your body with the nutrients it requires to function effectively. Your **alimentary canal** (or digestive tract) may be considered as your main digestive highway. It consists of a long tube with coils, large caverns and thin passageways. Other organs that provide chemicals to break down the food or absorb nutrients are attached to the alimentary canal. The alimentary canal begins at the mouth and ends at the anus, where waste products are removed. Excretion of waste products produced by the body's cells can also involve other organs, such as the skin, lungs and kidneys.

4.4.3 Digestive system — down we go...

Let's take a journey together to explore the various organs of your digestive system. What do they look like, where are they and what do they do?

Mouth

You ingest food, digest it, then egest it. The whole process of digestion starts with you taking food into your mouth. **Enzymes** (such as amylases) in your **saliva** are secreted by your **salivary glands** and begin the process of chemical digestion of some carbohydrates. Your teeth physically break down food in a process called mechanical digestion, then your tongue rolls the food into a slimy, slippery ball-shape called a **bolus**.

Look at those teeth!

In many vertebrates, mechanical digestion begins with the teeth. There are four main types of teeth in humans, each type with a different function and position in your mouth as shown below. Your teeth are your very own set of cutlery.

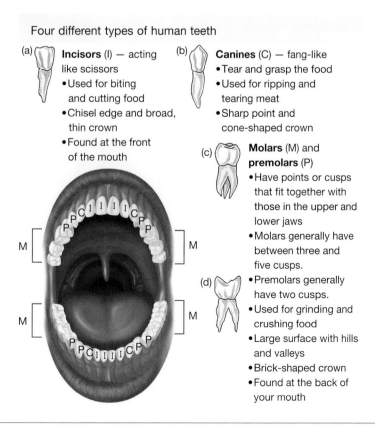

Four different types of human teeth

(a) **Incisors** (I) — acting like scissors
- Used for biting and cutting food
- Chisel edge and broad, thin crown
- Found at the front of the mouth

(b) **Canines** (C) — fang-like
- Tear and grasp the food
- Used for ripping and tearing meat
- Sharp point and cone-shaped crown

(c) **Molars** (M) and **premolars** (P)
- Have points or cusps that fit together with those in the upper and lower jaws
- Molars generally have between three and five cusps.
- Premolars generally have two cusps.

(d)
- Used for grinding and crushing food
- Large surface with hills and valleys
- Brick-shaped crown
- Found at the back of your mouth

Oesophagus to stomach

The bolus is then pushed through your **oesophagus** by muscular contractions known as **peristalsis**. From here it is transported to your **stomach** for temporary storage and further digestion.

Stomach to small intestine

Once the food gets from your stomach to your **small intestine**, more enzymes (including amylases, proteases and lipases) break it down into molecules that can be absorbed into your body. The **absorption** of these nutrient molecules occurs through finger-shaped **villi** in the small intestine. Villi are shaped like fingers to maximise surface area, which increases the efficiency of **nutrient** absorption into the surrounding capillaries. Once absorbed into the capillaries (of your circulatory system), these nutrients are transported to cells in the body that need them.

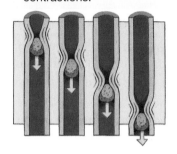

Partly digested food is forced along the oesophagus by peristalsis — a wave of involuntary muscular contractions.

4.4.4 The liver, pancreas and large intestine

Liver

Your liver is an extremely important organ with many key roles. One of these is the production of **bile**, which is transported to your **gall bladder** via the bile ducts to be stored until it is needed. Bile is transported from the gall bladder to the small intestine where it is involved in the mechanical digestion of **lipids** such as fats and oils.

Pancreas

Enzymes, such as **lipases**, **amylases** and **proteases**, which break down lipids, carbohydrates and proteins respectively, are made by the **pancreas** and secreted into the small intestine to chemically digest these components of food.

Large intestine

On its way through the digestive tract (alimentary canal), undigested food moves from the small intestine to the **colon** of the **large intestine**. It is here that water and any other required essential nutrients still remaining in the food mass may be absorbed into your body. **Vitamin D** manufactured by bacteria living within this part of the digestive system is also absorbed. Any undigested food, such as the **cellulose** cell walls of plants (which we refer to as fibre), also accumulates here and adds bulk to the undigested food mass.

The **rectum** is the final part of the large intestine where the faeces is stored before being excreted through the **anus** as waste.

The absorption of most nutrients into your body occurs in the ileum, the last section of the small intestine. The finger-like villi on its walls give it a large surface area that speeds up nutrient absorption. Many tiny blood vessels called capillaries transport the nutrients from the villi into your bloodstream. Undigested material continues on to the large intestine where water and vitamins may be removed, and then the remainder is pushed out through the anus as faeces.

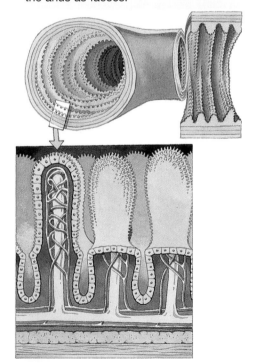

Fat stuff

Breaking down lipids, such as fats and oils, is hard work! Because lipids are insoluble in water, they tend to clump together into large blobs. Bile helps solve this problem. As half of the bile molecule is attracted to water and the other half attracted to lipids, it helps to **emulsify** or separate the lipids so the lipase enzymes can gain access to them and do their job. This is an example of mechanical digestion (bile) and chemical digestion (lipase) working together to get the job done!

4.4.5 Enzymes

Chemical digestion is usually assisted by compounds called enzymes that increase the rate of the chemical reactions. Without enzymes, a single meal could take many years to break down. Mechanical digestion increases the rate of chemical digestion because it increases the surface area of the food particles. This exposes more of the food surface to the digestive chemicals and enzymes.

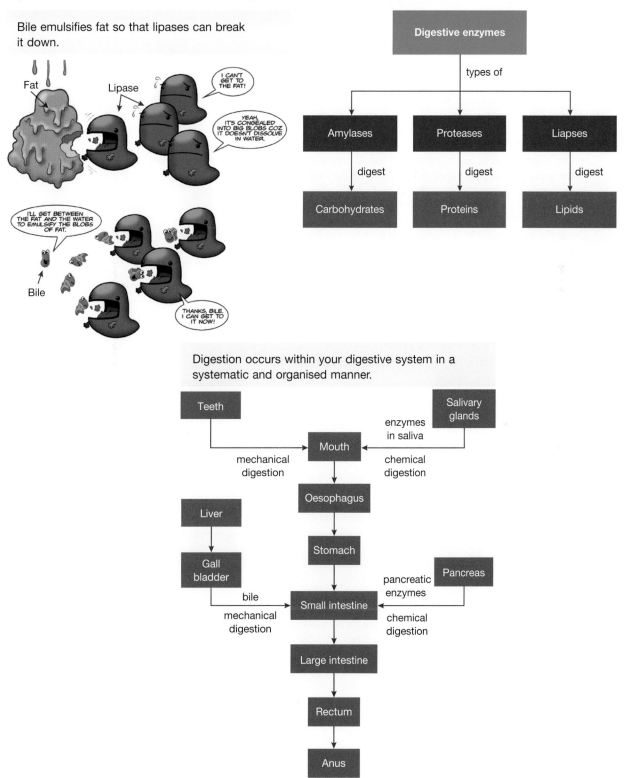

Bile emulsifies fat so that lipases can break it down.

Digestion occurs within your digestive system in a systematic and organised manner.

Chemical digestion begins in your mouth where enzymes in saliva begin to break down some of the carbohydrates in the food that you eat.

Not too hot!

Enzymes are made of protein. That is why it is important that they are not overheated. If they are too hot, they can become **denatured**. It's a bit like cooking meat — once they are denatured, they can't go back to how they were before, so they can't work as enzymes anymore. Different enzymes operate best within specific temperature ranges.

INVESTIGATION 4.2

Observing villi

AIM: To investigate the internal structure of the lining of the small intestine

Materials:
prepared slides of the walls of the small intestine
monocular light microscope

Method and results

- Use a light microscope to observe the prepared slide of the walls of the small intestine.
1. Draw a diagram of your observations. Record the magnification used, label a villus and use descriptive labels to record your detailed observations.

Discuss and explain

2. Describe the function of a villus. (Read through the information previously given in this section if you are unsure.)
3. With reference to your observations, suggest how the shape of a villus suits its function.

INVESTIGATION 4.3

Does temperature affect enzymes?

AIM: To investigate the effect of temperature on enzyme activity

Materials:
4 beakers   8 test tubes
milk   4 thermometers
fresh pineapple puree (Fresh pineapple can be pureed
using a food processor. If fresh pineapple is not available,
use junket powder or a junket tablet dissolved in
10 mL water.)

Method and results

1. Copy the table below and complete it with your results.

Water bath	Temp. of milk and pineapple mix (°C)	Time taken to set (min)

- Add water to the beakers so that they are two-thirds full. Use cold tap water and ice for beaker 1, cold tap water for beaker 2, hot tap water for beaker 3 and boiling water (from a kettle) for beaker 4. These are the 'water baths'.
- Half-fill four test tubes with milk and put one test tube in each water bath.
- Place one teaspoon of fresh pineapple puree (or 1 mL junket solution) in the other four test tubes. Put one of these test tubes in each water bath.
- Allow the test tubes to stand in the water baths for at least 5 minutes.
- For each water bath, pour the fresh pineapple puree into the milk and stir briefly.

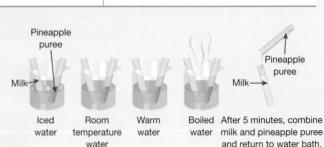

Iced water | Room temperature water | Warm water | Boiled water After 5 minutes, combine milk and pineapple puree and return to water bath.

2. Quickly record the temperature of the milk and pineapple mixture and then allow it to stand undisturbed. The mixture will eventually set. Record the time taken to set. If the milk has not set after 15 minutes, record the time as 15+.

Discuss and explain

3. Pineapple juice and junket contain an enzyme that causes a protein in milk (casein) to undergo a chemical reaction and change texture; that is why the milk sets. At what temperature did the enzyme work best? Explain.
4. Did the enzyme work well at very high temperatures? Explain your answer.
5. Which variables were controlled in this experiment?
6. Do you think that the same results would be obtained if tinned pineapple puree was used instead of fresh pineapple? Explain your answer.
7. Identify the strengths and limitations of this investigation, and suggest ways to improve it.
8. Propose a research question about enzymes that could be investigated.

4.4.6 'Ase' endings

There are specific enzymes for specific tasks. The substance that they break down is called a **substrate**. The resulting substance is referred to as the **product**. As the diagram at right shows, the enzyme remains unchanged at the end of the process.

Enzymes that break down carbohydrates, such as starch, into glucose are called amylases (this process is shown in the diagram on the bottom right). Those that break down fats and oils into fatty acids and glycerol are called lipases. Proteases are enzymes that break down proteins into amino acids.

Each enzyme has specific conditions in which it works. For example, amylases that break down carbohydrates in the stomach work in acidic environments, while those in the small intestine work best in alkaline conditions.

4.4.7 Personal explosions

Well, excuse me! Have you burped or passed wind recently? Have you had diarrhoea or vomited? These 'personal explosions' are related to the processing of nutrients by your body.

Burping, or belching, occurs when air is swallowed or sucked in. This may happen when you talk while you eat, eat or drink too quickly, or drink fizzy drinks (such as those with carbon dioxide gas dissolved in them). When you eat too fast and don't chew your

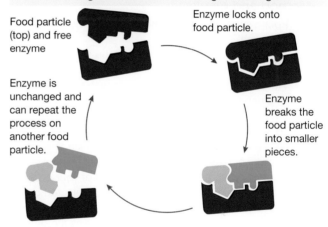

Enzymes speed up chemical reactions in the body, but are not changed so can be reused again and again.

Food particle (top) and free enzyme

Enzyme locks onto food particle.

Enzyme breaks the food particle into smaller pieces.

Enzyme is unchanged and can repeat the process on another food particle.

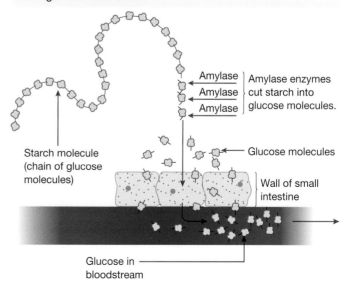

Amylases in the saliva and stomach break starch down into glucose molecules.

Starch molecule (chain of glucose molecules)

Amylase
Amylase
Amylase

Amylase enzymes cut starch into glucose molecules.

Glucose molecules

Wall of small intestine

Glucose in bloodstream

food enough, more acid can be produced in your stomach. When you burp, some of this acid can rise up into your oesophagus, resulting in a burning sensation called **heartburn**.

Flatulence refers to the release of gases when you 'pass wind' through your anus. These gases are produced by bacteria in your large intestine. The odour and composition of the gases depends on the foods you have eaten and the amount of air you have swallowed.

INVESTIGATION 4.4

Making a burp model

AIM: To construct a burp model and to design and construct a model that demonstrates the functioning of a process related to the digestive system

Materials:
vinegar
baking soda
medium/large balloon
funnel

Method and results

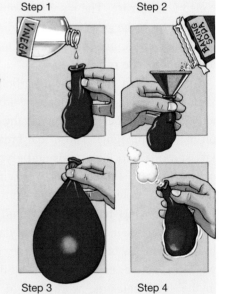

Step 1   Step 2

Step 3   Step 4

- Pour a small amount of vinegar into the bottom of the balloon 'stomach'.
- Add some baking soda to the balloon 'stomach' using a funnel.
- Using your fingers, pinch the balloon closed at its neck.
- Watch as your 'stomach' expands with gas.
- Unpinch the top of the balloon (or 'oesophagus/food tube') to release the gas (or burp).
- Try to make your model sound like the real thing!
1. Summarise your observations in a flow chart that includes labelled diagrams or digital/photographic images.
2. Select an organ belonging to an animal of your choice. Find out more about the structure and function of your selected organ and how it does its job. Summarise your findings.
   - Design and make a simple model (such as the one used for this experiment) to show how your selected organ achieves its function, or what happens when something goes wrong.
3. Summarise your design plans and labelled diagrams or digital images into an advertising brochure or digital multimedia advertisement.

Discuss and explain
4. Comment on the challenges you experienced during the design and construction of your model, and suggest ways that you could overcome these if you were to do it again.

Diarrhoea is the excessive discharge of watery faeces. It occurs when the muscles of the large intestine contract more quickly than normal, usually in an effort to rid your body of an infection. As a result, the undigested food moves through too rapidly for enough water to be absorbed into your body.

Green vomit? Messages from your stomach wall travel to the 'vomiting centre' of your brain, resulting in forceful ejection of your stomach contents (and occasionally also contents from your small intestine). **Vomiting** can be caused by eating or drinking too much, anxiety, infections or chemicals that irritate your stomach wall. If the vomit is green, it may be due to the colour of food ingested or the presence of bile, which is produced by your liver to help digest food in your small intestine.

4.4 Exercises: Understanding and inquiring

To answer questions online and to receive **immediate feedback** and **sample responses** for every question, go to your learnON title at www.jacplus.com.au. *Note:* Question numbers may vary slightly.

Remember

1. Identify the organs in the figure at right.
2. Identify four types of teeth and describe their functions.
3. State the name of the:
   (a) type of digestion that involves enzymes
   (b) enzymes that break down fats
   (c) enzymes that break down proteins
   (d) substances that enzymes act on.
4. Describe the process of peristalsis and suggest why it occurs.
5. Describe the relationship between:
   (a) teeth and mechanical digestion
   (b) the pancreas and the small intestine
   (c) the liver, gall bladder and the small intestine
   (d) the villi, small intestine and capillaries
   (e) bile, lipase and fats.
6. Describe what happens to enzymes when they get too hot.
7. Order the following organs into the correct sequence: stomach, large intestine, oesophagus, anus, mouth, small intestine.
8. Match the organ with its function in the table below.

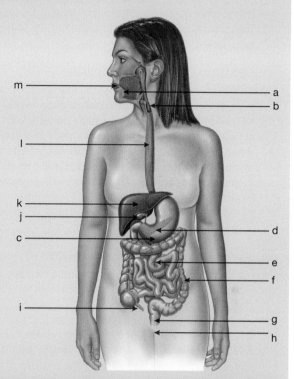

Organ		Function	
(a)	Gall bladder	A	Where the breakdown of starch and protein is finished and fat breakdown occurs
(b)	Large intestine	B	Temporary storage of food and where protein digestion begins
(c)	Liver	C	Tube which takes food from mouth to stomach
(d)	Oesophagus	D	Stores undigested food and waste while bacteria make some vitamins
(e)	Pancreas	E	Stores faeces
(f)	Rectum	F	Makes enzymes used in the small intestine
(g)	Small intestine	G	Makes bile, stores glycogen and breaks down toxins
(h)	Stomach	H	Stores bile made in the liver until needed in the small intestine

9. Explain why it is important to break down food that we eat.

Think

10. Which teeth are used to:
    (a) bite into a pear
    (b) crush and grind nuts?
11. Use Venn diagrams to compare:
    (a) mechanical and chemical digestion
    (b) lipases and proteases
    (c) small intestine and large intestine.
12. Suggest why it is necessary to drink fluids when you suffer from diarrhoea.

Think and discuss

13. State whether the following statements are true or false. Justify your response.
    (a) Mechanical digestion occurs when chemicals in your body react with food to break it down.
    (b) Ingestion involves taking food into your body, whereas digestion involves breaking food down.
    (c) Many enzymes have names that end with the suffix 'ase'.

(d) 'Bolus' is the term used to describe the muscular contractions that push food down your oesphagus to your stomach.

(e) Plant cell walls make up much of the fibre that accumulates in our large intestines.

(f) The process of denaturing enzymes kills them.

14. Suggest how you can still swallow food if you were positioned upside down.

15. Take a small piece of bread into your mouth. Although at first you don't taste much, after a while, it may taste sweet. Suggest reasons why.

Investigate, think and design

16. Design an investigation to test the following hypotheses:
    * Fresh pineapple results in a faster enzyme reaction than canned pineapple.
    * The length of time that pineapple puree is kept in ice affects the rate of enzyme reaction.
    * Different coloured junket tablets result in different rates of enzyme reaction.

17. When cows burp, they release methane gas into the air. This gas is believed to be one of the major causes of global warming. It has been suggested that cows could be responsible for about 20 per cent of the methane in the atmosphere. Research these claims.
    (a) On the basis of your findings, do you agree? Justify your response.
    (b) Design an experiment that could be used to test the claim that cows contribute to increased methane gas in the atmosphere.

Imagine, investigate and create

18. Cows have four stomachs. Find out the function of each stomach. Construct a model.

19. Research an enzyme of your digestive system. Find out what it does and where, then construct a poster or model to show how it works.

20. (a) Imagine that you are either a cheese and tomato sandwich or a hamburger.
    (b) List the ingredients of the food you chose in part (a).
    (c) Research what happens (and where) to each of these ingredients when eaten.
    (d) Construct a flowchart to show the process of digestion in the human body, including events and locations.
    (e) Use this information to write a story in either a cartoon or picture book format.
    (f) Convert your story into a play.
    (g) Perform your play to the class using animations, team members or puppets.

21. Use information in this section and other resources to relate structural features to the functions of the following parts of the digestive system.

Part of system	Structural features	Function
Oesophagus		
Stomach		
Small intestine		
Villi		
Large intestine		

22. Construct a 'working' model of the human digestive system.

23. Design and construct a 'trivial pursuit'–type board game about the digestive system. Include questions in three different categories. Create game pieces that are relevant to the digestive system.

24. Create a picture book or poster that helps explain to primary students how food is digested.

25. Construct a plasticine model (to scale) of the human digestive system, with each organ being a different colour.

26. Find out more about causes, symptoms, treatments and implications of (a) belching (b) flatulence or (c) vomiting. Create a cartoon or poster to summarise your findings.

learn**on** RESOURCES — ONLINE ONLY

Watch this eLesson: From dinner plate to sewerage system (eles-0056)

Complete this digital doc: Worksheet 4.1: Mechanical and chemical digestion (doc-18720)

Complete this digital doc: Worksheet 4.2: The digestive system (doc-18721)

4.5 Digestive endeavours

Science as a human endeavour

4.5.1 Do you look after your teeth?

When each part of the digestive system is healthy, your digestion works smoothly without you thinking about it. But things like tooth decay or being a coeliac can cause you problems.

Your teeth can decay when bacteria in your mouth turn sugar from your food into acid. This acid can 'eat' a hole in your tooth enamel and dentine. Once this hole reaches a nerve, you get a toothache. The illustration at right shows the structure of a tooth and where decay usually occurs — at the top of large back teeth and at the side where one tooth touches another.

Structure of a tooth showing where tooth decay occurs

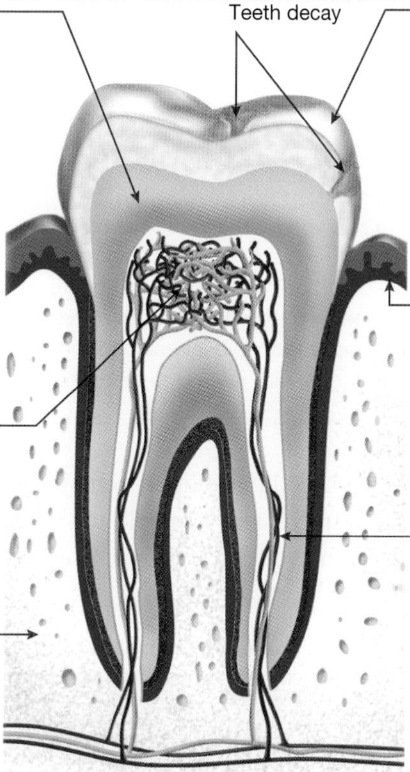

Dentine makes up most of the tooth. It is a bone-like material that gives the tooth its shape. Dentine is not strong and wears away if exposed.
Teeth decay
Enamel is the hardest substance in the body. It forms a coating over the exposed surface of the tooth.
Gum surrounds the tooth, stopping food particles getting into the root.
The **pulp** contains the nerves and blood vessels.
The **root canal** is the channel where the nerves and blood vessels go down into the jawbone.
Teeth are locked into the **bone** of the jaw.

If you don't clean your teeth regularly (at least once a day), they can become covered with a thin film of food, saliva and bacteria. This is called plaque. As this plaque rots, it causes your gums to swell and bleed. This is known as gum disease.

INVESTIGATION 4.5

How well do you brush your teeth?

AIM: To investigate the effectiveness of brushing teeth

Materials:

3 nutrient agar plates	*toothbrush*
3 small labels	*toothpaste*
3 cotton buds	*incubator set at 35 °C*
sticky tape	*dissection microscope*

Note: This is best done after lunch, prior to cleaning teeth.

Method and results

- Label the three agar plates as 'Dirty', 'Toothpaste' and 'Brush only'.

- Wipe a cotton bud carefully over your uncleaned teeth and gums.
- Gently wipe in a zigzag pattern over the agar plate labelled 'Dirty'.
- Replace the lid and seal around the edges with sticky tape.
- Repeat the steps above after brushing with no toothpaste and then with toothpaste.
- Incubate the agar plates for 24–48 hours.
- Observe the agar plates using the dissection microscope. Do not open the agar plates. Observe through the plastic dish.

1. Record your observations using diagrams with descriptive labels.

Discuss and explain

2. Which agar plate showed (a) the most microbial growth and (b) the least microbial growth? What does this suggest?
3. Suggest why we brush our teeth with toothpaste.
4. Suggest improvements to the design of this investigation.
5. Design an experiment that could investigate the effect of different types of toothpaste.
6. Propose a research question related to teeth that could be further investigated.

Our water supply and toothpaste often contain fluoride, which helps prevent tooth decay. Fluoride protects the enamel and helps repair or rebuild the enamel in your teeth. Damaged or missing teeth can make it difficult for you to chew your food properly and therefore may affect digestion of foods.

Oral surgery is one of the many tooth-related careers you can choose from.

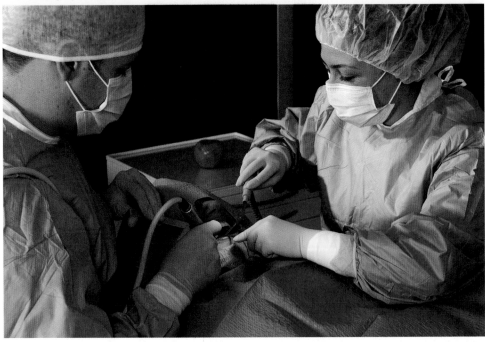

4.5.2 A future in teeth?

Dentistry is only one example of many different 'tooth pathway' careers that you may be aware of. Examples of other dental specialities include oral and maxillofacial surgeons, dental–maxillofacial radiologists, endodontists, oral physicians, oral pathologists, orthodontists, paediatric dentists, periodontists, prosthodontists, public health dentists and special needs dentists.

Missing a tooth? A synthetic replacement for a tooth root is used in a tooth implant.

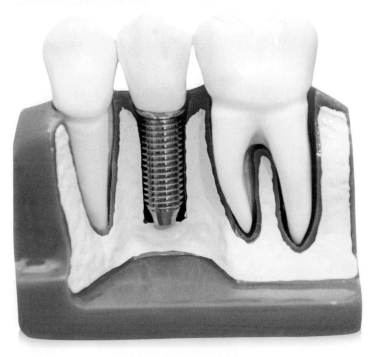

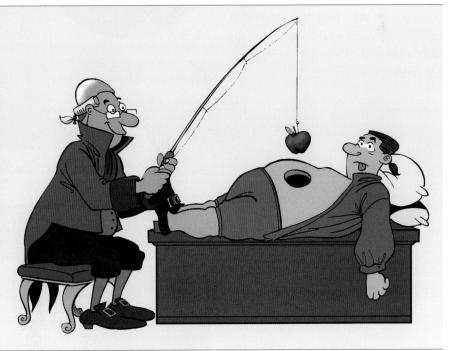

HOW ABOUT THAT!

Just to whet your appetite … have you heard the story of Alexis St Martin? In 1822, he was shot in the stomach at close range. Wrong place at the wrong time! His wound healed, but left an open hole that showed the inside of his stomach! By dangling food suspended on a silk thread, his doctor William Beaumont made some breakthrough discoveries on how our stomachs work.

4.5.3 Going up or down?

Gastroscopy and colonoscopy involve the use of endoscopes (from the Greek words meaning to 'see within').

A major advantage of these techniques is that they gather important information without needing to cut into the body.

Gastroscopy involves passing a long flexible endoscope through your mouth, down your oesophagus and into your stomach and duodenum. Although it takes only about five minutes, a lot of information can be gathered about the health of this part of your digestive system.

A colonoscopy enables the doctor to look directly at the lining of your colon or bowel. In this case the endoscope is inserted through your rectum. This procedure may take about thirty minutes. The results may be used to investigate abnormalities or detect the presence of colon polyps, which in some cases may turn into cancer. A colonoscopy can enable small polyps to be removed or a biopsy to be taken for further testing.

Gastroscopy

Endoscope

Normal

View through endoscope

H. pylori gastritis

Colonoscopy

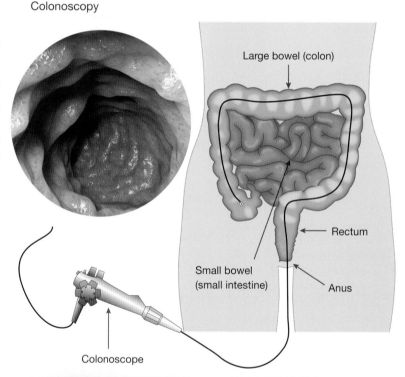

Large bowel (colon)

Small bowel (small intestine)

Rectum

Anus

Colonoscope

Australian scientists Dr Barry J. Marshall and Dr Robin Warren received the 2005 Nobel Prize in Medicine for their discovery that linked *Helicobacter pylori* bacteria to gastroduodenal disease. Their discovery dramatically improved the treatment of peptic ulcer disease.

Biopsies of (a) normal and (b) coeliac intestine

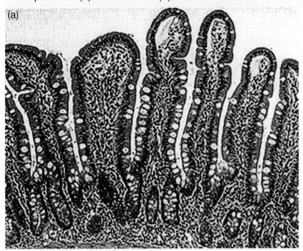

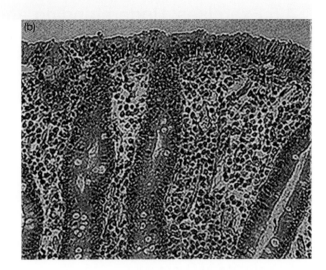

4.5.4 Villi alert!

Most nutrients are absorbed into your bloodstream when they get to the last section of your small intestine. The walls of this part of your intestine are lined with finger-like projections called villi.

The shape of villi increases the surface area through which nutrients can diffuse into tiny blood vessels called capillaries. Nutrients are then transported to other parts of your body.

Coeliac disease is an auto-immune disease — that is, one in which your body produces antibodies to attack your own tissues. In this case, it is the villi of the small intestine that are damaged. This interferes with the absorption of nutrients.

People with coeliac disease are intolerant to gluten. This protein is found in wheat, rye, barley and oats. Eating these foods triggers their immune system to damage the villi in their small intestine. Coeliac Australia refers to the condition as a 'hidden epidemic'. Coeliac disease affects approximately 1 in 70 people in Australia, with many (about 80 per cent) not even knowing that they have it. If left undiagnosed, more severe consequences, such as a variety of nutritional deficiencies, bowel cancer and osteoporosis, may result.

Australian researchers are attempting to develop a vaccine as a new treatment for coeliac disease. In 2009, Dr Robert Anderson and his team at the Walter and Eliza Hall Institute of Medical Research in Melbourne began the world's first trials of a coeliac vaccine. If this treatment is successful, it could mean the end of gluten-free diets for people with the condition.

Dr Robert Anderson and his team at the Walter and Eliza Hall Institute of Medical Research in Melbourne began the world's first trials of a coeliac vaccine.

4.5 Exercises: Understanding and inquiring

To answer questions online and to receive **immediate feedback** and **sample responses** for every question, go to your learnON title at www.jacplus.com.au. *Note:* Question numbers may vary slightly.

Remember

1. Outline the relationship between diet, coeliac disease and the digestive system.
2. Approximately how many people in Australia are affected by coeliac disease?
3. Use a Venn diagram to compare gastroscopy and colonoscopy.
4. Describe the discovery that led to two Australian scientists winning the 2005 Nobel Prize in Medicine.

Think and design

5. Design an investigation to test the following hypotheses:
   - Drinking fluoridated water reduces tooth decay.
   - Mouthwash prevents the growth of bacteria that cause tooth decay.
   - Drinking bottled water rather than tap water increases tooth decay.

Investigate and create

6. Select one of the 'tooth pathway' careers in the text 'A future in teeth'. Find out details of the training required and what a career in this pathway would entail. Present your findings in a brochure and include a section that describes what 'a day in the life of ...' this career would be like.
7. Imagine that you have invited two friends over for a sleepover. One of them has coeliac disease and the other is lactose intolerant.
   (a) Find out the cause and/or symptoms associated with each of these conditions.
   (b) Find out what sorts of foods you could offer your friends.
   (c) Design a dinner and breakfast menu that includes foods that each of your friends would be able to eat.
8. Research one of the following digestion-related diseases and report on the cause, symptoms, treatment or cure, possible consequences and current research.
   - Heartburn
   - Inflammatory bowel disease
   - Irritable bowel syndrome
   - Appendicitis
   - Constipation
   - Crohn's disease
   - Diverticulosis
   - Gallstones
   - Haemorrhoids
   - Pancreatitis
   - Peptic ulcer

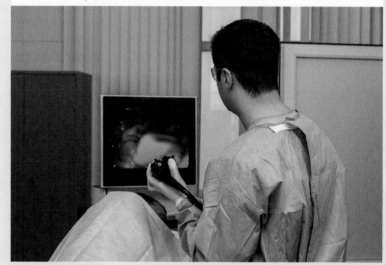

An endoscopist at work

Investigate and report

9. Find out more about one of the following digestion-related scientific careers and report your findings as a diary journal entry.
   - Gastroenterologist
   - Endoscopist
   - Colorectal surgeon
10. Recently, scientists have suggested a link between the presence of bacteria *Helicobacter pylori* and cancer protection. Find out more about this research and suggest possible implications that it may have.

4.6 Circulatory system — blood highways

4.6.1 What's in blood?

An average-sized human has about five litres of blood; that's about a bucketful. Blood is made up of red blood cells (**erythrocytes**), white blood cells (**leucocytes**), blood platelets and the straw-coloured fluid they all float in, called **plasma**.

Ready to carry!

Each drop of blood contains about 300 million **red blood cells**. These red blood cells travel around the body up to 300 000 times, or for about 120 days. After this they literally wear out and die. Each second you are manufacturing about 1.7 million replacement red blood cells in your bone marrow.

Red blood cells are red because they contain an iron-containing pigment called **haemoglobin**. Oxygen reacts with haemoglobin in red blood cells to form **oxyhaemoglobin**, which makes the blood an even brighter red. This change in colour intensity can indicate the amount of oxygen being transported in blood at a particular time.

Erythrocytes (red blood cells) and leucocytes (white blood cells)

HOW ABOUT THAT!

The amount of oxygen carried by haemoglobin varies with altitude. At sea level, about 100 per cent of haemoglobin combines with oxygen. At an altitude of 13 000 metres above sea level, however, only about 50–60 per cent of the haemoglobin combines with oxygen.

The shape and size of red blood cells makes them well suited to their function. Their small size allows them to fit inside tiny capillaries. When mature, red blood cells lack a nucleus, increasing space available to carry haemoglobin and hence oxygen. Their biconcave shape means that they have a large surface area for their size, which also assists in their important oxygen transporting role.

Fit to fight!

White blood cells contain a nucleus, and are larger and fewer in number than red blood cells. They are often referred to as the 'soldiers' in the blood as they are involved in fighting disease. Some white blood cells produce chemicals called antibodies; others engulf and 'eat' bacteria and other foreign matter. When you are ill or fighting an infection, the number of white blood cells in your blood increases for this reason.

Clot and cover...

If you cut yourself, you bleed. This is because a blood vessel has been cut. **Platelets** in the blood help it to clot and plug the damaged blood vessel. This seal prevents germs getting in.

HOW ABOUT THAT!

Insect blood looks a little like raw eggwhite, because it contains no pigment. The blood of crabs and crayfish, however, contains the pigment haemocyanin. This pigment has copper in it and is blue when combined with oxygen. This differs from haemoglobin in humans, which is red when combined with oxygen.

You have all of this in your blood.

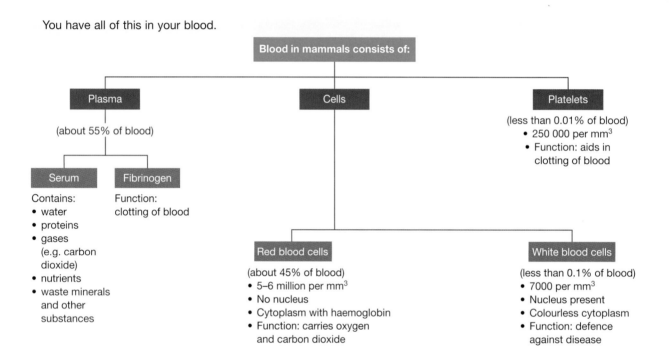

Blood in mammals consists of:

Plasma
(about 55% of blood)

Serum
Contains:
• water
• proteins
• gases
  (e.g. carbon
  dioxide)
• nutrients
• waste minerals
  and other
  substances

Fibrinogen
Function:
clotting of blood

Cells

Red blood cells
(about 45% of blood)
• 5–6 million per mm³
• No nucleus
• Cytoplasm with haemoglobin
• Function: carries oxygen
  and carbon dioxide

White blood cells
(less than 0.1% of blood)
• 7000 per mm³
• Nucleus present
• Colourless cytoplasm
• Function: defence
  against disease

Platelets
(less than 0.01% of blood)
• 250 000 per mm³
• Function: aids in
  clotting of blood

4.6.2 Mix and match?

How much do you know about the red stuff that flows throughout your body? Did you know that your blood might not mix too well with that of your friends? Blood can be grouped into eight types using the ABO system and the Rhesus (Rh) system. Your blood type is inherited from your parents.

These classification systems are based on whether particular chemicals (antigens) are present or absent on your red blood cells. If you are Rh-negative, you do not have the Rhesus factor on your red blood cells; if you do, you are Rh-positive.

The ABO system divides blood into groups A, B, AB and O. If you need a blood transfusion, it is very important to know your blood type and that of the donor because some blood types cannot be mixed. If the wrong types are mixed, the blood cells may clump together and cause fatal blockages in blood vessels.

How common is your blood?

Viewing blood cells

AIM: To observe blood cells under a light microscope

Materials:
prepared slide of blood smear
microscope

Method and results

- Place the prepared slide onto the microscope stage.
- Use low power to focus, then carefully adjust to high power.
- Find examples of red blood cells and white blood cells on the slide.

1. Draw diagrams of representative red blood cells and white blood cells. On your diagram, include descriptive labels and the magnification used.
2. Estimate (a) how many red blood cells could fit inside a white blood cell and (b) how many of each cell type could fit across the field of view.

Discuss and explain

3. Summarise the similarities and differences between the structures of red blood cells and white blood cells.

4.6.3 Connected pathways

Your **circulatory system** is responsible for transporting oxygen and nutrients to your body's cells, and wastes such as carbon dioxide away from them. This involves interactions between blood cells, blood vessels and your heart.

Blood vessels called **arteries** transport blood from your heart, whereas others, called **veins**, transport blood back to the heart. Materials are exchanged between blood and cells through tiny blood vessels called **capillaries** that are located between arteries and veins.

Arteries, veins and capillaries

Arteries have thick, elastic, muscular walls and carry blood under high pressure away from your heart. Veins have thinner walls and possess valves that prevent the blood from flowing backwards as they take blood to your heart.

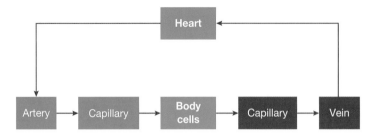

Veins have valves to ensure that blood flows in only one direction.

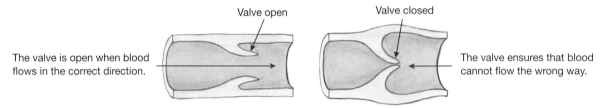

Valve open

Valve closed

The valve is open when blood flows in the correct direction.

The valve ensures that blood cannot flow the wrong way.

Capillaries are the most numerous and smallest blood vessels. Your body contains about 100 000 km of capillaries, which penetrate almost every tissue, so no cell is very far away from one. Capillaries are very important because they transport substances such as oxygen and nutrients to cells and remove wastes such as carbon dioxide.

Your circulatory system consists of your heart, blood vessels and blood. Arteries, capillaries and veins are the major types of blood vessels through which your blood travels.

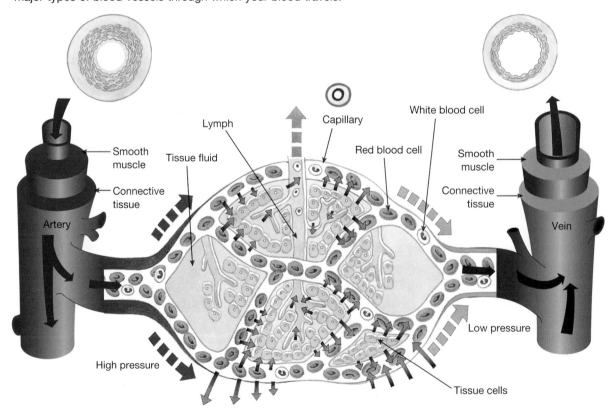

In the capillaries, oxygen diffuses out of the blood and waste produced by cells diffuses into the bloodstream.

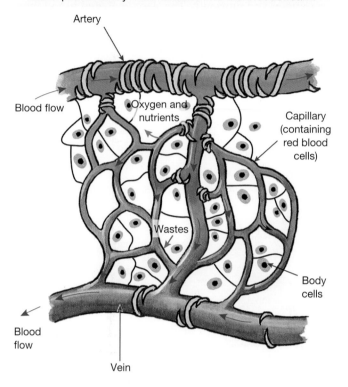

Artery

Blood flow

Oxygen and nutrients

Capillary (containing red blood cells)

Wastes

Body cells

Blood flow

Vein

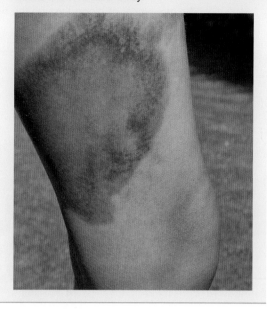
4.6.4 Have a heart

Often linked with emotions, love and courage, the heart has a special meaning for most of us. In a clinical sense, however, it is merely a pump about the size of your clenched fist.

Two pumps in one

To be more precise, the human heart is actually *two* pumps. One side contains **oxygenated blood** and the other deoxygenated. Your veins bring 'used' **deoxygenated blood** (stripped of oxygen and bluish in colour) back to your heart. All of these veins join up into a larger vein called the **vena cava**. Entering the top right chamber of your heart, blood is pumped into the bottom right chamber. It is then pumped out to your lungs

The heart is actually two pumps. One side pumps oxygenated blood and the other deoxygenated blood.

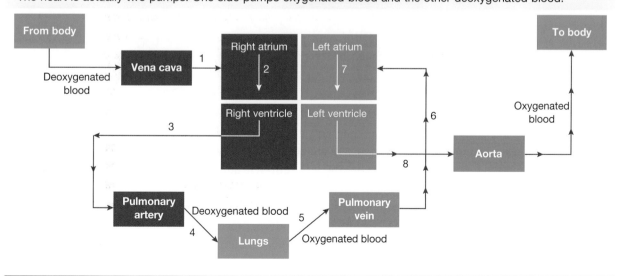

From body

Deoxygenated blood

Vena cava

Right atrium

Left atrium

1

2

7

Right ventricle

Left ventricle

3

6

8

To body

Oxygenated blood

Aorta

Pulmonary artery

Deoxygenated blood

4

Lungs

5

Oxygenated blood

Pulmonary vein

where it picks up oxygen and becomes oxygenated and more reddish in colour. It also loses some of its carbon dioxide. The oxygenated blood then returns to the left-hand side of your heart to be pumped out through arteries to your body tissues, where it delivers oxygen and nutrients. The deoxygenated blood then returns to the right-hand side of the heart for the cycle to be repeated.

The movement of blood through the heart

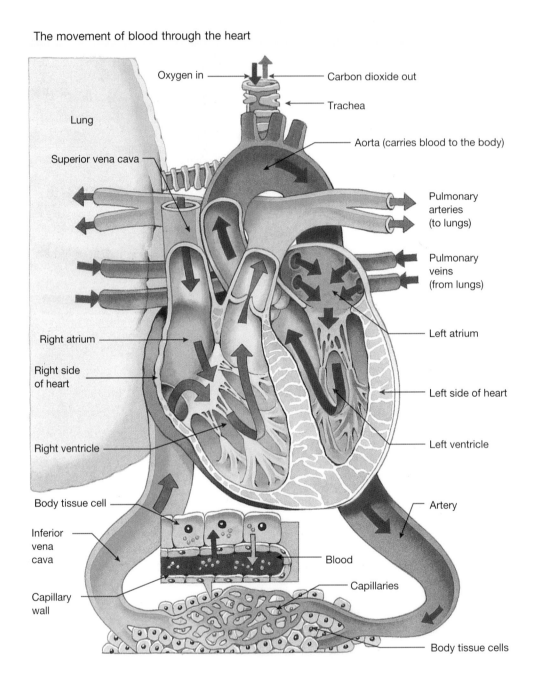

Four chambers

The human heart has four chambers. The upper two chambers are called the **left atrium** and **right atrium** (plural = atria), and the lower two chambers are the **left ventricle** and **right ventricle**. The two sides of the heart are different. The walls of the left side are thicker and more muscular because they need to have the power to force the blood from the heart to the rest of the body.

Flap-like structures attached to the heart walls, called **valves**, prevent the blood from flowing backwards and keep it going in one direction. If you listen to your heart beating, you will hear a **'lub dub'** sound.

The 'lub' sound is due to the valves between the ventricles and atria shutting. The 'dub' sound is due to the closing of the valves that separate the heart from the big blood vessels that lead to the lungs and the rest of the body.

4.6.5 Blood pressure

The heart's pumping action and the narrow size of the blood vessels result in a build-up of considerable pressure in the arteries. The force with which blood flows through the arteries is called **blood pressure**. It is affected by different activities and moods. It also goes up and down as the heart beats, being highest when the heart contracts (**systolic pressure**) and lowest when the heart relaxes (**diastolic pressure**). A person's blood pressure is expressed as a fraction. This fraction is the systolic pressure over the diastolic pressure, such as 120/70.

4.6.6 Keeping the pace

During each minute that you are sitting and reading this, about 5–7 litres of blood completes the entire circuit around your body and lungs. In a single day, your heart may beat about 100 000 times and pump about 7000 litres of blood around your body.

A normal human heart beats about 60–100 times a minute, this rate increasing during exercise or stress. With each **heartbeat**, a wave of pressure travels along the main arteries. If you put your finger on your skin just above the artery in your wrist, you can feel this **pulse** wave as a slight throb.

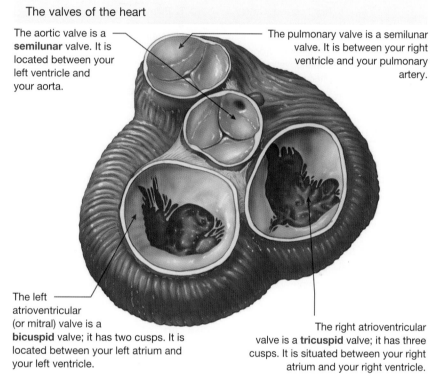

The valves of the heart

The aortic valve is a **semilunar** valve. It is located between your left ventricle and your aorta.

The pulmonary valve is a semilunar valve. It is between your right ventricle and your pulmonary artery.

The left atrioventricular (or mitral) valve is a **bicuspid** valve; it has two cusps. It is located between your left atrium and your left ventricle.

The right atrioventricular valve is a **tricuspid** valve; it has three cusps. It is situated between your right atrium and your right ventricle.

INVESTIGATION 4.7

Heart dissection

AIM: To observe the structure of a mammalian heart

Materials:
sheep's heart preferably with the blood
vessels still attached
dissecting instruments
dissecting board
newspaper or paper to cover
dissection board

Method and results

- Place newspaper on the dissection board, then place the heart on top of the paper.
- Use the diagram on the next page to identify the parts of the heart.
- Try to locate where blood enters and leaves the heart:
  (a) to and from the lungs
  (b) to and from the rest of the body.

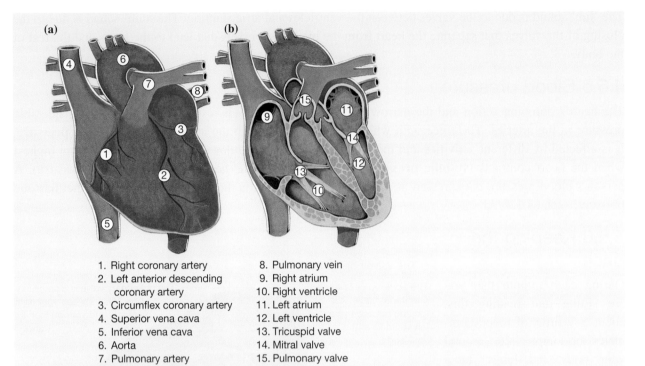

(a)   (b)

1. Right coronary artery
2. Left anterior descending coronary artery
3. Circumflex coronary artery
4. Superior vena cava
5. Inferior vena cava
6. Aorta
7. Pulmonary artery

8. Pulmonary vein
9. Right atrium
10. Right ventricle
11. Left atrium
12. Left ventricle
13. Tricuspid valve
14. Mitral valve
15. Pulmonary valve

1. Sketch and label the heart and use arrows to show the direction of blood flow.
   • Cut the heart in two so that both halves show the two sides of the heart (similar to the illustration on page 134).
2. In a diagram, record your observations of the thickness of the walls on the left side of the heart compared with the right side.
3. Suggest reasons for the differences observed.
   • Try to locate the valves in the heart.

Discuss and explain

4. Describe the valves and suggest their function.
5. Write a summary paragraph about the structure and function of the heart.

Your pulse rate immediately after exercise can be used as a guide to your physical fitness. The fitter you are, the less elevated your heart rate will be after vigorous exercise.

The regular rhythmic beating of the heart is maintained by electrical impulses from the heart's **pacemaker**, which is located in the wall of the right atrium. Some people with irregular heartbeats are fitted with artificial, electronic pacemakers to regulate the heart's actions and correct abnormal patterns.

Try clenching your fist every second for five minutes. Getting a little tired? The heart is made up of special muscle called **cardiac muscle**, which never tires. Imagine having a 'cramp' or 'stitch' in your heart after running to catch the bus! Owing to its unique electrical properties, heart muscle will continue to beat even if it is removed from the body. Scientists have shown that even tiny pieces of this muscle cut from the heart will continue to beat when they are placed in a test tube of warm salty solution.

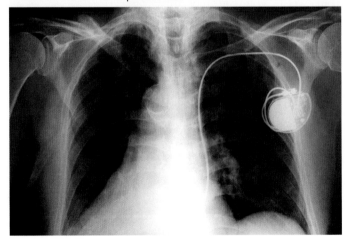

Person fitted with a pacemaker

4.6 Exercises: Understanding and inquiring

To answer questions online and to receive **immediate feedback** and **sample responses** for every question, go to your learnON title at www.jacplus.com.au. *Note:* Question numbers may vary slightly.

Remember

1. Match the circulatory system term with its description in the table below.

Term	Description
(a) Artery	A The bottom two chambers of the heart
(b) Atria	B Cell involved in transporting oxygen around the body
(c) Capillary	C Blood vessel that takes blood to the heart
(d) Heart	D Cell involved in protection against infection
(e) Red blood cell	E Blood vessel that takes blood away from the heart
(f) Vein	F The top two chambers of the heart
(g) Ventricles	G Organ that pumps blood around the body
(h) White blood cell	H Blood vessel that exchanges substances with cells

2. Outline what blood is, and what it does.
3. Name and describe the types of blood vessels in which blood travels around your body.
4. Compare red blood cells, white blood cells and blood platelets.
5. Describe the relationship between arteries, capillaries and veins.
6. What is the difference between:
   (a) the blood in the two sides of the heart
   (b) the structure of the two sides of the heart
   (c) systolic and diastolic pressure?
7. Explain why there are valves in the heart.
8. Outline what blood pressure is caused by.
9. (a) How many times does a normal human heart beat each minute?
   (b) Suggest what may cause the heart rate to increase.
   (c) Explain how the rhythmic beating of the heart is maintained.
10. What is unusual about cardiac muscle?

Think

11. Use information in this section and other resources to relate structural features to the functions of the following parts of the circulatory system.

Part of system	Structural features	Function
Arteries		
Veins		
Capillaries		
Red blood cells		

12. Think of other ways that information about the components of blood could be organised visually. Organise the material in one of these ways.

Think and discuss

13. (a) Some people have religious grounds for disagreeing with the use of blood transfusions. Imagine a four-year-old child with a life-threatening condition. Her parents will not allow her to have the blood transfusion that she needs. What should the doctors do? Discuss this with your team and report your decision to the class. If there are any differences of opinion, organise a class debate on the issue.

(b) Would your response be different if the child was 18 years old and wanted the blood transfusion, but her parents would not allow it?

14. A day after donating blood, a person finds that they have an infectious disease that can be transmitted by blood. What should they do? Discuss this with your team, giving reasons for your opinions.

Using data

15. Which blood type is the most common? Which is the least common?
16. Which blood group(s), A, B, AB or O, can be accepted by:
    (a) all blood groups
    (b) blood group AB
    (c) blood group A?

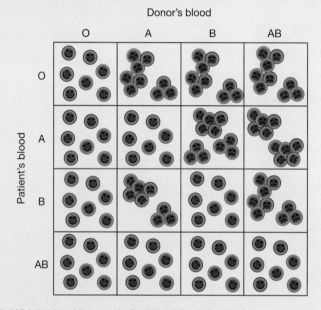

17. Which blood group, A, B, AB or O, can receive transfusions from all blood types?
18. Convert the information in the table above into a Venn diagram, target map or another visual thinking tool.
19. Observe the image at right of human blood cells.
    (a) Identify which are white blood cells and which are red blood cells.
    (b) Describe how you distinguished between the two types of blood cells.
    (c) Which are in the greatest abundance? Suggest a reason for this.

Investigate, think and discuss

20. The higher the altitude, the less oxygen there is in the air. Propose a reason people living at high altitudes usually have more red blood cells than people living at low altitudes.
21. Find out more about how blood circulates in insects and lobsters.
22. Construct a bar graph to show the proportions of the different parts of blood.

Human blood cells seen through a light microscope. The white blood cells are shown as purple, each with a nucleus.

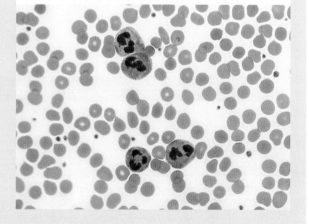

23. Find out what happens when people donate their blood at a blood bank. How often can you donate blood, how long does it take and how much blood do they take? Prepare a brochure, storyboard, PowerPoint presentation or cartoon to share your findings.

24. Use the internet to research waste removal and excretion for an animal of your choice. Present your findings in PowerPoint or as a poster.

25. Imagine that you have a friend who is anaemic. She is constantly tired and very pale.
    (a) Using the internet and other resources, find out what you could do to help her improve her health.
    (b) Report back to your team, sharing your ideas and any other relevant information. Have your team scribe summarise your ideas in a cluster map or mind map.
    (c) As a team, decide on a strategy for helping your anaemic friend.
    (d) Share your strategy with other teams as a mind map, flowchart, concept map or another visual tool.
    (e) Reflect on what you have learned during this activity. How might it influence your future behaviour or thinking? Could any of the strategies designed by the groups be used to solve any other problems? If so, which ones?

26. With a partner, construct a PMI chart for a law that makes it compulsory for everyone over 16 to donate blood at least once a year.

Investigate, think and create

27. (a) Copy the 'connected pathways' diagram on page 132 into your workbook.
    (b) Use a coloured pencil to show the path taken for a red blood cell to travel from the pulmonary vein to the pulmonary artery, if it goes via the intestines.

28. Mark the following sites (a, b, c, d) on your diagram. In which blood vessel(s) would you expect the highest:
    (a) blood pressure
    (b) blood glucose levels
    (c) blood carbon dioxide level
    (d) oxygen level?

29. List the following in the order that a red blood cell would reach them after leaving the aorta: pulmonary artery, left ventricle, right atrium, intestine, lung, pulmonary vein, left atrium, liver, right ventricle.

30. Convert your classroom or sports oval into a 'circulatory highway system'. Pretend to be a red blood cell and travel along the route it would take around the body.

31. Divide the class into groups, with each group making a model of a type of blood cell or vessel. Combine all of the models to convert the room into a circulatory system. In teams, write a play to use your circulatory system. Act it out for the class or get the class to act it out for you.

32. (a) Construct a model of the human heart using clay or bread dough.
    (b) Use information in this section and other resources to relate the structural differences between the atria and ventricles of the heart to their function.

33. Make up a rhyme or poem to describe the flow of blood through the heart. Use as many of the **bold blue** words in this section as possible.

34. Construct a 'working' model of a human heart that shows the movement of blood through the various chambers and blood vessels.

learn on RESOURCES — ONLINE ONLY

Watch this eLesson: Heart valve (eles-0858)

Try out this interactivity: Beat It! (int-0210)

Complete this digital doc: Worksheet 4.3: Blood and blood highways (doc-18717)

4.7 Transport technology

Science as a human endeavour

4.7.1 Scientific theories can change over time

Our understanding of the circulatory system has been built by scientists and physicians throughout human history. With new observations and evidence, some theories have been discarded and others developed or modified. New technologies have enabled new observations to be made, which have resulted in new ways of thinking about the structure and functioning of the human body.

Claudius Galen (*c*.129–*c*.199 AD)

For over a thousand years, the key training books used for doctors were based on the ideas of the Greek physician Claudius Galen. Galen's ideas were based on his observations of dissections of animals (other than humans). Galen described the human heart as being made up of two chambers and also being the source of the body's heat. He believed that blood was made by the liver and travelled to the right chamber of the heart and that the left chamber made 'vital spirits' which were then transported by arteries to body organs.

Andreas Vesalius (1514–1564)

Hundreds of years later another physician, Andreas Vesalius, began to transform medical knowledge — by questioning all previous theories. He believed that it was necessary to dissect bodies to find out how they worked. As the Church did not allow this, he took bones from graves and even stole a body from the gallows. His drawings showed the position and working of the muscles and organs in the body. Vesalius's observations proved that some of Galen's theories were wrong and he discovered anatomical structures previously unknown. His findings helped establish surgery as a separate medical profession.

Andreas Vesalius

A drawing by Vesalius

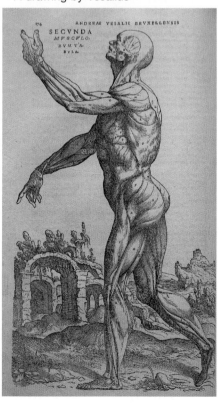

William Harvey (1578–1657)

Although Vesalius had assisted in revising the structure of the human heart, there was still confusion about its function. About 100 years later, William Harvey, an English physician, published his work on blood circulation which led to another change in how we think about the heart and our circulatory system.

William Harvey

4.7.2 Heart technology

Heart and blood vessel diseases are a key cause of death for many Australians. Medical research and new technologies strive to minimise the effects of diseases and disorders of the circulatory system.

Faulty heart and vein valves

The heart, like many other pumps, depends on a series of valves to work properly. These valves open and close to receive and discharge blood to and from the chambers of the heart. They also stop the blood from flowing backwards. If any of the four heart valves becomes faulty, the function of the heart may be impaired.

Veins throughout the body may also contain valves that keep the blood flowing in one direction. Defective valves in leg veins can cause blood to drain backwards, and to pool in the veins closest to the skin surface. These veins can become swollen, twisted and painful, and are called **varicose veins**.

A faulty heart valve may be replaced by an artificial valve. Why are the heart valves so important to the functioning of the heart?

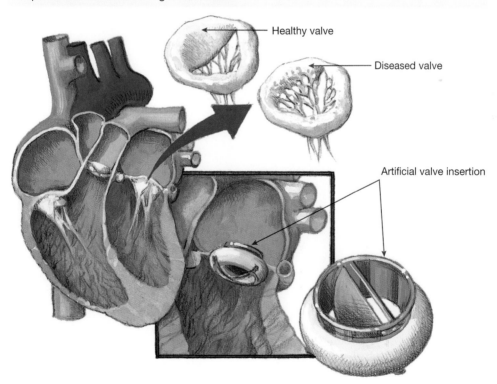

Healthy valve

Diseased valve

Artificial valve insertion

4.7.3 'If I only had a heart...'

The tin man from *The Wizard of Oz* would have been very happy with the development of an artificial heart. This mechanical device is made of titanium and plastic. Surgeons also implant a small electronic device in the abdominal wall to monitor and control the pumping speed of the heart. An external battery is strapped around the waist and can supply about 4–5 hours of power. An internal rechargeable battery is also implanted inside the wearer's abdomen. This is so they can be disconnected from the main battery for about 30–40 minutes for activities such as showering.

An artificial heart

4.7.4 A heart – but no pulse?

If only the left ventricle is damaged, and the rest of the heart is in good working order, a backup pump may be implanted alongside the heart. One model of these devices results in its wearers having a gentle whirr rather than a pulse. This is the sound of the propeller spun by a magnetic field to force a continuous stream of blood into the aorta.

Getting the beat!

An **electrocardiogram (ECG)** shows the electrical activity of a person's heart. ECG patterns are valuable in diagnosing heart disease and abnormalities.

Electrocardiograms

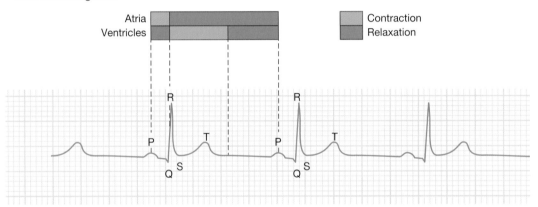

(a) Normal electrocardiogram

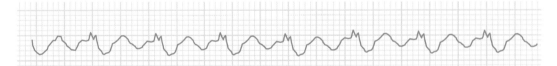

(b) Abnormal electrocardiogram

INVESTIGATION 4.8

Check your heart
AIM: To investigate the short-term effects of exercise on heart rate and blood pressure

Materials:
stopwatch
blood pressure monitor
optional: data logging or digital measuring devices

Method and results

1. Copy the table below into your workbook and enter your own results.

Test	Heart rate (bpm)	Blood pressure (mm Hg)
Before exercise		
After walking		
After running up stairs		

- Find your pulse, either on the inside of your wrist or in your neck (see the illustrations). Make sure you use two fingers, not your thumb, to find your pulse.

Two places where your pulse should be easy to find: (a) radial location (wrist) (b) carotid location (neck)

2. Measure and record your heart rate in beats per minute (bpm) by counting the number of times your heart beats in 15 seconds and then multiplying this number by 4.
3. Measure and record your blood pressure using the blood pressure monitor.
4. Go for a walk in the playground or around the school oval. Measure and record your heart rate and blood pressure again.
5. Run up and down a flight of stairs. Measure and record your heart rate and blood pressure again.

Discuss and explain

6. What effect does exercise have on heart rate and blood pressure?
7. Identify strengths and limitations of this investigation and suggest improvements.
8. Design and carry out an experiment to test the following hypothesis: 'There is a link between a person's resting heart rate and the number of hours the person spends exercising each week'.

4.7.5 Artificial blood

A current wave of interest in vampire movies and books has brought with it discussion about the merits of artificial blood sources. The interest in artificial blood, however, is not new; people have thought about its use in blood transfusions for hundreds of years. William Harvey's description in 1628 of how blood circulated through the body prompted a variety of unsuccessful investigations into the use of alternative fluid substitutes. Shortage of blood supplies during war and disease epidemics has fired up the quest for an artificial blood substitute. Currently, the two most promising red blood cell substitutes are haemoglobin-based oxygen carriers (HBOCs) and perfluorocarbon-based oxygen carriers (PFCs).

PFCs are usually white, whereas HBOCs are a very dark red. Although PFCs are entirely synthetic, HBOCs are made from sterilised haemoglobin. The haemoglobin may be from human or cow blood, human placentas or bacteria that have been genetically engineered to produce haemoglobin. As the haemoglobin doesn't have a cell membrane to protect it, various techniques such as cross-linking and polymerisation are used to make it less fragile. Some scientists are even investigating the idea of wrapping it in an artificial membrane.

4.7.6 Transplant pioneer

If your heart or lungs were not working properly and you had needed a heart or lung transplant in the 1980s, the doctor to see was Victor Chang.

Victor Chang was an Australian doctor who was awarded a Companion of the Order of Australia for his contribution to medicine. Dr Chang played an important role in establishing the heart transplant unit at St Vincent's Hospital in Sydney. He set up a team of 40 health professionals who were the finest in their field

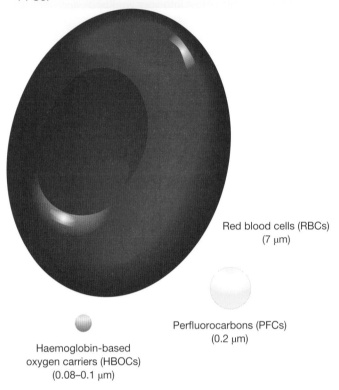

Human red blood cells are much larger than HBOCs and PFCs.

Red blood cells (RBCs) (7 μm)

Perfluorocarbons (PFCs) (0.2 μm)

Haemoglobin-based oxygen carriers (HBOCs) (0.08–0.1 μm)

and developed new procedures and techniques that led to an improved rate of success. Of his patients, 92 per cent were still alive one year after their heart or lung transplant operation and 85 per cent were still alive five years later.

The first heart transplant operation that Victor Chang carried out at St Vincent's Hospital was in 1984 on a young girl called Fiona Coote. Fiona is now an adult and, although she has since needed a second heart transplant, she owes her life to Dr Chang.

Dr Victor Chang also developed an artificial heart valve, called the St Vincent heart valve, and was working on developing an artificial heart. Unfortunately, his life was tragically cut short in 1991 when he was murdered by gunshot.

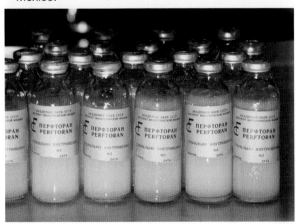

These vials contain the blood substitute perftoran. Perftoran is a perfluorocarbon-based oxygen carrier (PFC) that is currently used in Russia and Mexico.

The late Dr Victor Chang, pioneering heart transplant surgeon

In the future, will artificial blood vessels like this one be made by bacteria?

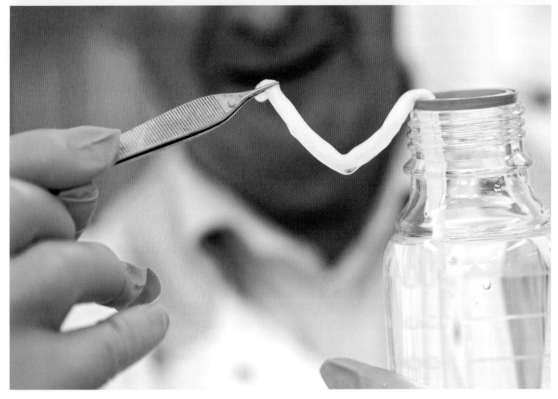

Artificial blood vessels?

Will the artificial blood vessels of the future be made by bacteria? Molecular biologist Helen Fink, working in Sweden, has suggested this may be the case. The cellulose produced by *Acetobacter xylinum* bacteria is strong enough to cope with blood pressure and function within our bodies, and could be used for artificial blood vessels in heart bypass operations in the future.

4.7 Exercises: Understanding and inquiring

To answer questions online and to receive **immediate feedback** and **sample responses** for every question, go to your learnON title at www.jacplus.com.au. *Note:* Question numbers may vary slightly.

Remember

1. What are varicose veins and what causes them?
2. What is an electrocardiogram and when is it useful?
3. Explain why valves are important to the functioning of the heart.
4. Describe how an ECG is used to detect heart abnormalities.
5. Outline how heart valves are similar to the valves in veins.
6. Construct a matrix table to show the differences between red blood cells, HBOCs and PFCs.

Think

7. Look at the electrocardiograms on page 142.
   (a) At 'P', are the muscle cells of the atria contracted or relaxed?
   (b) After the 'QRS' wave, is the ventricle relaxed or contracted?
   (c) How does the normal electrocardiogram differ from the abnormal electrocardiogram?
   (d) Suggest what might be wrong with the heart activity shown on the abnormal electrocardiogram.

Investigate

8. What are artificial hearts made of and how do they work?
9. How can blood loss cause death?
10. Use the internet to research one of the following: stroke, heart murmur, 'hole in the heart', atherosclerosis, angina, heart attack, arrhythmias, pericarditis, hypertension. Present your findings as a PowerPoint presentation or a poster.
11. Find out more about Dr Victor Chang and report your findings to others in the class.
12. There are a number of issues surrounding the development and use of artificial blood. Find out what these are and then construct a PMI chart as a summary. What is your opinion about artificial blood? Provide reasons for your opinion.
13. Find out and report on 'transport technology' research that Australian scientists are currently involved in.
14. (a) Which organs are most successfully transplanted into humans?
    (b) List sources of the organs for transplant and identify associated issues.
    (c) Describe how donors and organ recipients are matched.
    (d) Organ recipients can require specific treatment after the operation. Outline what this involves and why it is needed.
15. If you required a new heart, would you prefer an artificial one or one from a human or other natural source? Provide reasons for your response.
16. Outline your opinion on being an organ donor yourself.
17. Find out issues related to organ transplants and construct a PMI summary.
18. Research one of the following circulation topics and report your findings to the class: blood transfusions, rhesus babies, varicose veins, leukaemia, haemophilia, thrombosis, embolisms, aneurisms.
19. Plasma is the liquid part of blood. It can be used to make a number of products to help others in need. Report back on the uses of three of the following: Intragram, Anti D, Albumex 20, Factor VIII, Monofix, Prothrombinex, Thrombotrol VF, normal immunoglobulin, hyper-immunoglobulin.

Investigate, think and discuss

20. Dr Mary Kavurma and Dr Seana Gall are Tall Poppy Science Award winners. This award recognises young scientists who excel at research, leadership and communication. Dr Kavurma is a scientist at the University of New South Wales involved in research into atherosclerosis and cardiovascular disease. Dr Gall is based at the Menzies Research Institute, University of Tasmania, and her research field is cardiovascular epidemiology.
    (a) Find out more about their research and that of other scientists in this field of science.
    (b) Find out more about Australia's Tall Poppy Science Awards and other winners.
21. Find out more about Galen, Vesalius and Harvey, and their work and discoveries. Suggest how they were influenced by the times in which they lived. Why didn't they just accept the ideas of their times? Why did they ask questions? Propose a question or hypothesis that you may have asked if you lived in each of their times.
22. Find examples of how developments in imaging technologies have improved our understanding of the functions and interactions of our body systems. Share your findings with others.
23. Investigate how technologies using electromagnetic radiation are used in medicine; for example, in the detection and treatment of cancer of the circulatory or respiratory system.
24. (a) Identify issues associated with organ transplantation.
    (b) As a team, select one of these issues and find out why it is an issue.
    (c) What is your opinion on the issue?
    (d) Share your opinion and reasons for it with other members of your team.
    (e) Construct a team PMI on the organ transplant issue.

Investigate, think and create

25. In your team, design and perform an experiment to investigate the effect of different types of activities on your heart rate.
26. Doctors use a stethoscope to listen to heartbeats. Make and test your own stethoscope using rubber tubing and a plastic funnel.
27. What does a cardiologist do? Find an example of an Australian cardiologist and write a 'diary entry' for a day at work for them.

28. Find articles in the media that advertise foods or drinks that can reduce heart disease. In a team, research the claims and summarise your findings in a SWOT diagram. As a class, be involved in a debate that includes members from different interest groups or with different perspectives or biases.
29. Can diet, exercise or lifestyle choices change the chances of you having a heart attack? Research this question, summarising your findings in a PMI chart. On the basis of your data, what is your personal answer to this question? Give reasons for your opinion.
30. (a) Find examples of scientific research on the circulatory system.
    (b) Create a poster, PowerPoint presentation or podcast on the research and present your findings to the class.
31. (a) Use the internet to identify problems relating to the circulatory system.
    (b) Select one of these problems and construct a model or animation to demonstrate its effect on normal body function.

4.8 Respiratory system — breathe in, breathe out

4.8.1 Cells need energy!

Breathe in deeply… now breathe out. You have exchanged gases with your environment. You have supplied your body with some essential oxygen and removed some unwanted carbon dioxide. You do this about 15–20 times a minute without even having to think about it. Where does this oxygen go and where did the carbon dioxide come from?

Your cells need **oxygen** as it is essential for **cellular respiration**. This process involves breaking down **glucose** so that energy is released in a form that your cells can use. This reaction produces **carbon dioxide** as a waste product that needs to be removed.

$$\text{glucose} + \text{oxygen} \rightarrow \text{carbon dioxide} + \text{water} + \text{energy}$$

HOW ABOUT THAT!

Wrong way, turn back! There is a flap of tissue at the top of the trachea called the **epiglottis**. This tissue's job is to stop food 'going down the wrong way'. If food does go the wrong way, a cough moves the food back up, to either be removed or travel its correct pathway — down your oesophagus to your stomach.

Respiratory system

The main role of your **respiratory system** is to get oxygen into your body and carbon dioxide out. This occurs when you inhale (breathe in) and exhale (breathe out). The respiratory system is made up of your **trachea** (or windpipe) and your **lungs**.

Getting oxygen to your lungs

When you breathe in, air moves down your trachea, then down into one of two narrower tubes called **bronchi** (bronchus). After that, the air moves into smaller branching tubes called **bronchioles**, which end in tiny air sacs called **alveoli** (alveolus). It is at the alveoli that gases (such as oxygen and carbon dioxide) are exchanged between the respiratory system and the circulatory system.

(a) Organs of the respiratory system with (b) a portion of the lung expanded to show details

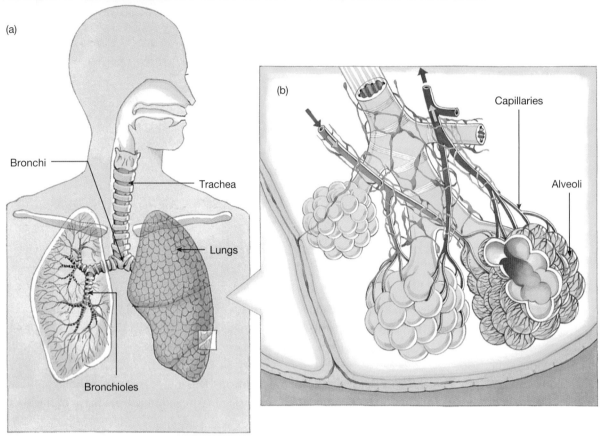

4.8.2 Working together to get oxygen from lungs TO cells

Your circulatory and respiratory systems work together to get oxygen to your cells. Once you have breathed in and oxygen has reached your alveoli, oxygen diffuses into red blood cells in capillaries that surround the alveoli.

The oxygen diffuses into the red blood cells because there is a higher concentration of oxygen inside the alveoli than inside the blood cells. Once inside the red blood cells, oxygen binds to haemoglobin to form oxyhaemoglobin. It is in this form that oxygen travels throughout your body. The blood that it travels in is referred to as oxygenated blood.

The pathway oxygen travels to your lungs when you inhale.

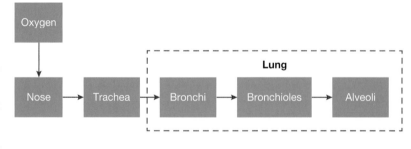

The oxygenated blood travels from your lungs via the **pulmonary vein** to the left atrium of your heart. From here, it travels to the left ventricle where it is pumped under high pressure to your body through a large artery called the **aorta**. The oxygenated blood is then transported to smaller vessels (**arterioles**) and finally to capillaries through which it diffuses into body cells for use in cellular respiration.

In an alveolus, oxygen diffuses into the blood and carbon dioxide diffuses out of the blood.

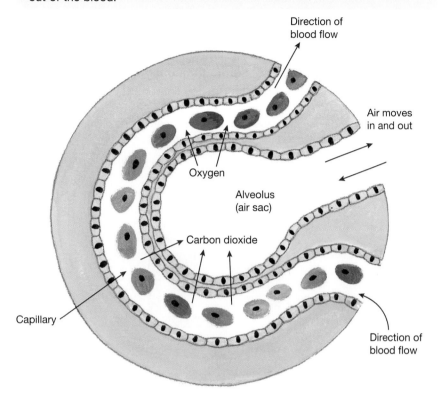

The pathway oxygen travels from your lungs to your body cells.

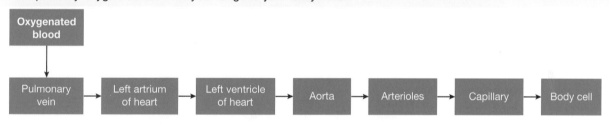

The pathway carbon dioxide travels from your body cells to your lungs.

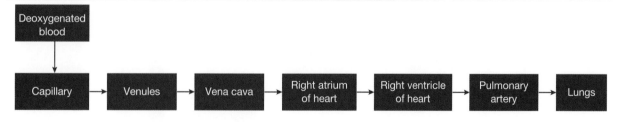

HOW ABOUT THAT!

Phew … garlic breath! Have you ever heard someone say this? Garlic or onion breath comes from further down than your mouth! It has travelled through a number of your body systems. After you have eaten food containing either of these, and it has been digested, it is absorbed through the walls of your intestines and then into your blood. When the smelly onion or garlic blood reaches your lungs through your circulatory system, you breathe out the smelly gas.

4.8.3 Working together to get carbon dioxide FROM cells

Carbon dioxide is a waste product of cellular respiration and needs to be removed from the cell. When carbon dioxide has diffused out of the cell into the capillary, the blood in the capillary is referred to as deoxygenated blood. This waste-carrying blood is transported from the capillaries to small veins (**venules**) to large veins called vena cava, then to the right atrium of your heart. From here it travels to the right ventricle where it is pumped to your lungs through the **pulmonary artery** (the only artery that does not contain oxygenated blood). This process is shown in the flowchart on the previous page.

Your respiratory and circulatory systems form connected highways that provide your cells with what they need and remove what they don't.

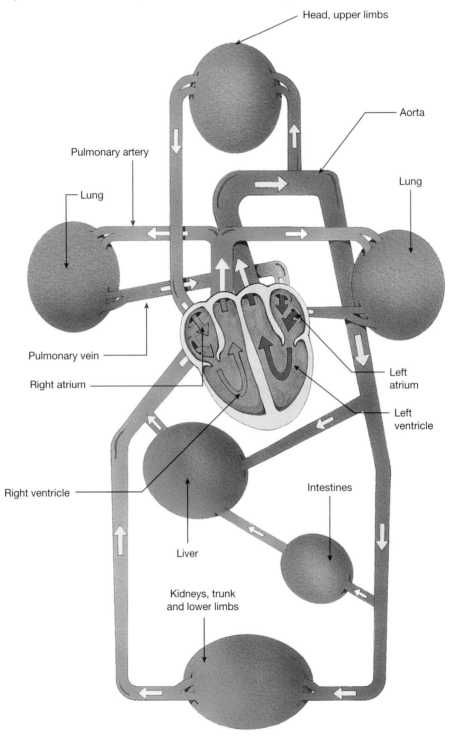

Exhaling carbon dioxide from lungs

Once the deoxygenated blood reaches the alveoli of the lungs, carbon dioxide diffuses out of the capillaries. This occurs because there is a higher concentration of carbon dioxide inside the capillaries than in the alveoli. Carbon dioxide is then transported into the bronchiole, then bronchi and trachea, until it finally exits through your nose (or mouth) when you exhale.

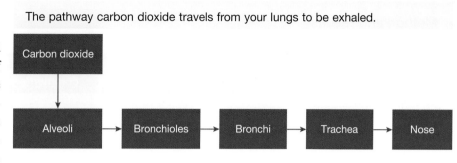

The pathway carbon dioxide travels from your lungs to be exhaled.

Carbon dioxide → Alveoli → Bronchioles → Bronchi → Trachea → Nose

4.8.4 Brain AND muscle?

The respiratory system also relies on organs from other systems. When you breathe in, a muscle beneath your rib cage called the **diaphragm** tightens. This allows the lungs to expand and air to be pulled into them. When you breathe out, the diaphragm relaxes, which reduces the size of the lungs and pushes air out. The largest volume of air that you can breathe in or out at one time is called your **vital capacity**.

Breathing involves muscle movements that are automatic and controlled by the respiratory centre in the brain.

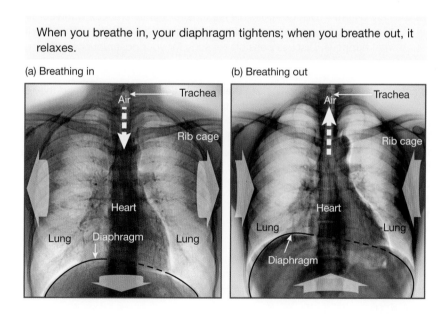

When you breathe in, your diaphragm tightens; when you breathe out, it relaxes.

(a) Breathing in

(b) Breathing out

INVESTIGATION 4.9

Hands on pluck
AIM: To investigate the trachea, lungs, heart and liver of a mammal

Materials:
sheep's pluck (heart and lungs) with part of the liver and trachea attached
newspaper and tray to place the pluck on
plastic disposable gloves
balloon pump and rubber tubing

Method and results

1. Copy the table below into your workbook.

Organ	Shape (sketch)	Approx. size (cm)	Colour	Texture	Other comments	System to which the organ belongs
Trachea						
Lung						
Heart						
Liver						

2. Carefully observe and record the shape, size, colour and texture of the sheep's trachea, lungs, heart and liver. Include notes on how they are connected. Can you see any blood vessels?
   - Push a piece of rubber tubing down the trachea to the lungs and use a balloon pump to blow some air into the trachea.

CAUTION

For hygiene reasons, it is not recommended that you use your mouth to blow into the tube inserted in the trachea.

3. Cut through the lung, heart and liver tissue. Make a record of your observations describing how they are similar and how they are different. Discuss possible reasons for the differences with your team members.
4. Using a scalpel or scissors, cut off a small piece of heart, lung and liver. Place each piece into a beaker of water and observe what happens. Discuss possible reasons for your observations with your team members.

Discuss and explain

5. Could you see any blood vessels? Try to find out their names and what sort of blood they carry.
6. Suggest why there are rings of cartilage around the trachea.
7. Suggest reasons for the differences in texture between the heart and lungs.
8. Suggest reasons for the differences in the shapes of the organs that you observed.
9. Comment on something that you learned or found particularly interesting from this investigation. Share your comment with others.
10. Research and report on the following points for each of the organs in this investigation:
    - its function and how it carries this out
    - the system to which it belongs
    - a disease relevant to it.

INVESTIGATION 4.10

Measuring your vital capacity
AIM: To investigate the vital capacity of lungs

Materials:
balloon
ruler

Method and results
- Blow up a balloon to about 20 cm in diameter two or three times to stretch it. Release the air each time.
- Take the biggest breath you can, then blow out all the air into the balloon. Tie up the end of the balloon to hold in your 'blown out' air.
1. Use a ruler to measure and record the diameter of the balloon as shown on the opposite page.
2. Use the table at the bottom of the opposite page to determine your approximate vital capacity in litres.

3. Release the air from the balloon and repeat your measurement of vital capacity three more times. Average your results to get your best estimate of the maximum 'blow-out' of your lungs.

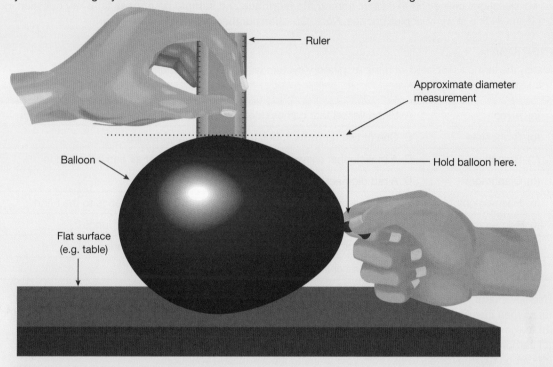

Ruler

Approximate diameter measurement

Balloon

Hold balloon here.

Flat surface (e.g. table)

4. Draw up a table with the following headings.

Name	Male or female?	Does this student play a wind instrument?	Lung capacity (L)

5. Collect results from all the students in your class and complete the table.
6. Calculate the average lung capacity for all the females and for all of the males in your class.
7. Calculate the average lung capacity for all students in your class who play a wind instrument.

Discuss and explain

8. Suggest why were you asked to stretch the balloon first.
9. Suggest why you measured your vital capacity four times.
10. With reference to your results, do females have a bigger or smaller lung capacity than males in your class?
11. Compare the average lung capacity for students who play a wind instrument with the average value for students who do not. Do your results suggest that playing a wind instrument has an effect on lung capacity? Explain.
12. Suggest another way of measuring the amount of air exhaled with each breath.

How to measure the diameter of the balloon

Balloon diameter (cm)	8	9	10	11	12	13	14	15	16	17	18	19	20	21
Approx. vital capacity (litres)	0.3	0.4	0.5	0.7	0.9	1.2	1.4	1.8	2.1	2.6	3.0	3.6	4.2	4.8

4.8 Exercises: Understanding and inquiring

To answer questions online and to receive **immediate feedback** and **sample responses** for every question, go to your learnON title at www.jacplus.com.au. *Note:* Question numbers may vary slightly.

Remember

1. Match the terms associated with the respiratory system with their description in the table below.

Term	Description
(a) Alveoli	A Blood vessel that carries deoxygenated blood from the heart to the lungs
(b) Bronchiole	B One of two narrower tubes that leads off the trachea
(c) Bronchus	C A muscle that allows the lungs to expand so that air can be pulled in
(d) Diaphragm	D A red pigment that binds to oxygen
(e) Haemoglobin	E Blood vessel that carries oxygenated blood from the lungs to the heart
(f) Pulmonary artery	F Tube through which air moves from your mouth to your lungs
(g) Pulmonary vein	G Tiny air sac through which oxygen diffuses into capillaries
(h) Trachea	H Small branching tube with alveoli at its end

2. Use flowcharts to identify the pathway that:
   (a) oxygen travels to get from the air outside your body to the alveoli of your lungs
   (b) oxygen travels to get from your lungs to your body cells
   (c) carbon dioxide travels to get from your body cells to the alveoli of your lungs
   (d) carbon dioxide travels to get from your lungs to the air outside your body.

Using data

3. The table below shows approximate percentages of various gases breathed in and breathed out.

What goes in and what comes out

Gas	Oxygen (%)	Carbon dioxide (%)	Water vapour (%)	Nitrogen (%)
Air breathed in	21	0.04	1	78
Air breathed out	15	4	5	76

   (a) (i) Compare the percentage of oxygen breathed in to that breathed out.
       (ii) Suggest a reason for this pattern.
   (b) (i) Compare the percentage of carbon dioxide breathed in to that breathed out.
       (ii) Suggest a reason for this pattern.
   (c) The percentages in the table can vary in different weather conditions and at different heights above sea level. Research these variations and the possible implications this may have on humans.

Think and discuss

4. Describe how oxygen gets from the alveoli of your lungs into blood cells in your capillaries.
5. Differentiate between the terms 'cellular respiration', 'respiratory system' and 'breathing'.

Think and investigate

6. Use information in this section and other resources to relate structural features to the functions of the following parts of the respiratory system.

Part of system	Structural features	Function
Trachea		
Alveoli		
Lungs		
Capillaries		

7. Some people describe the structure of the lungs as an upside-down hollowed-out tree. Which parts of the lungs might be the following parts?
   (a) Trunk   (b) Branches   (c) Twigs   (d) Leaves
8. Give reasons for the following pieces of advice.
   (a) It is better to breathe through your nose than your mouth.
   (b) You should blow your nose when you have a cold rather than sniff it back.
   (c) You should not talk while you are eating or drinking.
9. Find out what a spirometer is.
10. Did you know that mountain climbers often find it difficult to breathe? Some wear oxygen tanks to allow them to climb very high mountains. Research the effects of high altitude on breathing and report your findings.
11. Some singers can hold a musical note for a very long time — investigate what muscles and techniques they use to be able to do this.

Think and create

12. Construct a model lung as shown in the diagram below. You can use the following items:
    - two clear 1-litre plastic bottles with tops
    - four balloons
    - two plastic drinking straws
    - rubber bands or very sticky tape
    - plasticine or Blu-Tack
    - scissors.
13. Carefully observe the lung model shown below.

A model lung. When the rubber sheet at the bottom is pulled down, the pressure inside the jar drops and air is sucked into the balloon. The balloon inflates (blows up).

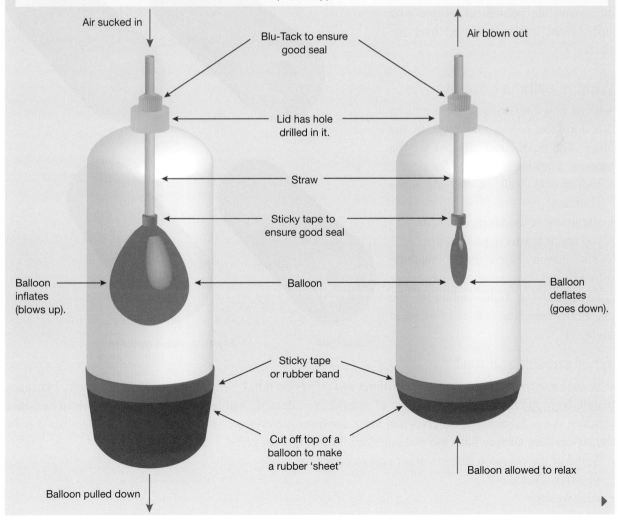

Air sucked in

Blu-Tack to ensure good seal

Air blown out

Lid has hole drilled in it.

Straw

Sticky tape to ensure good seal

Balloon inflates (blows up).

Balloon

Balloon deflates (goes down).

Sticky tape or rubber band

Cut off top of a balloon to make a rubber 'sheet'

Balloon allowed to relax

Balloon pulled down

(a) Identify which body parts are represented in the model by the:
   (i) straw
   (ii) rubber sheet at the bottom of the bottle
   (iii) balloon connected to the straw
   (iv) plastic bottle.
(b) Pull the rubber sheet at the bottom of your model downwards. Record your observations. Release the rubber sheet. Record your observations. Discuss how your observations relate to how a lung works. Suggest how the model could be improved.

learn on RESOURCES — ONLINE ONLY

📄 **Complete this digital doc:** Worksheet 4.4: Breathing — constructing a report (doc-18718)

4.9 Short of breath?

Science as a human endeavour

4.9.1 Asthma

If you do not suffer from **asthma**, it is very likely that you know someone who does. Asthma is a very common condition and the number of people who suffer from it has increased over the years.

What is asthma?

Asthma is a narrowing of the air pipes that join the mouth and nose to the lungs. The pipes most affected are the bronchi. They become narrower as:
- the muscle wall of the air pipes contracts
- the lining of the air pipes swells
- too much mucus is produced.

The narrow pipes make breathing difficult and can result in wheezing, coughing and a tight feeling in the chest. The coughing is usually worse at night.

Asthma is a narrowing of the air pipes.

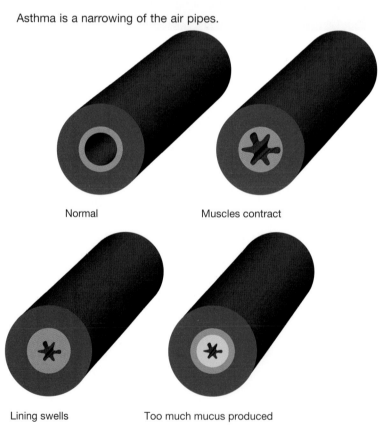

Normal   Muscles contract

Lining swells   Too much mucus produced

What causes asthma?

It is not known why some people get asthma and others do not. It seems that it can be inherited, but many people from families without a history of asthma are affected. Asthma is certainly the result of sensitive airways. An asthma attack occurs when those sensitive airways are 'triggered'. If the sufferer has a cold, the airways are already inflamed and are more likely to be triggered.

Some of the common triggers of an asthma attack are:
- vigorous exercise
- cold weather

- cigarette smoke
- dust and dust mites
- moulds
- pollen
- air pollution
- some foods and food additives
- some animals.

Not all asthma sufferers are affected by the same triggers. Some people suffer attacks only as a result of exercise. Others might be affected by any one or more of the triggers. It is important that those who get asthma try to find out what triggers the attacks. Many of the triggers can be avoided.

4.9.2 Controlling the triggers

The best way to control asthma is to avoid the triggers. While this is not always possible, it is worthwhile to recognise the triggers, so that you can minimise them.

Pollen and moulds

Pollen from some grasses and trees is very light and becomes airborne on even slightly windy days. The inhaling of pollen can be reduced by avoiding outdoor activities and keeping windows and doors closed on breezy spring days. Moulds live in warm, humid conditions and thrive in bathrooms, kitchens and bedrooms. Their spores are easily breathed in, triggering attacks in some asthma sufferers. Moulds can be reduced by airing the house regularly.

Air pollution

Those asthma sufferers whose attacks are triggered by air pollution are warned to remain indoors as much as possible and avoid vigorous activity on smoggy days. If tobacco smoke is a trigger, the cigarette smoke of others needs to be avoided.

Dust mites

Dust mites are a common trigger of asthma attacks. Dust mites are microscopic animals that live in their thousands in warm, moist and dark places like doonas, sheets, pillows, carpets and curtains. Dust mite droppings float in the air and are easily inhaled.

A house dust mite

Since you share so much of yourself and where you live with this fellow Australian, you should probably know its name. It is the most common dust mite (a relative of spiders and ticks) in Australia, *Dermatophagoides pteronyssinus*. The good news is that it is half a millimetre long and doesn't bite. The bad news is that there may be thousands of them living in your pillow, each defecating about 20 faecal pellets a day, reproducing (each female laying about 30 eggs in her lifetime), dying and decomposing. The fact that dust mites mate for 24 hours at a time (perhaps because their penis is only about as wide as their sperm) may make this particularly disturbing!

Our skin scales are the main food source for these dust mites, so wherever we are, they are. Dr Janet Rimmer (a respiratory physician and Director of the National Asthma Council Australia) also suggests that, of the 45 per cent of Australians who are affected by allergies, about 80 per cent are allergic to dust mites. But not all researchers have bad news about dust mites. Dr Matthew Colloff, a CSIRO researcher, has found them so interesting that he wrote a book (called *Dust mites*) about them.

Even the cleanest house has dust mites, but their numbers can be reduced by:

- exposing your mattress to the sun, because dust mites are susceptible to drying out
- washing bedding materials and bedclothes with tea-tree or eucalyptus oil or in hot water (above 55°C)
- removing soft toys that collect dust or hot washing them weekly
- regularly vacuuming curtains and carpets
- airing the bedroom by keeping doors and windows open
- replacing carpets with hard flooring.

HOW ABOUT THAT!

Dust mites thrive best in bedding and carpets because these contain plenty of dead human skin cells. Humans shed a complete layer of dead skin cells every month. That amounts to about 1 kilogram of skin cells each year. In fact, most of the dust in your house consists of dead skin cells.

4.9.3 Asthma medication

Asthma medications can be divided into two main groups: preventers and relievers. Preventers make the lining of the airways less sensitive and therefore less likely to be triggered. Relievers open up the airways once an attack has commenced. Most asthma medications are applied with inhalers or 'puffers', which direct the medication straight into the air tubes for fast action. Severe attacks of asthma require other drugs and sometimes extra oxygen needs to be supplied.

4.9.4 Up in smoke

Asthma is not the only condition that can interfere with your lungs functioning as they should. Some human activities can damage not only your lungs, but also those of others around you. Smoking is one example of such an activity. About 15 000 Australians die each year as a result of diseases caused by smoking. Smoking is actually the largest preventable cause of death and disease in Australia.

4.9.5 Just one cigarette

There are clearly many long-term effects of smoking. However, the diagram on the opposite page shows what happens to you after smoking just one cigarette.

There are some more obvious effects such as bad breath, body odour and watery eyes. After several cigarettes, your teeth and fingers become stained. Your sense of taste is reduced. Even your stomach is affected as acid levels increase.

Smoking and your lungs

Lung cancer is the most well-known disease caused by smoking. Chemicals that cause cancer are called **carcinogens**. Cigarette tobacco contains a number of carcinogens. The chemicals in cigarettes also clog up the fine hairs in your air tubes with a mixture of mucus and foreign chemicals.

Cough it up

Coughing is the body's way of trying to clear the air tubes. However, not all of the clogging can be cleared by coughing. A dirty mixture remains in the air tubes, causing swelling, making them sensitive and slowing down the passage of air. Eventually, the sticky mixture sinks down into the lungs, where it blocks some of the pathways to the alveoli, where freshly breathed air should deliver oxygen to the blood.

The diseases caused by this blocking process are called chronic obstructive pulmonary diseases, or COPD. **Emphysema** is the worst of these diseases and results in the eventual destruction of the alveoli.

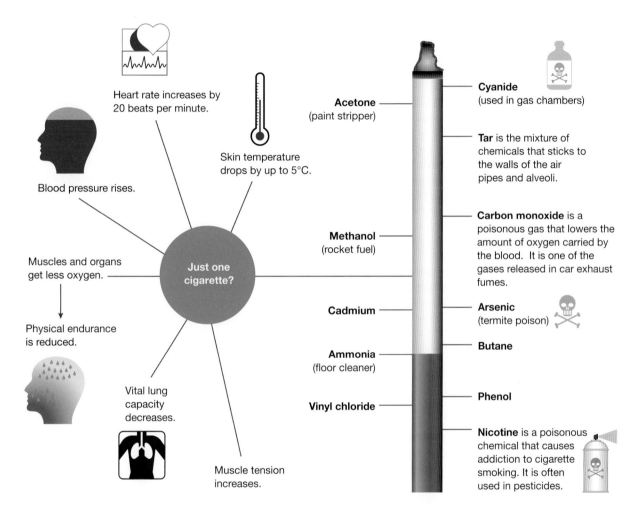

4.9.6 Professor Robyn O'Hehir

BSc, MBBS (Hons I), FRACP, PhD, FRCP, FRCPath

1. **What is your current science-related title?**
   I am a Professor of Medicine, with particular responsibilities for allergy, clinical immunology and respiratory medicine, at Monash University, Melbourne. I am also the Director of the Department of Allergy, Immunology and Respiratory Medicine at the Alfred Hospital in Melbourne.

2. **What field of science are you in?**
   Allergy, cellular immunology and respiratory medicine. I was appointed to the first Chair in Allergy and Clinical Immunology in Australia.

3. **Describe some science that you are involved in at the moment.**

Millions of people around the world suffer from allergies. I am sure you know several friends who have asthma or hay fever, or you may even have them yourself. Asthma and hay fever are usually triggered by proteins called allergens, from house dust mites or grass pollens. Allergies to peanuts and shellfish are less common but often more serious, because they can trigger life-threatening allergic reactions called anaphylaxis. Allergies are caused by reactions between white blood cells ('T cells') and environmental proteins that are usually harmless. My research group is trying to find ways to damp down the allergic T-cell responses.

Allergen immunotherapy (allergy shots) is the only treatment that can prevent allergic diseases, but currently it can't be used for peanut allergies, even though this is one of the most serious allergens. To develop a safe and effective vaccine against peanut allergies, we are identifying parts of critical peanut proteins that can build up tolerance in allergic patients without risking anaphylaxis.

4. **What do you enjoy about being a scientist?**

I enjoy the fact that my research not only is laboratory-based, exploring novel methods for switching off allergic responses, but also lets me see patients and train other doctors in how to do research from bench to bedside to the community. I head an active clinical department, still carry out clinics with patients, and am actively engaged in national and international tests of new preventions and treatments for allergies. My combined research and clinical duties allow translation of our research findings into better clinical practice.

5. **What triggered your interest in science?**

I decided to specialise in allergy and respiratory medicine, focusing on asthma, following my experiences as a young trainee physician at the Alfred Hospital in Melbourne. Asthma was a huge problem in Australia at that time, and many times I resuscitated young adults in the hospital emergency room — and I watched them return, with appropriate medication and careful education, to confident, full lives. Some remain my patients today. The ability to dissect underlying mechanisms of disease and then work towards new therapeutics and practices to benefit patients is a great excitement and honour. The diversity of patients and their needs ensures that every day is quite different.

6. **Do you have any other comments that may be of interest to Year 8 Science students?**

A career in science combined with medicine may take a bit longer in terms of training, but it gives you a fantastic ability to do interesting work that is intellectually demanding and also involves working with lots of people who need your help. I am very glad that I chose a career in science and medicine.

4.9 Exercises: Understanding and inquiring

To answer questions online and to receive **immediate feedback** and **sample responses** for every question, go to your learnON title at www.jacplus.com.au. *Note:* Question numbers may vary slightly.

Remember

1. What happens to the air pipes to the lungs during an asthma attack to make breathing difficult?
2. Why is an asthma attack more likely to be triggered in a person with a cold?
3. What is an asthma trigger?
4. What are the two major types of asthma medication and how are they different from each other?

Think and discuss

5. (a) In your team, brainstorm ideas about the common triggers of asthma that can be controlled. Summarise your discussion in a bubble map.
   (b) Construct a table similar to the one below.
   (c) Again as a team, add suggestions to your table of ways that the trigger could be controlled.

Trigger	How the trigger can be controlled
Moulds	Air the house regularly.

6. (a) If you suffer from asthma, prepare a talk for the rest of your class explaining:
   (i) what it is
   (ii) how it affects you
   (iii) how you control it or try to prevent attacks.
   (b) If you do not suffer from asthma, write a set of at least five questions that you could ask an asthma sufferer in an interview. If possible, conduct the interview and record the answers in writing, or as audio or video.

Think

7. Draw up a two-column table. The first column should be headed 'Reasons for smoking'; the second column should be headed 'Reasons for not smoking'. With at least one other person, complete the table. Then compare your table with others. You might be able to construct a large table for the whole class.

8. Smoking-related diseases cost taxpayers many millions of dollars because hospitals are mostly paid for by governments. Write down your opinion of each of the proposals below. Give reasons for your opinion.
   (a) The cost of hospital treatment for diseases caused by smoking should be paid for by the patient because it was their fault that they got sick.
   (b) Cigarettes should cost more. The extra money made from them could then be given to hospitals to help pay for treating smoking-related diseases.
   (c) Cigarette companies who make profits from smoking should be made to pay for hospital treatment of patients with diseases caused by smoking.

With over 4000 chemicals in each cigarette, smoking can lead to any of these conditions and effects.

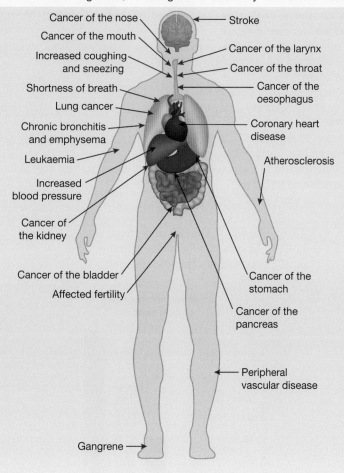

Investigate and discuss

9. Propose a series of questions to find out more about each of the areas below, investigate them, and then share your findings with others.
    (a) Allergies
    (b) Asthma
    (c) Anaphylaxis
    (d) Allergen immunotherapy
    (e) Clinical immunology and respiratory medicine
10. (a) Find out more about Anaphylaxis Australia Inc.
    (b) Outline the topics covered in a first aid course for management of anaphylaxis.
    (c) What is an EpiPen and how is it used?

Percentage of adult Australians who smoke

Year	1945	1964	1969	1974	1976	1980	1983	1986	1989	1992	1998	2001	2004
Males (%)	72	58	45	41	40	40	37	33	30	28	29	28	26
Females (%)	26	28	28	29	31	31	30	28	27	24	24	21	20

Analyse and evaluate

11. The table above shows how the popularity of smoking has changed over the past 70 years or so.
    (a) Draw a line graph of the data in the table. Use 'Year' on the x-axis and '% of adult Australians who smoke' on the y-axis. Draw lines for males and females in different colours.
    (b) Why do you think that the percentage of females who smoke has changed little while the percentage of males who smoke has declined greatly?
    (c) Use dotted lines to show your prediction of the trends up to the year 2020. What percentage of males and females do you predict will be smoking in 2020?
12. Study graph 1 on the below.
    (a) Copy and complete the following statements.
        (i) People who smoke 10 cigarettes a day are _____ times more likely to develop lung cancer than non-smokers.
        (ii) People who smoke 30 cigarettes a day are _____ times more likely to develop lung cancer than people who smoke 10 cigarettes a day.
    (b) If a packet of cigarettes costs $15 and contains 20 cigarettes, calculate how much a person smoking 40 cigarettes a day spends on smoking:
        (i) each day   (ii) each week   (iii) each year.

Graph 1: The risk of getting lung cancer increases with the number of cigarettes smoked daily.

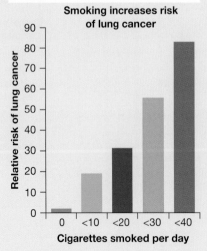

13. Study graph 2 on the right.
    (a) Describe how the incidence of lung cancer deaths changed between 1900 and 1980.
    (b) Identify when the number of male smokers peaked.
    (c) Identify when the number of deaths from lung cancer peaked.
    (d) Explain why there is a 20-year gap between the two numbers.
    (e) The graph shows data for male smokers only. Predict when the number of cases of lung cancer deaths in women peaked (use the graph you drew for question 11 to answer this).

Investigate and create

14. Draw a poster that sends one single important message about smoking.
15. Design an experiment that would investigate the effect of passive smoking on heart rates. Investigate whether any relevant research has been performed in this area. If so, share it with others in your class.
16. (a) Research the structure and function of an alveolus.
    (b) Suggest how the structure of an alveolus is related to its function.
    (c) Suggest how smoking affects the ability of an alveolus to perform its function.

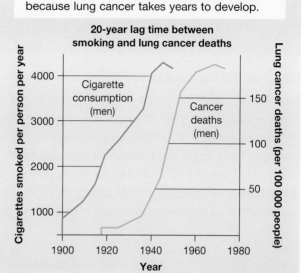

Graph 2: The number of deaths from lung cancer has risen as cigarette consumption has increased but there is a 20-year lag time because lung cancer takes years to develop.

4.10 The excretory system

Being alive requires energy and nutrients. It also results in the production of wastes that need to be removed.

4.10.1 Excretory system

Excretion is any process that gets rid of unwanted products or waste from the body. The main organs involved in human excretion are your **skin**, lungs, **liver** and **kidneys**. Your skin excretes salts and water as sweat, and your lungs excrete carbon dioxide (produced by cellular respiration) when you breathe out.

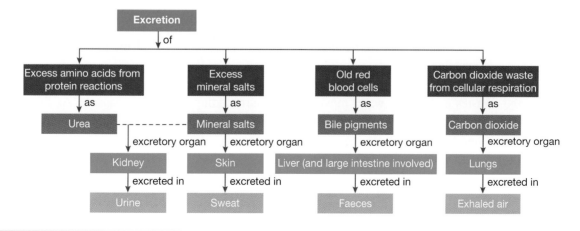

Your liver is involved in breaking down toxins for excretion, and your kidneys are involved in excreting the unused waste products of chemical reactions (e.g. urea) and any other chemicals that may be in excess (including water) so that a balance within our blood is maintained.

4.10.2 Kidneys

If you put your hands on your hips, your kidneys are close to where your thumbs are. You have two of these reddish-brown, bean-shaped organs. Without them you would survive only a few days.

Organs, tubes and urine

Your kidneys play a key role in filtering your blood and keeping the concentration of various chemicals and water within appropriate levels. Each of your kidneys is made up of about one million **nephrons**. These tiny structures filter your blood, removing waste products and chemicals that may be in excess. Chemicals that your body needs are reabsorbed into capillaries. The fluid remaining in your nephrons travels through tubes called **ureters** to your **bladder** for temporary storage. As it fills, your bladder expands like a balloon. It can hold about 400 mL of this watery fluid which contains unwanted substances called **urine**. **Urination** occurs when urine moves from your bladder through a tube called the **urethra** and out of your body.

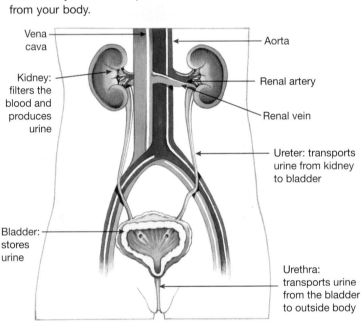

Your kidneys have an important role in the excretion of wastes from your body.

Vena cava

Aorta

Kidney: filters the blood and produces urine

Renal artery

Renal vein

Ureter: transports urine from kidney to bladder

Bladder: stores urine

Urethra: transports urine from the bladder to outside body

Nephrons — how their structure suits their function

Each nephron is made up of a long tubule (very fine tube) that forms a cuplike structure at one end called the **Bowman's capsule**. This structure surrounds a cluster of capillaries called the **glomerulus** (from an ancient Greek word meaning 'filter').

Blood containing wastes travels to the glomerulus within each nephron in your kidneys, where the blood is filtered. Wastes and excess water move into the surrounding Bowman's capsule. As this 'waste' fluid moves along the tubules, any useful substances are reabsorbed back into capillaries that are 'twisted' around the tubules, and hence back into circulation. The remaining fluid becomes urine, which eventually travels in your ureters to your bladder prior to urination.

Each of your kidneys is made up of about a million nephrons.

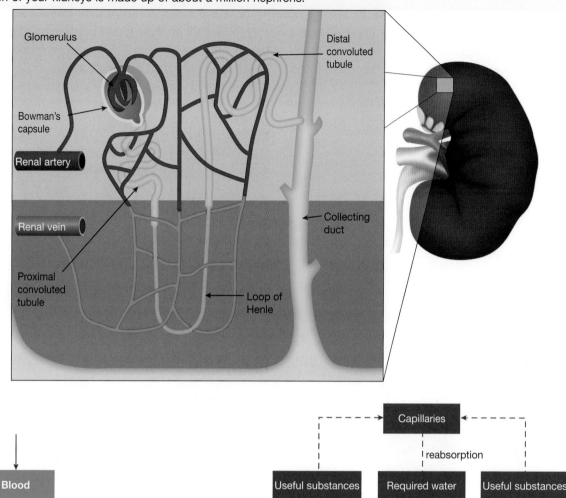

4.10.3 Have a drink!

Both blood and urine are mostly made up of water. Water is very important because it assists in the transportation of nutrients within and between the cells of the body. The concentration of substances in blood is also influenced by the amount of water in it.

Water helps the kidneys do their job because it dilutes toxic substances and absorbs waste products so that they can be transportation out of the body. If you drink a lot of water, more will be absorbed from your large intestines and your kidneys will produce a greater volume of dilute urine. If you do not consume enough liquid, you will urinate less and produce more concentrated urine.

4.10.4 Haemodialysis

People with kidney disease may not be able to remove the waste materials from their blood effectively. They may need to be linked up to a machine that does this job for them; their blood is passed along a tube that lets wastes, such as urea, pass out of it. However, useful substances, such as glucose, proteins and red blood cells, stay in the tube and are kept in the blood. This process is called **haemodialysis**.

Haemodialysis

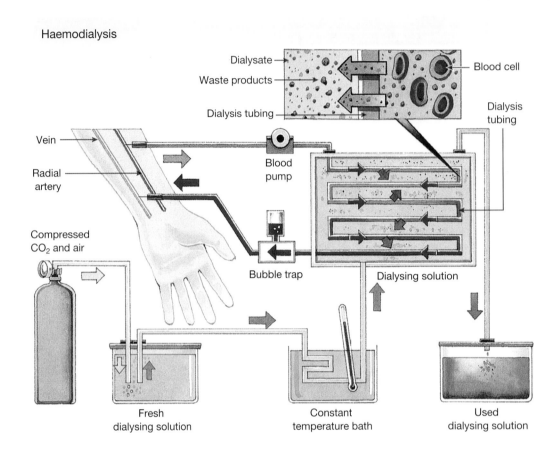

4.10.5 Liver

Livers are busy places!

Over a litre of blood passes through your liver each minute. Your liver is like a chemical factory, with more than 500 different functions. Some of these include sorting, storing and changing digested food. The liver removes fats and oils from the blood and modifies them before they are sent to the body's fat deposits for storage. It also helps get rid of excess protein, which can form toxic compounds dangerous to the body. The liver converts these waste products of protein reactions into urea, which travels in the blood to the kidneys for excretion. It also changes other dangerous or poisonous substances so that they are no long harmful to the body. Your liver is an organ that you cannot live without.

Too much alcohol?

The liver is also involved in breaking down alcohol. Alcohol is converted into a substance called acetaldehyde, then to acetate and finally into carbon dioxide and water. The carbon dioxide is transported from the liver to the lungs, and then exhaled out of the body. The water may be removed as vapour in breath, sweat on skin or as urine.

Alcohol can also affect the amount of urine produced by the kidneys. Reabsorption of water may be reduced in the kidneys, resulting in the production of more urine. Increased urination can result in dehydration and consequently impair other body functions.

HOW ABOUT THAT!

The human kidneys remove excess salt from the blood to help keep levels constant. Different types of animals have other ways of removing excess salt from their bodies. Turtles, for example, have salt-secreting glands behind their eyes. Hence you may see a turtle 'shedding tears'. On the other hand, penguins and some other seabirds, such as the Southern Giant Petrel, may appear to have runny noses because that is where their salt-secreting glands are located.

4.10 Exercises: Understanding and inquiring

To answer questions online and to receive **immediate feedback** and **sample responses** for every question, go to your learnON title at www.jacplus.com.au. *Note:* Question numbers may vary slightly.

Remember

1. Match the terms associated with the excretory system with their description in the table below.

Term	Description
(a) Bladder	(A) Watery fluid produced by kidneys that contains unwanted substances
(b) Kidney	(B) Transports urine from bladder to outside body
(c) Ureter	(C) Stores urine
(d) Urethra	(D) When urine moves from the bladder, through the urethra and out of the body
(e) Urination	(E) Transports urine from kidneys to bladder
(f) Urine	(F) Filters the blood and produces urine

2. Define the term 'excretion'.
3. Draw and label a diagram of the kidneys showing the following attachments: renal arteries, renal veins, ureters, bladder.
4. Outline what happens when you drink a lot of water.
5. Describe one way in which excess salt is removed from your body.
6. Explain how haemodialysis assists people with kidney disease.
7. Describe the relationship between:
   (a) a kidney and a nephron
   (b) kidneys and urine
   (c) alcohol, lungs and kidneys.

8. Distinguish between:
   (a) the ureter and the urethra
   (b) a Bowman's capsule and a glomerulus
   (c) the bladder and the kidney.

Think and discuss

9. Carefully observe the haemodialysis diagram on page 166. Suggest reasons the following are included in the process:
   (a) blood pump
   (b) bubble trap
   (c) constant temperature bath.
10. Suggest what you would expect to find in used dialysing solution.
11. Suggest why red blood cells don't pass through the dialysis tubing.
12. Use a Venn diagram to compare haemodialysis with real kidneys.

Analyse and evaluate

13. Use the table below and the other information in this section to answer the following questions.
    (a) Draw two bar graphs to show the quantity of water, proteins, glucose, salt and urea in blood and in urine.
    (b) Which substance is in the greatest quantity? Suggest a reason for this.
    (c) Which substances are found only in blood?
    (d) Which substances are found in urine in a greater quantity than in blood? Suggest a reason for this.
    (e) When would the amount of these substances in the urine become greater or less than in the blood?

Substance	Quantity (%)	
	In blood	In urine
Water	92	95
Proteins	7	0
Glucose	0.1	0
Chloride (salt)	0.37	0.6
Urea	0.03	2

Investigate and create

14. Research and report on one of these conditions: urinary incontinence, kidney stones, kidney transplants, cystitis, blood in urine, proteinuria, nephritis.
15. Find out and report on:
    (a) the differences between the urethra in human males and females
    (b) why pregnant women often need to urinate more frequently
    (c) how the prostate gland in males may affect urination in later life
    (d) which foods can change the colour or volume of urine
    (e) which tests use urine in the medical diagnosis of diseases.
16. Find out more about nephrons and how they work. Construct a model of a nephron that shows how it is linked to blood vessels and how urine gets to your bladder from it.
17. Research the nephrons of animals that live in different environments (for example, deserts, oceans and rivers). Comment on similarities and differences in their structure. Suggest reasons for the differences.

learn on RESOURCES – ONLINE ONLY

Complete this digital doc: Worksheet 4.6: Removing waste from the blood (doc-18722)

4.11 Musculoskeletal system — keeping in shape

4.11.1 Bones

Your **musculoskeletal system** consists of your **skeletal system** (bones and joints) and your **skeletal muscle system**. Working together, these two systems protect your internal organs, maintain posture, produce blood cells, store minerals and enable your body to move.

Did you know that an adult human **skeleton** contains over 200 separate **bones**? Without a skeleton we may resemble jelly-like blobs! Not only does it provide a structure for our muscles to attach to, allowing us to move, but it also provides support, protects vital organs (e.g. brain and heart) and forms a frame that gives our body shape.

Your bones have many different shapes and sizes, depending on the job that they have to do. They can be short, thick, round or flat. The longest bone in your body is the femur, or thigh bone. The smallest bone, the stages, or stirrup bone is in your ear and is about 3 mm in length. A feature that all bones have in common, however, is that they are all light and strong. Why do you think they share this feature?

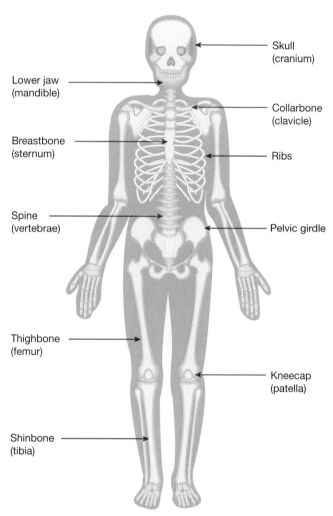

It's hard being a bone

Bones are alive. If bones were not alive, how would you grow taller? How would a broken arm or leg mend? Bones contain living cells and need a blood supply to provide them with oxygen and other nutrients.

Not only are bones busy providing you with support and movement, they are also busy inside. Bones contain soft tissue called **bone marrow**. This is very important because it is where blood cells are made.

Throughout the first twenty years of your life, most of the soft and rubbery **cartilage** that made up your skeleton is gradually replaced with bone. Your trachea, nose and ears, however, are made mostly of cartilage and the ends of your bones remain covered in cartilage.

Compact bones, such as the long bone of your femur, have a strong and hard outer layer that contains **calcium** and **phosphorus**. This is why you need an adequate supply of these minerals. Investigation 4.11 shows what could happen to your bones without a supply of these important **minerals**.

The hardening of your bones as you get older is called **ossification**. After ossification, the bone is made up of about 70 per cent non-living matter and 30 per cent living matter. As you get even older, your bones may get dry and **brittle**, which is why older people break their bones more easily.

4.11.2 Joints

A **joint** is the region where two bones meet. Your knees and elbows are examples of joints. Bones at a joint are held together by bundles of strong fibres called **ligaments**. The cartilage that covers the end of each bone is itself covered with a liquid called **synovial fluid**. The cartilage and synovial fluid work together to stop bones from scraping against each other.

The region where bones meet is called a joint. Cartilage and synovial fluid stop bones scraping against each other.

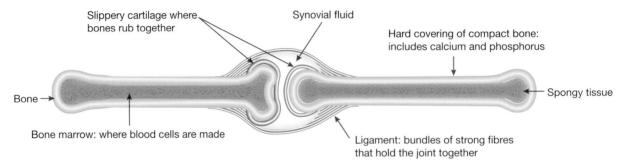

Slippery cartilage where bones rub together

Synovial fluid

Hard covering of compact bone: includes calcium and phosphorus

Bone →

Spongy tissue

Bone marrow: where blood cells are made

Ligament: bundles of strong fibres that hold the joint together

Most joints allow your bones to move. The amount and direction of movement allowed depends on the type of joint. Twist your neck. The joint between your skull and spine is a **pivot joint**, which allows this twisting type of movement (figure a). Bend your elbow. Your elbows and knees are **hinge joints**, like those of a door. They allow movement in only one direction (figure b). Roll your shoulder. Your hip and shoulder joints are **ball and socket joints**, allowing movement in many directions (figure c).

Some joints, such as those that join the plates in your skull, do not move. These are called **immovable joints**. While not allowing movement, these joints provide a thin layer of soft tissue between bones. Their job is to absorb enough energy from a severe knock to prevent the bone from breaking.

The plates of the skull are immovable joints

Different types of joints: (a) pivot joint (b) hinge joint (c) ball and socket joint

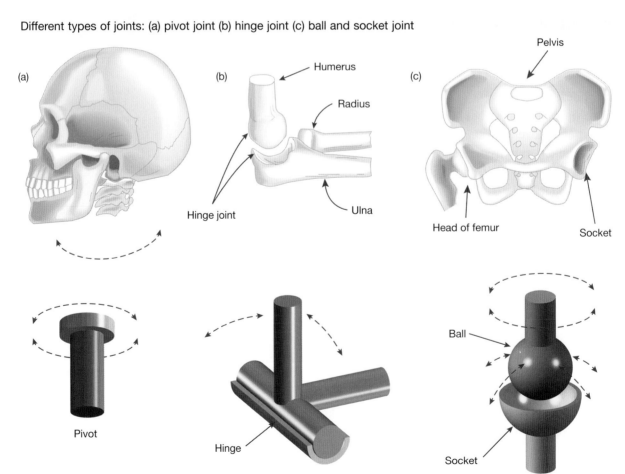

(a)

(b)

Humerus

Radius

Hinge joint

Ulna

(c)

Pelvis

Head of femur

Socket

Pivot

Hinge

Ball

Socket

4.11.3 Muscles

Muscles are tough and elastic fibres. The movement of muscles is controlled by the brain, which sends signals through your nerves. Muscles such as those that make your heart pump and those that control your breathing rate are called **involuntary muscles** — they work without you having to think about it. The muscles that are connected to bones are called **voluntary muscles** because you have to choose to use them.

All pull, no push!

Muscles are connected to the bones of your skeleton by bundles of tough figures called **tendons**. Muscles pull on bones by contracting or shortening. Muscles never push.

4.11.4 Broken bones

Breaks and fractures

When a bone breaks, the ends of the bone need to be put back into place (set) so that they can grow together. If a bone is shattered into several pieces, it is sometimes possible to use pins or wire to hold the pieces in place while the bone heals. A **greenstick fracture** occurs when the bone cracks but does not break. This type of **fracture** is common in children because their bones are more flexible.

When your biceps contract, your arm bends upwards. When your triceps contract, your arm straightens.

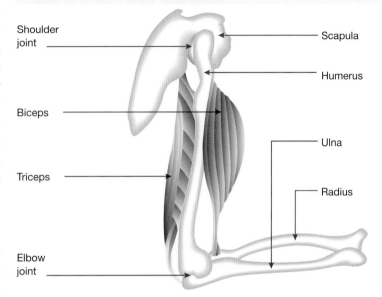

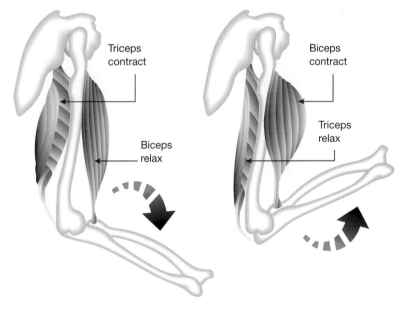

On the mend?

New technologies are being researched and developed to help fix broken bones. Some of these involve special cells called **stem cells**, while others involve the use of special 'glues' that hold bones together and aid the healing process. Scientists at CSIRO are currently working on a liquid gel called NovoSorb that glues the fractured bone together so that it is supported while it heals. As this gel degrades naturally, it does not require follow-up surgery to remove pins as is needed with older technologies.

Osteoporosis

Osteoporosis is a loss of bone mass that causes bones to become lighter, more fragile and easily broken. It occurs in middle-aged or elderly men and women. In Australia, about 60 per cent of women and about 30 per cent of men are affected by osteoporosis at some stage. It is believed to be caused by lack of calcium in the diet.

In your teenage years, you can help protect yourself from getting osteoporosis later in life by having a healthy diet and exercising. Your diet should include dairy products such as milk, cheese and yoghurt and other foods high in calcium. Such a diet will help ensure that your bone mass is adequate as an adult.

4.11.5 Ouch! Torn or swollen?

Sprains

Sprains occur when ligaments joining bones at a joint are torn or stretched. Sprains usually happen when you fall onto a joint, such as an elbow or an ankle, and twist it.

Arthritis

Arthritis is a swelling of the joints that makes movement difficult. Osteoarthritis occurs mainly in elderly people and is caused by wear and tear of the joints. The cartilage gradually breaks down, thus allowing bare bones to grate against each other instead of sliding or turning smoothly. Rheumatoid arthritis is a swelling of the tissue between the joints. The swelling causes the joints to slip out of place, which then causes great pain and deformities.

Tennis elbow

Tennis elbow occurs when the lining of the elbow joint swells and produces too much synovial fluid. The joint becomes swollen and painful. This occurs when the joint is used a lot and is most common in tennis players.

Torn hamstrings are a common but painful sporting injury.

Torn hamstrings

Torn hamstrings are a common sporting injury. The hamstring muscle joins the pelvis to the bottom of the knee joint, running along the back of the thigh. It controls the bending of the knee and straightening of the hips. A sudden start or turn in sport often stretches the hamstring muscle too far. It tears, causing great pain. Cold and unprepared muscles are more likely to tear. Proper warming up before strenuous sporting activity is one way to reduce the chances of tearing a muscle.

INVESTIGATION 4.11

Rubbery bones

AIM: To investigate the effect of calcium and phosphorus deficiency on bone

Background information

Vinegar is an acid that dissolves minerals such as calcium and phosphorus, removing them from bones.

Materials:

2 chicken or turkey bones   water
2 jars (or beakers)   vinegar

Method

- Clean the two chicken or turkey bones and leave them to dry overnight.
- Observe the bones and then place one bone in a jar of vinegar and the other in a jar of water.

- Allow the bones to soak for at least three days. Then remove the bones and observe any changes.
- Return the bones to their previous jars for another week, then remove and observe any further changes in the bones. Can you tie either bone into a knot?

Results

1. Construct a table to record your observations.
2. Record your observations of the bone:
   (a) before placing them in the solutions
   (b) after soaking for three days
   (c) a week after your observation in part (b).

Discuss and explain

3. Suggest a reason for the inclusion of the jar of water in the investigation.
4. Describe changes that you observed for each bone.
5. Provide reasons for the changes that you observed.
6. Relate this investigation to your bones and your diet.
7. Propose a conclusion for your findings.

INVESTIGATION 4.12

Chicken wing dissection

AIM: To investigate the structure of a chicken wing

Background information

Take special care when using scissors and scalpels.

Materials:
chicken wing
dissection tray or board
newspaper
disposable gloves
scalpel
scissors

Method

- Sketch one of the joints in the chicken wing. Label the bones, the tendons and the muscles. Show clearly where the muscle inserts (attaches to the bones). Use arrows to show how the bones move when the muscle is shortened.
- Feel the cartilage with a gloved hand. Does the cartilage feel rough or slippery? Why does it need to be slippery?
- Is cartilage harder or softer than bone?

Results

1. Sketch one of the joints in the chicken wing. Label the bones, the tendons and the muscles. Show clearly where the muscle inserts (attaches to the bones). Use arrows to show how the bones move when the muscle is shortened.
2. Feel the cartilage with a gloved hand. Does the cartilage feel rough or slippery? Why does it need to be slippery?

Discuss and explain

3. Describe differences that you observed between cartilage, tendons, muscles and bones.
4. Relate this investigation to your own joints, muscles, tendons and bones.
5. Propose a conclusion for your findings.

4.11 Exercises: Understanding and inquiring

To answer questions online and to receive **immediate feedback** and **sample responses** for every question, go to your learnON title at www.jacplus.com.au. *Note:* Question numbers may vary slightly.

Remember

1. Match the common name with the scientific term in the table below.

Common name	Scientific term
(a) Breastbone	(A) Mandible
(b) Kneecap	(B) Cranium
(c) Lower jaw	(C) Vertebrae
(d) Shinbone	(D) Tibia
(e) Skull	(E) Sternum
(f) Spine	(F) Femur
(g) Thighbone	(G) Patella

2. Describe the job done by each of the following parts of a joint.
   (a) Ligament
   (b) Cartilage
   (c) Synovial fluid
3. Some joints are referred to as immovable joints. What is the use of having joints that don't move?
4. Write down an example of each of the following types of joint.
   (a) Hinge
   (b) Ball and socket
   (c) Pivot
   (d) Immovable
5. Ligaments and tendons are bundles of tough fibres. What is the major difference between a ligament and a tendon?
6. Describe the action of the biceps and triceps muscles as you bend your elbow to raise your forearm.

Think and investigate

7. Your musculoskeletal system consists of your skeletal system (bones and joints) and your skeletal muscle system (voluntary or striated muscle). Working together, these two systems protect your internal organs, maintain posture, produce blood cells, store minerals and enable your body to move.

    Use information in this section and other resources to relate structural features to the functions of the following parts of these systems.

Part of system	Structural features	Function
Bones		
Cartilage		
Joints		
Skeletal muscles		

8. Find out more about the structure and function of skeletal, smooth and cardiac muscle tissue.
9. What is dietary rickets and how is it caused?

Identify the incomplete skeletons and name the missing parts.

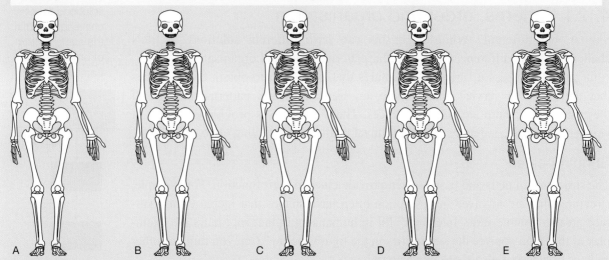

A   B   C   D   E

10. Look carefully at each of the skeletons above. Three of them are incomplete. Identify the incomplete skeletons and name the missing parts.
11. Apart from warming up just before a game, how do the best basketball and netball players reduce the likelihood of torn muscles and tendons?
12. What would happen if the cartilage in your knee joint wore out?
13. Research and report on one of the following science careers: orthopaedic surgeon, physiologist, physiotherapist, occupational therapist, rheumatologist, fitness trainer.

Investigate and discuss

14. (a) In teams of four, use a 'lucky dip' system to match each member of the team with one of the following topics.
    • Deviated septum
    • Broken nose
    • Nose cancer
    • Rhinoplasty
    (b) Each member should investigate their topic and report their findings to their team.
    (c) As a team, construct a cluster map or mind map to summarise the key points of your team findings.
    (d) As a team, select one of the four topics for further research. Brainstorm questions for your selected topic and research these questions.
    (e) Report your team's findings to the class.

Create

15. Make a skeleton mobile to hang from the ceiling.
    (a) Trace the skeleton diagram on page 169 (or a larger one from another resource), colour it and cut it into a number of sections.
    (b) Paste each section onto cardboard and thread the sections together to make a skeleton mobile.
16. Use a cut-out human skeleton to make a new, imaginary animal by rearranging the bones. Suggest a name and describe the lifestyle of your animal.
17. Make a working model of an arm to show how the biceps and triceps work. Use the illustration on page 171 as a guide. Materials you might use include icy-pole sticks or stiff cardboard (for bones), split pins (for ligaments), string or rubber bands (for muscles), polystyrene foam (for cartilage) and glue. Draw a labelled diagram of your model.

RESOURCES — ONLINE ONLY

📄 **Complete this digital doc:** Worksheet 4.7: Bones, joints and muscles (doc-18716)

4.12 Same job, different path

4.12.1 Patterns, order and organisation

Similar, but different? While organisms can have different solutions to life's challenges, these differences share similar patterns, order and organisation.

Organisms possess a variety of structures that help them to obtain the resources that they need to survive. While there are similarities and patterns in some of these structures, there are also differences. These differences provide examples of wonderful, creative solutions to the continual challenge of staying alive.

Shaping clues

The structures of cells and tissues often provide clues to their function. For example, structures that are involved in absorption often have shapes that increase their surface area to volume ratio. Intestinal villi in humans and plant root hairs are examples of this. Can you see the similarities in the figures below? Can you think of other cells or tissues that also share this pattern?

Organisms have different solutions to life's challenges within the same pattern of organisational framework.

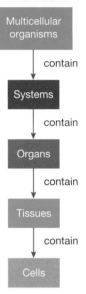

Multicellular organisms

↓ contain

Systems

↓ contain

Organs

↓ contain

Tissues

↓ contain

Cells

Similar structures? What might their function be?

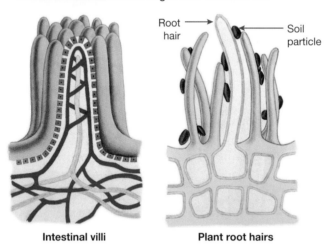

Intestinal villi   **Plant root hairs**

4.12.2 Respiratory routes

glucose + oxygen → carbon dioxide + water + energy

Cellular respiration is essential for life. Organisms require a supply of oxygen and a way to remove the carbon dioxide that is produced as waste. Although this gaseous exchange is essential, different types of organisms achieve it in different ways.

Unicellular organisms are small enough that gases such as oxygen and carbon dioxide can simply diffuse in and out of their cell. Likewise, some very thin multicellular organisms have many of their cells in direct contact with their environment. These organisms rely simply on diffusion for their exchange of gases. Flatworms, for example, do not need a respiratory system, as they use their whole body surface to obtain the oxygen they require from the water in which they live.

Flatworm

Notice any similarities or differences in these gas exchange surfaces?

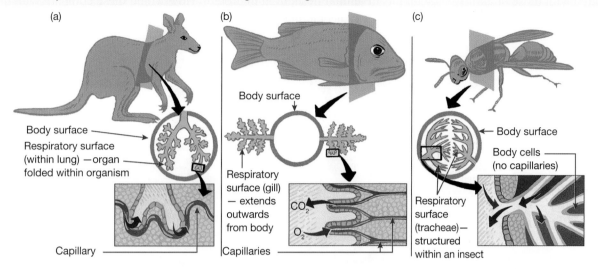

(a) Body surface — Respiratory surface (within lung) —organ folded within organism — Capillary

(b) Body surface — Respiratory surface (gill) — extends outwards from body — CO_2 — O_2 — Capillaries

(c) Body surface — Body cells (no capillaries) — Respiratory surface (tracheae)— structured within an insect

Some other small animals, such as worms living on land, can exchange gases through their mucus-covered skin. Oxygen from the air dissolves in the mucus, while carbon dioxide seeps out. Tiny blood vessels in their skin transport the gases to and from the rest of the worm's body.

Other animals may have specialised gas exchange organs. Three main kinds of these organs are lungs in mammals and amphibians, gills in fish, and tracheae in insects. Examine the figure above to compare the structure of these organs. How are they similar? How are they different?

4.12.3 What do they eat?

Animals can be classified on the basis of their diet. **Herbivores** eat plants; **carnivores** eat other animals; **omnivores** eat both plants and animals. An animal's teeth can provide hints to the types of food that it eats. Observe the teeth in the skulls of the vertebrates shown at right. Based on your observations, predict the types of food that they may eat.

Wombats are herbivores. They have large incisors for biting and cutting, but no canines. They also have large premolars and molars because the fibrous plant materials they eat need a lot of grinding. Tasmanian devils are carnivores. Because their prey is alive and moving, they possess large canines for stabbing and holding on to it. Their incisors are used for tearing meat. The molars and premolars in carnivores have cutting

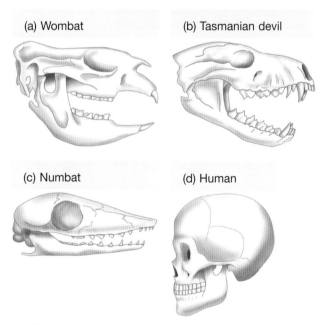

(a) Wombat

(b) Tasmanian devil

(c) Numbat

(d) Human

edges. Insectivores are carnivores that eat only insects. Their teeth are small and pointed so that they can crush the exoskeleton of the insect, which they then swallow whole. Even if you are a vegetarian, as a species, humans are considered to be omnivores. We possess all of the different types of teeth needed to break down both meat (from animal tissue) and plants.

4.12.4 Digestive differences

Although most vertebrates possess a digestive system that has a similar pattern, order and organisation, there may be differences that are related to nutritional needs and diet. Consider, for example, differences in the digestive systems of herbivores with diets that are high in plant material with lots of cellulose compared with those of carnivores with lots of animal flesh, high in protein. How would these compare with the digestive system of an organism that ate only nectar and pollen?

Notice any similarities or differences in these digestive systems?

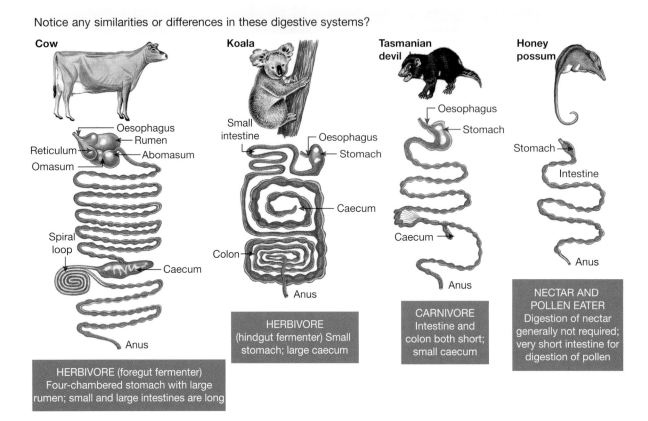

Cow

Oesophagus
Rumen
Reticulum
Abomasum
Omasum

Spiral loop

Caecum

Anus

HERBIVORE (foregut fermenter)
Four-chambered stomach with large
rumen; small and large intestines are long

Koala

Small intestine
Oesophagus
Stomach

Caecum

Colon

Anus

HERBIVORE
(hindgut fermenter) Small
stomach; large caecum

Tasmanian devil

Oesophagus
Stomach

Caecum

Anus

CARNIVORE
Intestine and
colon both short;
small caecum

Honey possum

Stomach
Intestine

Anus

NECTAR AND
POLLEN EATER
Digestion of nectar
generally not required;
very short intestine for
digestion of pollen

4.12.5 Heart count?

Two, three or four? Not all animals have a four-chambered heart like you. Fish have a heart with two chambers and blood passes through the heart only once each time around the body. The hearts of amphibians and most reptiles are three chambered and allow oxygenated and deoxygenated blood to mix. Birds and mammals are similar to amphibians and most reptiles in that blood flows through the heart twice in each circulatory trip, but they possess a heart with four chambers that does not allow the mixing of blood. What do they share? How are they different?

Around we go ... but which route do we take?

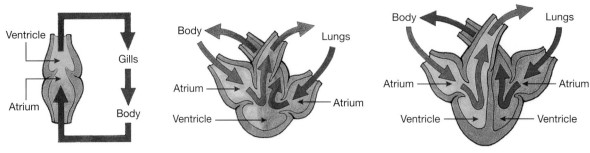

Ventricle
Gills
Atrium
Body

Body
Lungs
Atrium
Atrium
Ventricle

Body
Lungs
Atrium
Atrium
Ventricle
Ventricle

(a) A fish heart has two chambers. Note that blood passes through the heart only once for every circulation within the body.

(b) Amphibians and most reptiles have a three-chambered heart. Oxygenated and non-oxygenated bloods mix in the single ventricle as blood flows through the heart twice for every circulation within the body.

(c) Birds and mammals have a four-chambered heart and blood flows through the heart twice for every circulation within the body.

4.12.6 Throwing out the trash!

Different types of fish, living in different environments, can also differ in how they maintain their salt balance.

Proteins are involved in a variety of different chemical reactions that keep animals alive. **Ammonia** is formed when proteins break down. Ammonia is toxic to cells and requires either lots of water to release it into, or conversion into a less toxic form (such as urea or uric acid). Conversion into other forms costs the animal energy. Whichever form these **nitrogenous wastes** are in, they need to be removed from the animal's body.

Different types of animals use different strategies to remove nitrogenous wastes. This is linked to the amount of water available in the environments in which they live. Fish, for example, have a ready supply of water, so most fish release their nitrogenous waste as ammonia. The main nitrogenous waste excreted by humans is **urea**. Uric acid requires the least water for excretion. Insects, spiders and birds excrete their wastes as **uric acid**.

Have you noticed the pattern? The environment in which an organism lives, in this case the amount of water available, can play a role in how different species have evolved different strategies to the same problem of removal of their wastes.

Saltwater fish, such as snapper, drink sea water constantly and produce a small volume of urine. Freshwater fish, such as Murray cod, however, rarely drink, but make lots of urine.

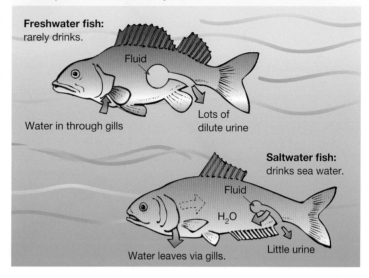

4.12.7 Inside or out?

Are you wearing your skeleton on the inside or outside? Vertebrates (such as humans) have an internal backbone. They also have an internal skeleton called an **endoskeleton**. You can read more about our endoskeleton on the previous pages. Invertebrates, however, do not possess a backbone. Some invertebrates, such as grasshoppers and ants, have their skeleton on the outside of their bodies. Their external skeleton is called an **exoskeleton**. Other invertebrates, such as worms and jellyfish, do not have any skeleton at all.

Due to their different body structures, invertebrates can use their muscles in different ways to achieve movement. Worms and slugs, for example, can stretch and shorten muscles in certain parts of their body to bring about movement. Even though the muscles of jellyfish have no bones or other hard parts to attach to, they propel themselves through water by pumping water into their body cavities and releasing it suddenly. The muscles in insects, such as grasshoppers, are attached to their exoskeleton. They can extend their legs by contracting the extensor muscles and relaxing their flexor muscles.

Birds produce nitrogenous waste in a form of solid uric acid. They store this in their bodies without diluting it with water and then excrete it with their faeces. Animals living in dry environments, such as insects and snakes, also excrete their wastes in this form to conserve water.

The muscles in insects are attached to the exoskeleton, the outer covering of the body. This grasshopper can extend its leg by contracting the extensor muscle and relaxing the flexor muscle.

Joint

Exoskeleton (cuticle)

Extensor muscle

Flexor muscle

INVESTIGATION 4.13

Inside or out?

AIM: To use models to investigate the differences between how muscles join to bones in animals with endoskeletons and exoskeletons

Materials:
2 cardboard tubes, each at least 30 cm long
sticky tape
rubber bands
large nail or other pointed object

Method

- Cut each cardboard tube into two pieces about 15 cm long.
- Using the nail, make two holes on opposite sides of each tube. These should be about 5 cm from one end of each piece.
- Label two pieces 'Endo A' and 'Endo B' and the other two pieces 'Exo A' and 'Exo B'.
- Tape Endo A and Endo B together on one side, so that they form a hinge at the ends with the small holes.
- Cut two rubber bands and thread the cut ends through the holes from the outside.
- Tie knots so that the rubber bands can't pull back through the holes.
- Tape Exo A and Exo B together in the same way as Endo A and Endo B.
- Cut another two rubber bands and thread the cut ends through the holes so that they run *inside* the tube.
- Make sure that they are stretched very tightly, and then tie knots on the outside of the tubes.

The rubber bands are like the muscles in your arm. They are attached to the bones on either side of your elbow. The arm bends at the joint when the muscle contracts.

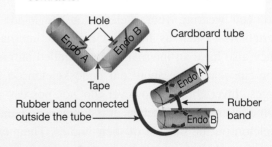

The rubber bands are like the muscles in an insect's limb. When a muscle contracts, the joint on which it operates straightens.

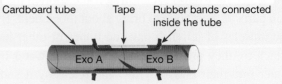

Results

1. When one rubber band contracts, what happens to the one on the opposite side?
2. Draw sketches of each and record your observations when the joint is moved.

Discuss and explain

3. Describe how the two skeletons are different.
4. Identify the strengths and limitations of the method and how you could improve it.
5. Research and construct models that demonstrate how two different types of organisms have developed different strategies to solve the same 'problem' that is related to their survival.

4.12 Exercises: Understanding and inquiring

To answer questions online and to receive **immediate feedback** and **sample responses** for every question, go to your learnON title at www.jacplus.com.au. *Note:* Question numbers may vary slightly.

Remember

1. Place the following terms in order of simplest to most complex: cell, organ, system, multicellular organism, tissue.
2. Provide an example of how structure can give clues about function.
3. Write the word equation for cellular respiration.
4. Describe two key functions of gaseous exchange.
5. Suggest why there are differences between herbivores and carnivores in the structures of their digestive systems.
6. Construct a table to summarise similarities and differences between the hearts and circulation of fish, amphibians and mammals.
7. Name two organs belonging to each of the following systems.
   (a) Respiratory system
   (b) Circulatory system
   (c) Excretory system
8. Construct a matrix table to compare the diets and teeth of herbivores, carnivores, omnivores and insectivores.

Think

9. Why don't herbivores have canine teeth?
10. How do we know what different types of dinosaurs ate, even though they haven't existed for about 65 million years?

Investigate and discuss

11. (a) In teams of three, use a 'lucky dip' system to match each member of the team with one of these organisms: earthworm, grasshopper, fish.
    (b) Each member should investigate the respiratory system of their matched organism and report their findings to their team.
    (c) As a team, construct a cluster map that summarises the key details and features of the respiratory systems of each of these organisms.
    (d) As a team, construct a matrix table that summarises the similarities and differences between the respiratory systems of humans, earthworms, grasshoppers and fish.
12. Select an animal of your choice.
    (a) Find out how it:
        (i) detects stimuli
        (ii) supports itself and moves
        (iii) takes in oxygen and removes carbon dioxide
        (iv) conducts messages from one part of its body to another.
    (b) Construct a model or animation to demonstrate one of the functions above.
13. There are differences in the form in which groups of animals excrete nitrogenous wastes. Find out the differences between humans, freshwater fish, saltwater fish and insects, Communicate your findings using models or in a puppet play, documentary or animation.

Investigate, design and create

14. Research and prepare a poster on the hearts of different types of animals.
15. (a) Outline the key differences between the structures of the digestive systems of a cow, a koala, a Tasmanian devil and a honey possum.
    (b) Suggest reasons for the differences.
16. Complete the following table.

Feature	Mammal	Fish
Number of chambers in the heart		
Times blood travels through heart in each circulatory trip		
Name of gaseous exchange respiratory organ		

17. Find out about the different tissues and systems that exist in plants. Present your information, using diagrams and lots of colour, on a poster or PowerPoint presentation. Be a creative as you can.
18. (a) Select one of the systems in the mind map on page 108 and find out more about what it does and how it works.
    (b) Use your findings to write a brief play that other students in the class can act out.
19. In a small team, formulate scientific questions about how the structure of the heart, kidney or lungs is related to the function of that organ. Research these questions and present your findings to the class.
20. Select an organ (for example, heart, lungs or stomach) and find out the answers to the following questions.
    (a) What is the function of the organ?
    (b) Which system does it belong to?
    (c) What other organs are in the same system?
21. Design and construct a model of one of the following systems: respiratory, excretory, reproductive, digestive.
22. Some animals that live in water, such as sea anemones, have a digestive sac which acts both as a mouth and an anus.
    (a) Find out more about the digestive system of sea anemones and construct a model to demonstrate your understanding of how it works.
    (b) Describe how this digestive system is similar to that of humans and how it is different.
    (c) Suggest reasons for the differences.

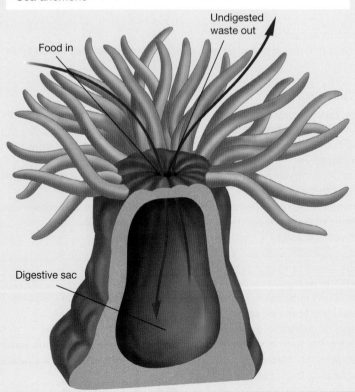

Sea anemone

Undigested waste out

Food in

Digestive sac

4.13 Flowcharts and cycle maps

4.13.1 Flowcharts and cycle maps

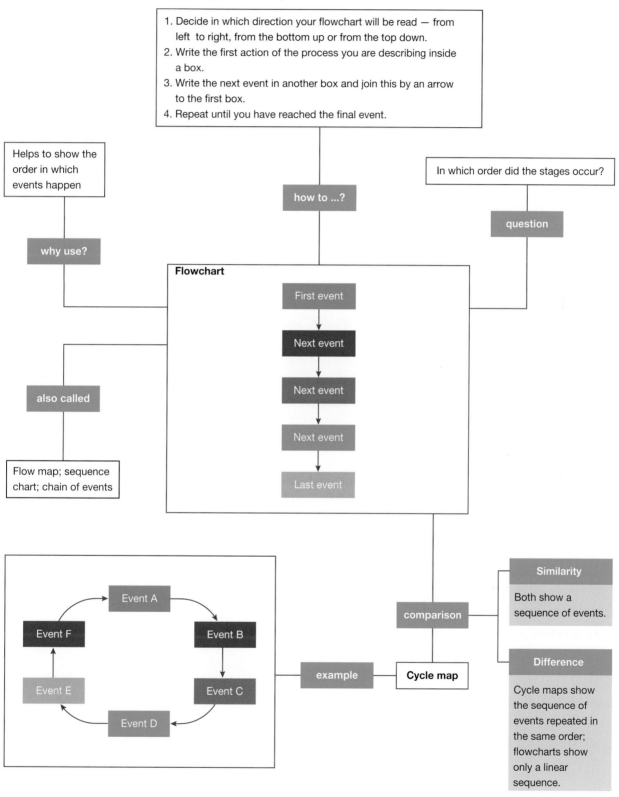

1. Decide in which direction your flowchart will be read — from left to right, from the bottom up or from the top down.
2. Write the first action of the process you are describing inside a box.
3. Write the next event in another box and join this by an arrow to the first box.
4. Repeat until you have reached the final event.

Helps to show the order in which events happen

In which order did the stages occur?

how to ...?

why use?

question

Flowchart

First event

Next event

Next event

Next event

Last event

also called

Flow map; sequence chart; chain of events

Event A

Event F

Event B

Event E

Event C

Event D

example

Cycle map

comparison

Similarity

Both show a sequence of events.

Difference

Cycle maps show the sequence of events repeated in the same order; flowcharts show only a linear sequence.

4.14 Review

4.14.1 Study checklist

Multicellular organisms

- state the relationship between cells, tissues, organs, systems and multicellular organisms
- describe the key functions of the following body tissues: connective, epithelial, skeletal, blood, muscle, nerve
- identify the organs and overall function of a system of a multicellular organism
- describe the structure of each organ in a system and relate its function to the overall function of the system
- compare similar systems in different organisms

Musculoskeletal system

- describe the structure of a human bone
- describe the relationship between bones, joints, ligaments and muscles

Circulatory system

- identify the components that make up blood
- compare red blood cells and white blood cells
- state the relationship between blood, heart, arteries, veins and capillaries
- give examples of technology related to the human heart

Respiratory system

- use a flowchart to describe the relationship between the trachea, alveoli, lungs, capillaries, oxygen and carbon dioxide

Digestive system

- sequence the structures of the digestive system and state the function of each
- describe how the structure of the teeth, oesophagus and villi in the small intestine assist their function
- describe how the tongue, gall bladder, pancreas and liver are involved in the digestive process
- outline the importance of peristalsis
- distinguish between the processes of ingestion, mechanical digestion, chemical digestion, absorption of nutrients, assimilation and egestion

Excretory system

- explain how wastes are removed from the human body

Individual pathways

ACTIVITY 4.1	ACTIVITY 4.2	ACTIVITY 4.3
Systems	Investigating systems	Analysing systems
doc-6081	doc-6082	doc-6083

learnon ONLINE ONLY

4.13 Review 1: Looking back

To answer questions online and to receive **immediate feedback** and **sample responses** for every question, go to your learnON title at www.jacplus.com.au. *Note:* Question numbers may vary slightly.

1. In this chapter you have seen the importance of your transport systems.
   (a) With your team, create mind or concept maps that summarise what you know about the following.
   - Blood
   - Blood vessels
   - Heart
   - Lungs
   - Kidneys
   - Liver
   (b) For each of the six parts of your body listed above, brainstorm in your team as many questions as you can. Then select one question for each body part to do your own research. Report your findings back to the group.
   (c) Summarise your group findings in a creative way — this could be as a cartoon, song, poem, play, model or simulation.

2. (a) In a pair, select one of the statements from the cartoon below and continue the conversation, taking turns asking relevant questions and answering them.
   (b) Find a new partner and repeat activity (a) with the same statement or another one.
   (c) In a team, create at least five other statements that could have been made on:
      (i) animal transport systems
      (ii) transport technology
      (iii) transport system disorders or diseases.

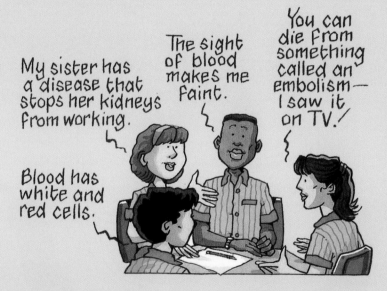

3. In a team, brainstorm ideas to come up with the following for transport systems.
   (a) Words for an alphabet key
   (b) Five 'what if' keys

4. Use your six thinking hats (see section 2.2 in your learnON title at www.jacplus.com.au) for three of the following issues or statements.
   (a) Drinking of any alcohol in Australia should be illegal.
   (b) Smoking in public should be punishable by a 10-year prison sentence.
   (c) Donating blood at least four times a year should be compulsory for all over the age of 16.
   (d) Only people under the age of 40 should be allowed to have a heart transplant.
   (e) Smokers should not be allowed to have surgery.
   (f) Blood transfusions should be illegal.
   (g) Everyone should have the right to a blood transfusion.
   (h) Organ donation should be compulsory.
   (i) Overweight people should not be allowed to have surgery on their circulatory system.

5. Write a story that tells of the life of a red blood cell.
6. Make a copy of the diagram below for your workbook.
   (a) Label the lettered parts (A to J) of the human circulatory system and blood vessels on your diagram.
   (b) Use a red pencil to colour in the blood vessels with oxygenated blood, and a blue pencil for those with deoxygenated blood.
   (c) State whether the blood in the following blood vessels is deoxygenated or oxygenated.
       (i) Aorta
       (ii) Pulmonary artery
       (iii) Pulmonary vein
       (iv) Vena cava
       (v) Carotid arteries
   (d) Draw up a table that shows the differences in structure and function of the arteries, veins and capillaries.

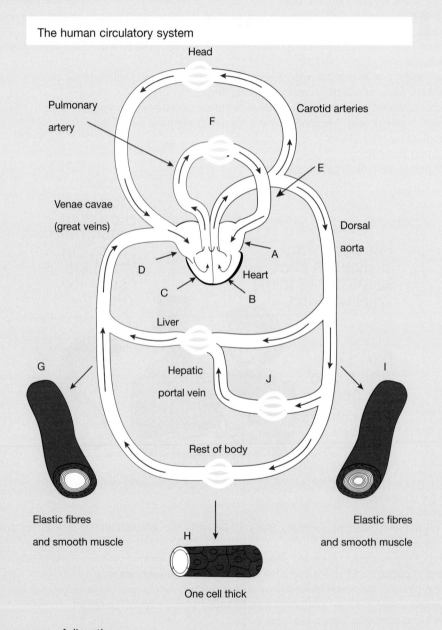

The human circulatory system

7. Outline the purpose of digestion.
8. Construct a Venn diagram to show similarities and differences between mechanical and chemical digestion.

9. Label the following diagram of the human digestive tract.

The human digestive tract

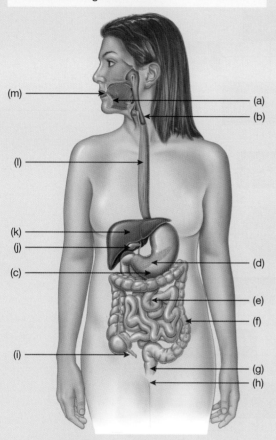

(m)
(a)
(b)
(l)
(k)
(j)
(c)
(d)
(e)
(f)
(i)
(g)
(h)

10. (a) Using the links and ideas in the two visual thinking maps on this page (as well as your own), construct a board for the game 'Nutridigest', which you will create.
(b) In a team of four, brainstorm as many questions as you can that could be placed on each of the squares. Creatively write these onto your Nutridigest cards.
(c) As a team, discuss the rules for your Nutridigest game. Write down those you agree on.
(d) Make a brochure that explains how to play your game.
(e) Trial playing the game with your team.
(f) Make any alterations to your game that you think would improve it.
(g) Play the game that has been created by another team.

What happens to food in each part of the digestive system?

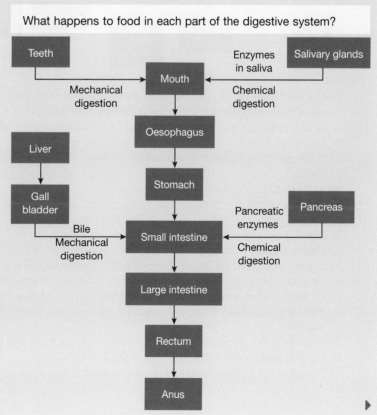

Teeth → Mouth ← Enzymes in saliva ← Salivary glands
Mechanical digestion / Chemical digestion

Mouth → Oesophagus → Stomach

Liver → Gall bladder → Bile / Mechanical digestion → Small intestine

Pancreas → Pancreatic enzymes → Small intestine / Chemical digestion

Small intestine → Large intestine → Rectum → Anus

What other points about enzymes could you add to this map? Can you suggest any more links between points already on the map?

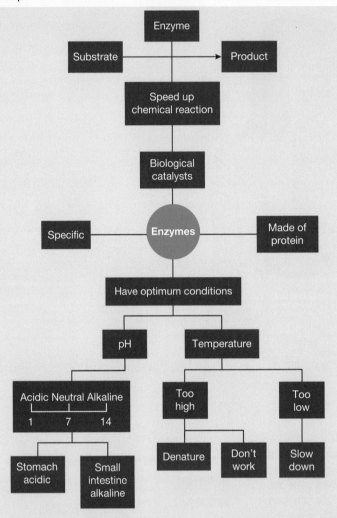

(h) In a team of eight, discuss the good things (strengths) and not-so-good things (limitations) of each game. Also, suggest ways that they could be improved in the next 'edition'.

11. (a) Have a lucky dip with names of parts of the digestive system and three different nutrients inside.
    (b) As a class, make your selections from the lucky dip.
    (c) Think about the function of your selection. Also, think about what sorts of actions or sounds it might have.
    (d) As a class, act out your roles in digestion.

12. (a) Construct a mind or concept map to summarise what you have learned during your study of 'living connections'.
    (b) Share your map with others in your team.
    (c) Create another mind or concept map that incorporates the learning of all of your team members.

13. As a team, create a song, poem, cartoon or play about something that you have learned.

14. (a) Discuss in your team what you have learned or found interesting in this chapter.
    (b) On sticky notes, write down other questions that you or your team may have about areas related to those in this chapter. Place the questions on a class question gallery board with those of other class members.
    (c) Once all of the questions are on the board, organise them into groups or themes.
    (d) In pairs, select one of the questions in the gallery.
    (e) Research the question and report your findings to the class.

15. The process of replacing oxygen with carbon dioxide is called gas exchange. Some animals exchange gases through lungs or gills. Some other animals exchange gases though their skin, and yet others through rows of air holes along both sides of their bodies.

    Investigate how animals and insects exchange gases to create a mind map and answer the following questions.
    (a) Construct a mind map, poster or PowerPoint presentation that summarises your findings on how at least five different animals achieve their exchange of gases.
    (b) Why do frogs need lungs when they can exchange gases through their skin?
    (c) Why can't fish survive in the air?
    (d) Why can't humans breathe under water without air tanks?
    (e) How is gas exchange in insects similar to that in humans?
    (f) What are two major differences between gas exchange in insects and humans?
16. (a) In teams, construct mind maps about each of the circulatory, respiratory and excretory systems.
    (b) Use the information in these mind maps to construct matrix grids to summarise the key information for each system.
17. (a) Name the organs in each of the following systems.
        (i) Circulatory system
        (ii) Respiratory system
        (iii) Excretory system
    (b) Create a word-find with these terms and then swap it with another team member to see how many words each of you can find.
18. Carefully observe the diagrams of the digestive systems of the animals below. Construct a matrix table that shows the similarities and differences between birds, earthworms, insects and humans.

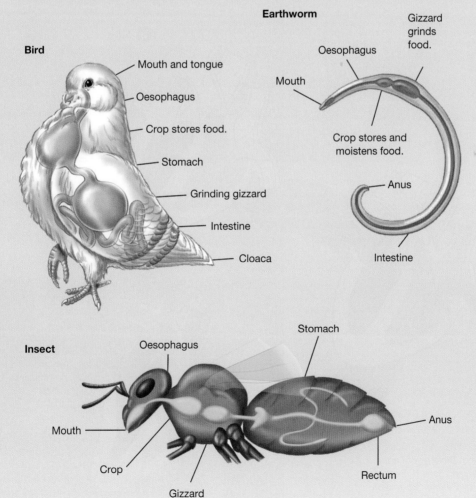

Earthworm

Gizzard grinds food.

Oesophagus

Mouth

Crop stores and moistens food.

Anus

Intestine

Bird

Mouth and tongue

Oesophagus

Crop stores food.

Stomach

Grinding gizzard

Intestine

Cloaca

Insect

Oesophagus

Stomach

Mouth

Crop

Gizzard

Anus

Rectum

Use section 4.13 Flowcharts and cycle maps.

19. (a) Read through the information in section 4.6 to refresh your memory on the structure and function of your heart.
    (b) Use a flowchart to show the movement of blood through your body using the following labels: left atrium, right atrium, right ventricle, left ventricle, pulmonary artery, pulmonary vein, lungs, aorta, vena cava, from body, to body.
20. (a) Use the cardiac cycle diagram below to answer the following questions.
    (i) In which stage do the atria contract?
    (ii) In which stage do both the atria and ventricles relax?
    (b) Construct flowcharts to help you summarise the information in the diagram.
21. Systole is the contraction of your heart muscle and diastole is the relaxation of your heart muscle. What do you think the following mean?
    (a) Atrial systole
    (b) Ventricular systole
    (c) Atrial diastole
    (d) Ventricular diastole
22. Construct a cycle map to outline the over all movement of blood through the heart.
    (a) Use this information to design a working model of a human heart.
    (b) Use a flowchart to plan the construction of you're heart model.
23. (a) Find out more about Leonardo da Vinci's models of the human heart.
    (b) Use a flowchart to map changes in scientific ideas about the heart throughout history.

One complete cardiac cycle can take about 0.8 seconds in an adult human with a pulse of about 75 beats per minutes.

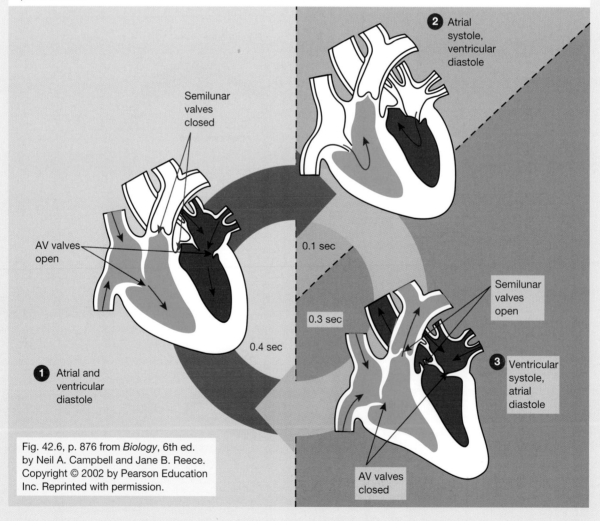

Semilunar valves closed

Atrial systole, ventricular diastole

AV valves open

0.1 sec

Semilunar valves open

0.3 sec

0.4 sec

1 Atrial and ventricular diastole

3 Ventricular systole, atrial diastole

AV valves closed

Fig. 42.6, p. 876 from *Biology*, 6th ed. by Neil A. Campbell and Jane B. Reece. Copyright © 2002 by Pearson Education Inc. Reprinted with permission.

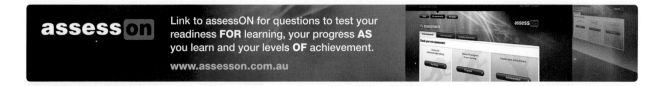

TOPIC 5
Systems — reproduction

5.1 Overview

Sex is fascinating. It has to be! It is the basic foundation for the continuation of life for most organisms on Earth. It can also be dangerous, desperate and competitive, as many insects and spiders would agree. The changes that occur at puberty in humans can be scary and exciting. But they all have the same purpose. They are the means by which you become an adult with the potential of passing on your genetic information to your offspring. It is all a part of the cycle of life. Around and around we go …

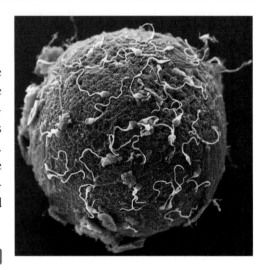

5.1.1 Think about reproduction **assesson**

- How fast can a sperm swim?
- Why is a dinner date a bad idea for a male red-back spider?
- What do a clitoris and a penis have in common?
- Can sperm build up and burst your testicles?
- If females have about 400 000 eggs at birth, is there a possibility that they can have that many children?
- How big is an 8-week-old fetus?
- Which contraceptives work the best?
- What's a 'test-tube baby'?
- Which foods contain the swollen ovaries of plants?
- Which vertebrates were the first to have a penis?
- Do 'virgin births' really exist?
- What is a hermaphrodite?

Numerous **videos** and **interactivities** are embedded just where you need them, at the point of learning, in your learnON title at www.jacplus.com.au. They will help you to learn the content and concepts covered in this topic.

5.1.2 Your quest
Perhaps you've noticed you're changing

During puberty, your body goes through some interesting, exciting and, sometimes, scary changes.

Investigate

1. In teams, find out the answers to the five questions asked by the students shown below.
2. Share and discuss your findings with those of other teams.

Do girls have different tubes for urinating and having babies?

What's the difference between fertilisation, intercourse, conception and contraception?

Why are some bellybuttons 'innies' and some 'outies'?

What are gonads? Do I have them?

When do boys start to make sperm?

INVESTIGATION 5.1

What's happening?
AIM: To increase awareness of changes during puberty

Method and results

1. In teams of two or more, sketch a figure of a girl and a boy on separate sheets of paper.
2. Add labels to show the changes for each during puberty.
   • Compare and discuss your figures with those of other teams in the class.
3. Make any changes or additions you wish to your diagram.

Discuss and explain

4. (a) As a team, suggest changes and possible additions to your diagrams.
   (b) As a class, collate examples of changes and possible additions to your diagrams.
5. (a) Suggest reasons why these changes occur.
   (b) As a class, suggest reasons why these changes may occur.

The timeline of our lives

AIM: To construct an individual human growth and development timeline

Method and results

1. Construct a timeline to show how you have changed over time: from the ages of 6 months, 2 years, 5 years, 8 years and 11 years to now. You may use photographs, cartoon sketches or plasticine models.
   - As we are all individuals, our timelines vary. Some of this variation is due to our genes that are inherited; some is due to our environments and lifestyles. Variation is very important for the survival of the species. Compare your timeline with those of others in your class.
2. Record the similarities and differences.

Discuss and explain

3. Pose questions prompted by your observations and reflections.

5.2 Private parts

5.2.1 Gamete factories …

Although testes and ovaries may look different, these two organs have the same job. They both make gametes.

Sperm are made in the **testes** of a male when he is sexually mature. Testes hang from the body within the **scrotum** to maintain sperm at a temperature of about 3 °C below that of the rest of the body. This temperature difference is essential for successful sperm production. Tight underwear or jeans can increase the temperature of the testes and so increase the number of damaged sperm produced.

Parts of the male reproductive system

Part	Function
Testes	Produce sperm cells
Scrotum	Where the testes are located. Keeps the testes at a slightly lower temperature than body temperature.
Vas deferens	The tube through which sperm cells travel from the testes to the penis
Prostate gland	Secretes some of the liquid that is added to sperm cells to form semen. The fluid secreted by the prostate gland is alkaline and contains many chemicals including enzymes. It plays an important role in keeping sperm cells alive once they enter the female reproductive system.
Seminal vesicle	Also contributes some of the liquid that makes up semen. The fluid produced by the seminal vesicle contains proteins, enzymes, sugar, vitamin C and other substances. The sugar provides a source of energy for sperm cells.
Urethra	The tube inside the penis through which semen leaves the male's body
Penis	The penis swells during sexual arousal. Semen containing sperm cells is ejaculated (released) from the penis into the vagina.

The male reproductive system

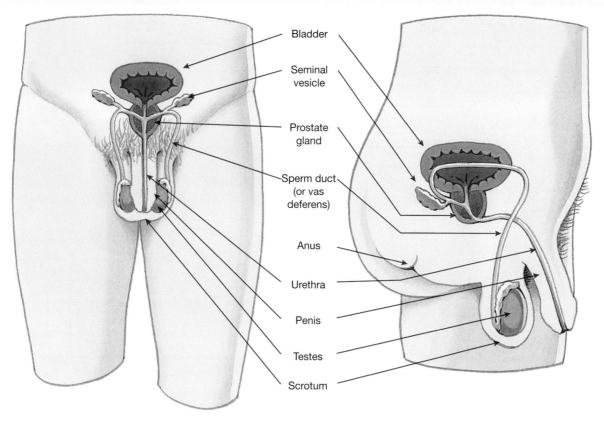

Bladder

Seminal vesicle

Prostate gland

Sperm duct (or vas deferens)

Anus

Urethra

Penis

Testes

Scrotum

HOW ABOUT THAT!

Amazing sperm

Sperm cells are less than half a millimetre long. Viewed through a microscope lens, spermatozoa (sperm, for short) remind you of tadpoles — a big head and a thin, whippy tail. They form in the testes, but only when the temperature is just right — a few degrees lower than body temperature. This is where the scrotum — a natural thermostat — does its job. It shrivels and scrunches up closer to the body when you are cold (keeping sperm warmer) and hangs away from your body when you are hot (cooling them down).

A sperm cell under a microscope

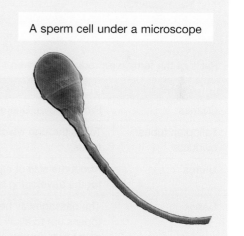

5.2.2 The female reproductive system

Ova (the female's sex cells) are made in the ovaries. Females are born with about 400 000 eggs, or ova, in their ovaries. These eggs are in sacs called **follicles**. Usually only one ovum (plural = ova) is ripened and released into the **fallopian tube** (or oviduct) each month, once the female is sexually mature.

Fallopian tubes are about the diameter of a human hair. They form a tunnel in which the sperm and ovum meet and are hence the site of **fertilisation**. Damage to these tubes can prevent the sperm and egg from meeting.

If the fertilised egg does not move down into the uterus, but remains in the tube, an ectopic pregnancy may result.

The female reproductive system

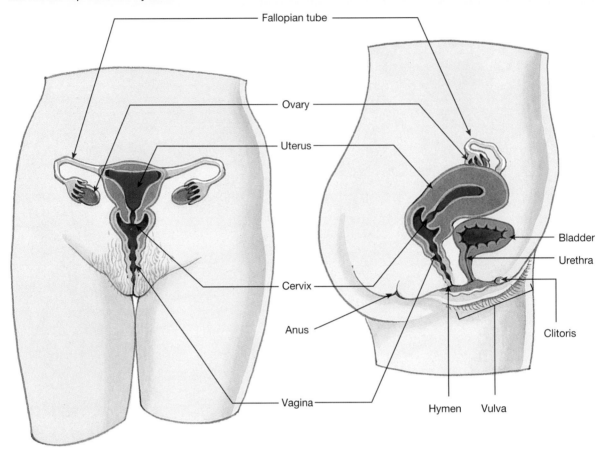

Parts of the female reproductive system

Part	Function
Ovaries	Produce ova. One egg is produced about once a month from one of the ovaries.
Fallopian tubes/ oviducts	Tubes through which ova must travel to reach the uterus. Fertilisation occurs in the fallopian tubes.
Uterus	About the size of a pear when not pregant. It is where the embryo implants and is 'home' for the developing baby. The uterus lining is called the endometrium.
Cervix	The passageway between the vagina and the uterus. During childbirth the cervix dilates (opens up) to allow the baby to come out of the uterus. A 'Pap smear' involves scraping some cells from the cervix's lining to check for any pre-cancerous changes.
Vagina	Elastic tube that connects the uterus to the outside world through which semen enters the female's reproductive system. During birth, babies are pushed out of the uterus and pass through the vagina to enter the world.
Clitoris	Swells during sexual arousal and becomes highly sensitive when erect

HOW ABOUT THAT!

Did you know that not all plants or animals have separate sexes? Some invertebrates are both male and female at once. These interesting combinations are called hermaphrodites. This enables an individual to achieve greater reproductive efficiency than if it was just the one sex.

Snails have been around for 600 million years and have developed intriguing methods of reproduction. Each snail has an organ called an ovotestis, which makes both sperm and eggs, and a single tube to carry both the sperm and the eggs.

After a complex courtship in which hermaphrodite snails rear up, each pressing its muscular foot against its partner, and stroking each other with their tentacles, they simultaneously insert their sex organ into the other's body. In this manner, each snail gives sperm to the other and each has its eggs fertilised.

5.2 Exercises: Understanding and inquiring

To answer questions online and to receive **immediate feedback** and **sample responses** for every question, go to your learnON title at www.jacplus.com.au. *Note:* Question numbers may vary slightly.

Remember

1. Put the following words into sentences:
   (a) testes, scrotum, sperm
   (b) ovaries, ova, follicles
   (c) vas deferens, fallopian tube, ovaries, testes
   (d) ovum, fallopian tube, sperm, fertilisation, ovary.
2. Explain why tight underwear is not recommended for males.
3. What is an ectopic pregnancy?
4. Draw a table as shown below. Classify the following organs and list them in the correct column of your table.
   fallopian tube, penis, urethra, testes, prostate gland, bladder, uterus, seminal vesicle, ovary, vas deferens, scrotum, cervix, vagina

Found in males only	Found in females only	Found in both males and females

5. Match each organ with its function. There may be more than one organ with the same function.

Organ	Function
a. Seminal vesicle	A. Produces gametes
b. Ovary	B. Where the baby grows and develops
c. Scrotum	C. Where fertilisation occurs
d. Testes	D. Keeps the testes slightly cooler than the rest of the body
e. Prostate gland	E. The passageway between the vagina and the uterus
f. Uterus	F. Produces some of the liquid found in semen
g. Cervix	
h. Fallopian tubes	

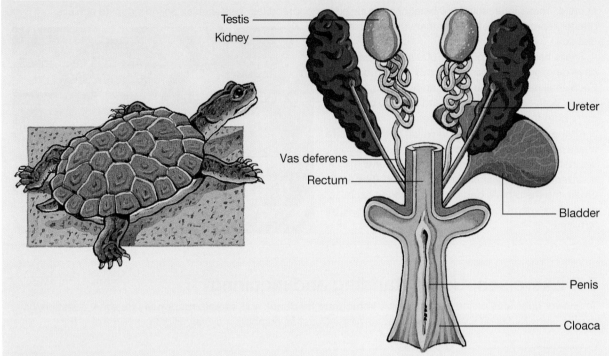

The reproductive system of a male turtle

Testis
Kidney
Ureter
Vas deferens
Rectum
Bladder
Penis
Cloaca

Think and discuss

6. Construct Venn diagrams to compare the following pairs:
   (a) ovaries and testes
   (b) vas deferens and fallopian tubes
   (c) penis and clitoris
   (d) seminal vesicles and prostate gland
   (e) uterus and vagina.

Investigate, think and discuss

7. Suggest why women with blocked fallopian tubes are unable to have babies.
8. Carefully observe the male and female reproductive figures on pages 195 and 196.
   (a) Identify three organs for each gender and state their function.
   (b) Suggest how the structure of each organ suits it for its function.
9. What is a hermaphrodite? Suggest possible advantages and disadvantages of this condition.
10. Male turtles have quite a bit in common with a human male. Carefully observe the figure above showing the reproductive system of a male turtle. Construct a Venn diagram that identifies the similarities and differences between the two systems.

Investigate and create

11. (a) Construct a model of the reproductive system of:
    (i) a human
    (ii) another animal of your choice.
    (b) Describe how they are similar and how they are different.
    (c) Design and create a third model that has the best features of each.
12. Imagine that you are a sperm or an egg. Find out more about how you are produced, stored and used in sexual reproduction.
    (a) Write a dramatic story about your life.
    (b) Construct puppets, animated cartoons or 'fancy dress' actors to recreate your story as a play.
13. Find out more about sperm and ova and represent your findings in the following formats.
    (a) Venn diagram
    (b) PMI chart
    (c) Flowchart

14. Find out more about the structure and function of the human male and female reproductive systems, and design one of the following.
    • A Trivial Pursuit–style game with questions on each system, a board, dice and a rule book
    • An 'information wheel' made of two pieces of circular card connected with a paper fastener ('foldback pin') in the centre. (For an example of how to create an information wheel, see page 204.

learn on RESOURCES – ONLINE ONLY

📄 **Complete this digital doc:** Worksheet 5.1: Human reproduction (doc-18737)

5.3 Why the changes?

5.3.1 Changes can be very exciting

Perhaps you have noticed that you are changing. Have you noticed any hairs in places where they weren't before, changes in your body shape or height, changes in your interests …?

Changes that you may currently be aware of are indications that you are becoming an adult. These physical changes are called **puberty**. The term 'puberty' comes from the Latin word *pubertas*, which means adulthood.

The main purpose of the changes that occur in puberty is to enable you to start producing children. Your sex organs grow and develop. Males begin to produce sperm and females begin to develop the ova they were born with. When combined, these gametes can produce babies.

We are all different

It's okay for the changes to occur at different times and at different rates, because everyone is different. We are all individuals. People reach puberty at different ages. Girls reach puberty between the ages of 8 and 17 years, and boys between 10 and 18 years. A message from your brain to your sex glands triggers all of these changes. When these glands get the message, they produce substances called **hormones**. These hormones travel around your bloodstream and trigger lots of changes.

Both boys and girls experience a growth spurt around the time of puberty.

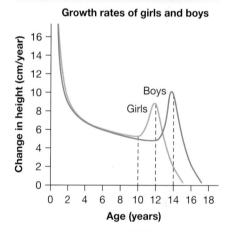

The average age at menarche (the first menstrual period) has decreased over the last two centuries.

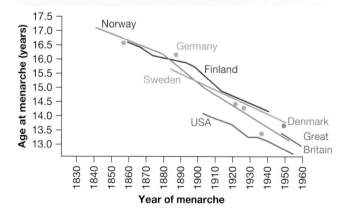

5.3.2 Some common questions

Q: *Why am I getting more pimples?*
A: The sex hormones (testosterone, for example) make the glands in the skin produce extra oils. This can cause the pores in the skin to become blocked and may result in pimples.

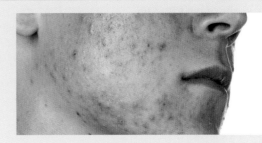

Q: *What causes the changes in a boy's voice during puberty?*
A: The voice box, or larynx, gets bigger, making the voice deeper. Males develop a larger larynx so they develop deeper voices. The squeak often heard during this time of change is due to the muscles of the larynx getting momentarily out of control. Although female voices can also change, the final result is not usually so noticeable.

Q: *In which new places am I likely to grow hair during puberty?*
A: If you are female, it's likely to be around your pubic region and under your arms. Males may notice an increase in these regions and also on their legs, arms, faces and chests.

Q: *What are wet dreams?*
A: These dreams happen to a boy during sleep and result in ejaculation of semen. They are quite normal (although not all boys have them) and are an indication that you are becoming sexually mature.

Q: *What is a breast bud?*
A: This is a little button of tissue just under the nipple from which the breast develops. Sometimes boys also get breast buds, but in their case they go away and do not develop.

HOW ABOUT THAT!

Sperm by the millions

The average amount of semen produced during an ejaculation (about a teaspoonful) contains about 200–500 million sperm cells! You might think it would take a long time for the testes to make 400 million sperm. Not so. Some 200 million sperm cells are manufactured each day by a fertile adult male. That's around 73 billion sperm cells in a year!

Q: *What role does the male hormone testosterone play in puberty?*
A: Testosterone:
  • is needed for sperm production
  • not only enlarges the penis, testes and scrotum, but also increases their sensitivity
  • increases body muscle bulk and promotes growth
  • deepens the voice
  • stimulates growth of body and facial hair
  • increases interest in sex.

Q: *Can sperm build up and burst your testes?*
A: No, unused sperm are stored for a while and then reabsorbed into your body. New sperm replace them.

Q: *At what age do females get periods and how long do they last?*

A: Most girls experience their first period between the ages of 11 and 14, although others may have it up to two years earlier or later. The first menstrual period is called **menarche**.

A period generally lasts about four to six days, with varying amounts of discharge over this time. The discharge contains cells from the lining of the uterus that was built up in preparation for a baby. A **menstrual cycle** is the time from the first day of one period to the first day of the next. It is usually about 28 days, but this varies in different women. It is quite common for a menstrual cycle to be irregular at first — until the body settles into its own pattern.

Menstrual cycle — what's happening?

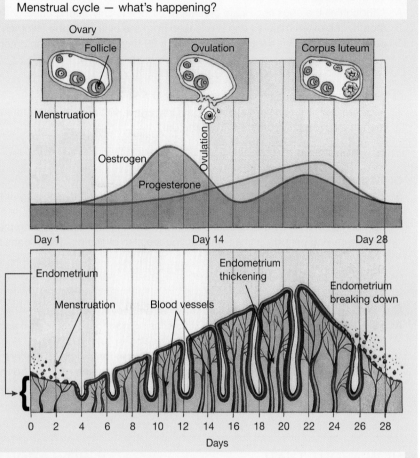

Solving the period puzzle …

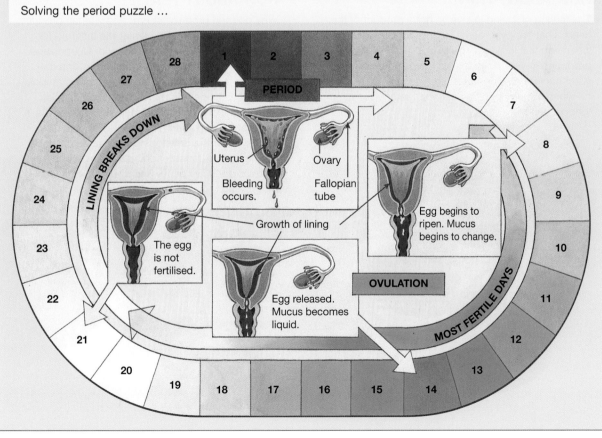

Q: *What is ovulation?*

A: Although all the eggs or ova are present at birth, it is only during puberty that they begin to ripen. Each month, one ovum matures and is released from the ovary into the fallopian tube. The release of the ovum is called ovulation. This continues about every 28 days until the woman enters menopause.

Q: *How many eggs are girls born with and about how many develop?*

A: Girls are born with about 400 000 eggs. Only about 400 of these will mature and be released during her menstrual cycles; the others will not mature.

Q: *What are periods?*

A: Each month, the lining of the uterus prepares itself for the fertilised egg. If the egg is not fertilised, the uterus lining is shed through the vagina. This monthly discharge, or shedding, is called **menstruation**, or a period. Sometimes there may be some discomfort or abdominal cramps during the first few days of your period. If you are having periods, it is a sign that you are also ovulating. This means that you are physically able to get pregnant and have a baby.

Q: *FSH, LH, oestrogen and progesterone: What are they?*

A: FSH (follicle-stimulating hormone), LH (luteinising hormone), oestrogen and progesterone are all hormones. FSH and LH are hormones that make ova develop in girls' ovaries and begin sperm production in boys' testes. Oestrogen and progesterone are produced in the ovaries. They are the most important female hormones and are involved in changes in the lining of the uterus. The hormones control the menstrual cycle as shown in the diagram on previous page.

HOW ABOUT THAT!

The feeling of 'being in love' is not a product of the heart. That happy, dreamy feeling experienced when you 'fall in love' is partly due to a chemical produced in your brain, called phenylethylamine. As chocolate also contains phenylethylamine, it is no wonder many people describe themselves as 'chocolate lovers'.

HOW ABOUT THAT!

Why do men produce sperm and women produce eggs? Dr Josephine Bowles at the Institute for Molecular Bioscience, University of Queensland, has been researching this question — and she has found an answer. A substance called retinoic acid, a relative of vitamin A, triggers a special type of cell division called meiosis. This results in the production of female gametes or ova. Cells in the testes of the developing male fetus produce a protein that degrades this substance, and so meiosis does not occur and sperm are produced rather than ova.

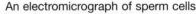
An electromicrograph of sperm cells

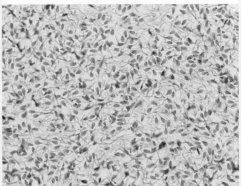

5.3 Exercises: Understanding and inquiring

To answer questions online and to receive **immediate feedback** and **sample responses** for every question, go to your learnON title at www.jacplus.com.au. *Note:* Question numbers may vary slightly.

Remember

1. Use a table to give definitions of the following terms: puberty, menstruation, ovulation, menarche, menstrual cycle, hormones, testosterone, breast buds.
2. Which hormones are responsible for triggering sperm production in males and ovum development in females?

Think and reason

3. Use the illustration of menstruation and other information in this section to answer the following questions.
   (a) What is ovulation?
   (b) On what day in a 28-day cycle is ovulation likely to occur?
   (c) At which time in the cycle is sperm most likely to meet (and fertilise) an egg?
4. Describe the changes in oestrogen and progesterone levels throughout the menstrual cycle.
   (a) Which hormone is found in the highest concentration just before ovulation?
   (b) Which hormone is found in the highest concentration when the uterine lining is thickest?
   (c) At which time would the lining provide the best 'home' for a fertilised egg?
5. Translate the graph on page 201 into a 28-day calendar.

Think and discuss

6. Suggest why the hips become 'fleshier' and the pelvic bones widen in females during puberty.
7. After puberty, the testes continue making sperm for the rest of a man's life. How is this different from gamete production in a woman? What are the consequences of this?
8. If a female has menstrual cycles, is she potentially able to have babies? Explain.
9. Why aren't all menstrual cycles, penises and breasts the same?

Investigate

10. Write down ten questions you have about puberty or reproduction. Use a variety of texts and resources to find the answers. Report the findings to your friends.

Create

11. Construct a crossword with your own clues and answers from information found in this section.
12. Write and act out a play to demonstrate the menstrual cycle. Include the following stages: egg ripening, ovulation, movement of the unfertilised egg into the fallopian tube and through the uterus, and both the egg and the uterine lining being shed through the vagina.
13. Get everyone in the class to write down two to five questions about the changes that occur during puberty and reproduction in a 'Dear Ethel' magazine format. After your teacher has collated these, select at least two to investigate. Report your findings by compiling a class 'Dear Ethel' magazine.
14. Copy and complete the table below. Use this table to help you construct an 'information wheel' about the human menstrual cycle.

Day in cycle	Key events	Hormonal changes	Possible images to include
1			
4			
8			
12			
14			
18			
21			
25			
28			

This is an example of how you might design your information wheel on the menstrual cycle, but there are many others. In this example, the bottom circle of the card is larger than the top piece. You may decide to reverse this or have both pieces the same size and add holes in the top piece to see the days of the timeline.

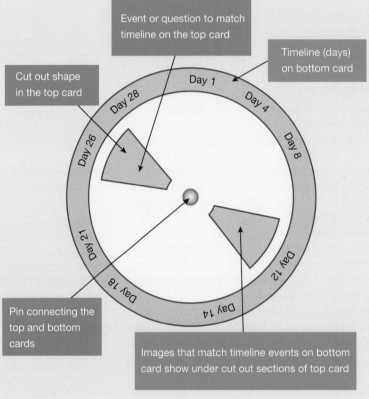

Event or question to match timeline on the top card

Timeline (days) on bottom card

Cut out shape in the top card

Pin connecting the top and bottom cards

Images that match timeline events on bottom card show under cut out sections of top card

Day 1
Day 28
Day 26
Day 21
Day 18
Day 14
Day 12
Day 8
Day 4

5.4 Getting together

5.4.1 Snap!

Let's talk about sex. Sexual reproduction is not something to be embarrassed about, but something incredible and fascinating.

In **sexual reproduction** offspring result from the joining of a male reproductive cell and a female reproductive cell. These reproductive cells are called **gametes** and are made in the reproductive organs. In animals, male gametes are called **sperm** and female gametes are called **ova** or egg cells.

Reproductive systems are designed to bring the male and female gametes together. The joining of sperm and egg cells is called fertilisation. This process mixes the genetic material from the nucleus of each parent and results in the formation of a **zygote**.

Meeting outside …

In some animals, especially those that live or breed in water such as fish and amphibians, fertilisation occurs *outside* the female's body. This is called **external fertilisation**. In this situation, the female releases her unfertilised eggs into the water to be fertilised by the male's sperm, which are also released into the water.

Meeting inside …

In animals that live and breed on land, **internal fertilisation** occurs. This keeps the gametes inside the body so there is no chance of dehydration occurring. In this situation, sperm are introduced into the female by a process called **copulation** (or **sexual intercourse**).

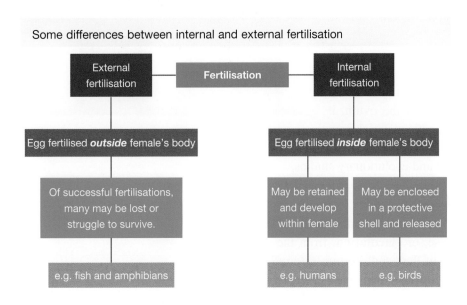

Some differences between internal and external fertilisation

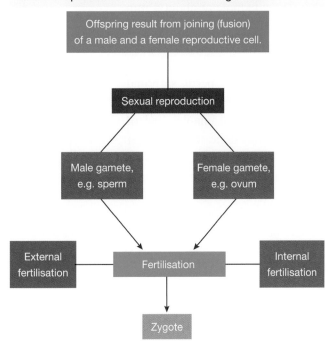

Sexual reproduction involves fusion of gametes.

5.4.2 Ova

Like sperm, ova are produced by a special type of cell division called **meiosis**. Unlike sperm, however, the ova that will be released throughout the female's reproductive years are already present at her birth. This brings differences in terms of **epigenetics** — an exciting new branch of science that involves studying the effect of our experiences on the expression of our genetic information.

Although the resulting zygote will contain a mixture of the genetic material (**nuclear DNA**) from the nucleus of both the sperm and the ovum, it will contain the genetic material from mitochondria (**mitochondrial DNA** or **mtDNA**) only from the mother's ovum. mtDNA forms the basis of many new genetic and evolutionary studies.

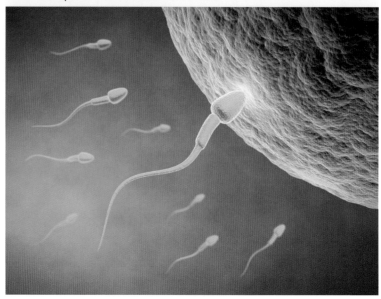
Human sperm cells surround an ovum.

5.4.3 Sperm

In humans, the sperm are mixed with a nutrient-rich fluid before their release, or **ejaculation**, from the male's penis. This combination of fluid and sperm is referred to as **semen**. The sperm make up less than one per cent of the semen. An ejaculation may release about 400 million sperm. Sperm are made up of three distinctive parts: the head, the mid-section and the tail. Swishing its tail, and powered by the mid-section, the sperm swims like a tadpole at about 4 mm per hour to reach its goal — the egg.

5.4.4 Success, at last!

Upon reaching a ripe egg, hundreds of sperm surround it and try to break through. The sperm release an enzyme that dissolves the covering of the egg and enables one of them to penetrate it. Once a sperm has done this, no other sperm can get through. The successful sperm then sheds its body and tail, and the head containing the nucleus continues to move towards the nucleus of the ovum to join with it.

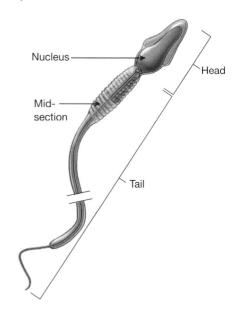
Sperm cell

Nucleus

Head

Mid-section

Tail

Because the nucleus contains the genetic information from the parents, this is the point at which the information is mixed. Once this has occurred, the resulting zygote begins its growth into a new individual. The amazing thing about all of this is that we all started this way. Wow! Imagine having been both a sperm and an egg cell. Isn't that incredible?

Sexual intercourse — getting the sperm to the egg

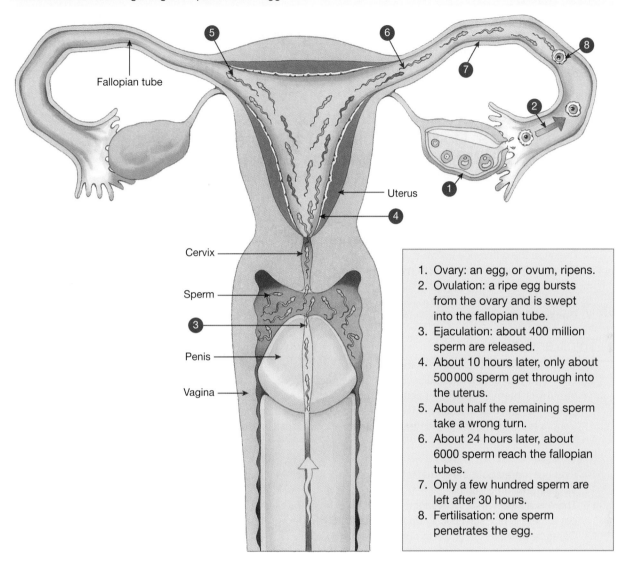

Fallopian tube

Uterus

Cervix

Sperm

Penis

Vagina

1. Ovary: an egg, or ovum, ripens.
2. Ovulation: a ripe egg bursts from the ovary and is swept into the fallopian tube.
3. Ejaculation: about 400 million sperm are released.
4. About 10 hours later, only about 500 000 sperm get through into the uterus.
5. About half the remaining sperm take a wrong turn.
6. About 24 hours later, about 6000 sperm reach the fallopian tubes.
7. Only a few hundred sperm are left after 30 hours.
8. Fertilisation: one sperm penetrates the egg.

5.4.5 Two or more?

Sometimes in the very early stages of division following fertilisation the embryo splits in two, so that two identical offspring are produced. This happens in the case of **identical twins**. They will always be the same gender as they both have the same genetic make-up.

Usually, only one ovum is released at a time. However, if several are released, twins or more can result from fertilisation by different sperm. In this case, the babies are not identical because they have different genetic combinations. These are called **fraternal twins**.

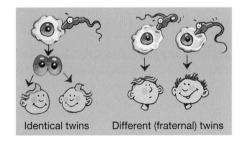

Identical twins   Different (fraternal) twins

The use of fertility drugs and treatments has resulted in an increase in the number of multiple births. This is because fertility treatments can affect ovulation, so that more than one egg is released at a time. Some of these drugs can increase the chance of twins by 25 times and of triplets up to 350 times!

5.4 Exercises: Understanding and inquiring

To answer questions online and to receive **immediate feedback** and **sample responses** for every question, go to your learnON title at www.jacplus.com.au. *Note:* Question numbers may vary slightly.

Remember

1. Define the term 'sexual reproduction'.
2. Put the following words into a sentence: gametes, sperm, ovum, zygote, fertilisation.
3. Describe the difference between external and internal fertilisation.
4. Draw a labelled diagram of a sperm.
5. Describe the events that take place when the sperm reach a ripe ovum.
6. Explain the difference between identical and fraternal twins.
7. Construct a flowchart that includes ovulation, ejaculation and fertilisation.

Think and discuss

8. Construct a Venn diagram to compare sperm and egg cells (ova).
9. With a partner or in a team, discuss the following questions.
   (a) Why is internal fertilisation generally more efficient than external fertilisation?
   (b) Why doesn't fertilisation occur each time a couple have sexual intercourse?
   Send a team member to other teams to share your discussion points.

Investigate

10. Find out more about the gametes of at least four different animals. Display your findings either as models or as diagrams on a poster.
11. There is a theory that, by wearing tight jeans, human males may affect the development of their sperm. Find out if there is any scientific evidence to support this theory.
12. Carefully observe the graph at right on the incidence of multiple births in Australia.
    (a) Suggest reasons for any patterns in the graph.
    (b) Suggest what the graph would look like if this year's data were added. Provide supporting information for your suggestion.
13. Research further into either epigenetics or mitochondrial DNA and share your findings with others.

Investigate and create

14. Write a story, play or poem about the successful sperm, from ejaculation to when it fertilises the ovum.
15. Design a board game that incorporates information about sperm and eggs.
16. Draw a descriptive timeline that includes: ovulation, ejaculation, sexual intercourse, the various stages of the sperm's travels through the female's reproductive tract, and fertilisation.

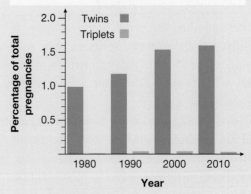

Incidence of multiple births in Australia since 1980 (as a percentage of total number of pregnancies)

Source: Based on ABS data.

5.5 Making babies

5.5.1 The first eight weeks

Conception occurs when the egg cell and sperm unite to form a zygote. When the zygote has divided into many more cells, it is known as an **embryo**. About ten days after fertilisation, the embryo completely embeds itself in the uterus lining (endometrium). This process is called **implantation**.

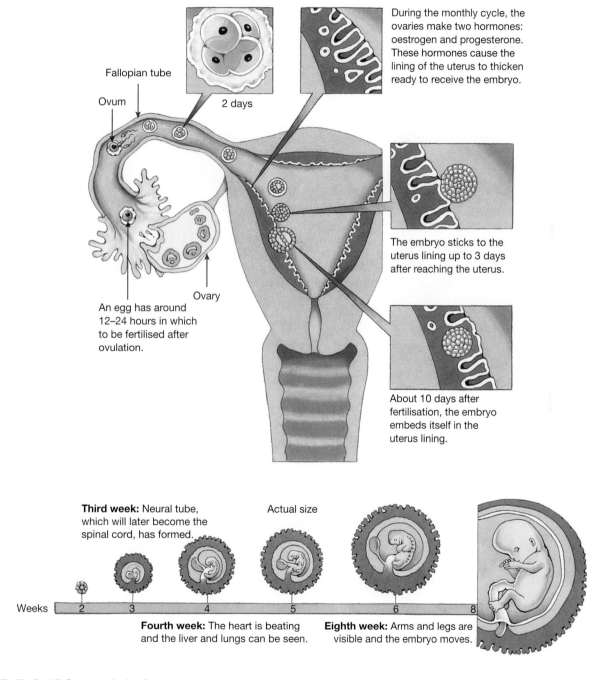

Fallopian tube

Ovum

2 days

During the monthly cycle, the ovaries make two hormones: oestrogen and progesterone. These hormones cause the lining of the uterus to thicken ready to receive the embryo.

The embryo sticks to the uterus lining up to 3 days after reaching the uterus.

Ovary

An egg has around 12–24 hours in which to be fertilised after ovulation.

About 10 days after fertilisation, the embryo embeds itself in the uterus lining.

Third week: Neural tube, which will later become the spinal cord, has formed.

Actual size

Weeks   2   3   4   5   6   8

Fourth week: The heart is beating and the liver and lungs can be seen.

Eighth week: Arms and legs are visible and the embryo moves.

5.5.2 'After eight'

In humans, at about eight weeks, when the embryo has developed a distinct head, arms and legs, it is called a **fetus**. The fetus obtains its nutrients and oxygen through a special organ called the **placenta**. This organ is connected to the mother's blood vessels through the uterus. The placenta also absorbs fetal waste products and acts as a barrier against harmful substances. The unborn child continues to develop inside a sac that is filled with fluid (called amniotic fluid) for the rest of its time within the uterus. The total time spent in the uterus is often called the **gestation period**. In humans, this is usually about 40 weeks. If a baby is born before 37 weeks, it is called **premature** and usually requires extra care and assistance.

Approximate size of a fetus at different stages of development

Development (weeks)	Length (cm)	Mass (g)
8	3	2
12	7.5	18
16	16	140
40	51	3400

A human embryo at 32 days

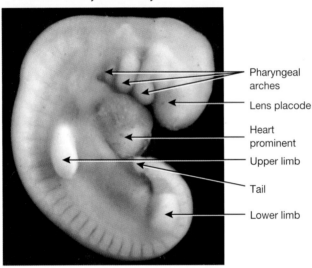

- Pharyngeal arches
- Lens placode
- Heart prominent
- Upper limb
- Tail
- Lower limb

A human embryo at 52 days

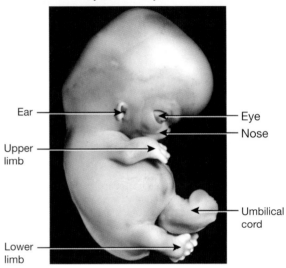

- Ear
- Eye
- Nose
- Upper limb
- Lower limb
- Umbilical cord

Ready for birth — the baby at 40 weeks' gestation

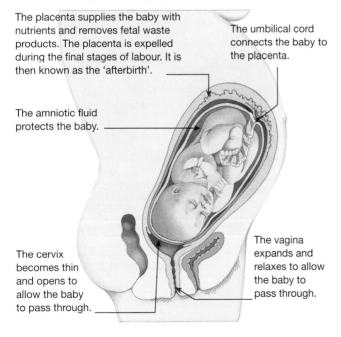

The placenta supplies the baby with nutrients and removes fetal waste products. The placenta is expelled during the final stages of labour. It is then known as the 'afterbirth'.

The umbilical cord connects the baby to the placenta.

The amniotic fluid protects the baby.

The cervix becomes thin and opens to allow the baby to pass through.

The vagina expands and relaxes to allow the baby to pass through.

5.5.3 Giving birth

Three stages are involved in giving birth to a baby. Giving birth is referred to as **labour** because it can be a lot of hard work for the mother. During the first stage, the cervix gradually widens. In the second stage, the woman feels a strong urge to push with each contraction of the uterus. During this stage the baby is born through the vagina, or birth canal. Usually the baby is born head first. Sometimes the baby is born bottom or feet first; this is referred to as a **breech** birth and is often more difficult. The third stage lasts from the baby's delivery until the placenta is delivered.

In some cases, the baby or mother need extra assistance. A **caesarean** may be performed in which doctors surgically remove the baby by cutting through the mother's abdomen to her uterus.

(a) First stage
Uterus begins to contract at regular intervals that get closer and closer together. These contractions begin pushing down on the baby. At some point, the sac holding the amniotic fluid breaks; the fluid leaks out of the mother's vagina.

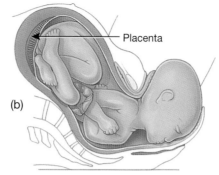
(b) Placenta

As contractions continue, the cervix stretches open, until it is about 10 cm wide. This stage can last for many hours, especially for first-time mothers

(b) Second stage
The mother gets a fierce urge to push (a bit like with a bowel motion) every time the uterus contracts. Bit by bit, this pushes the baby further down the vagina (birth canal).

(c) Third stage
The placenta is delivered after the baby is born. By this stage of the pregnancy it is a flattish, dinner-plate-shaped organ that looks a bit like a large piece of liver.

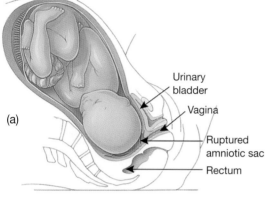
(a) Urinary bladder / Vagina / Ruptured amniotic sac / Rectum

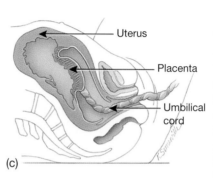
(c) Uterus / Placenta / Umbilical cord

HOW ABOUT THAT!

Oxytocin — the 'trust' hormone
Hormones can have a variety of effects on our bodies. **Oxytocin** is an example of a hormone that not only has the potential to change how we feel, but also has important reproductive roles. This hormone causes the uterus to contract during childbirth and has a key role in breastfeeding. When a baby suckles on the mother's nipple, oxytocin is released in the mother, triggering the 'let down' response in which milk is released for the baby.

Oxytocin is also thought to be involved in the promotion of trust, love, empathy and social recognition. It has been described as the 'cuddle chemical', as it is released when mothers cuddle their babies. The release of oxytocin may assist in the formation of bonds not only between mothers and their babies, but also between people in close relationships.

Cuddling and breastfeeding can result in the release of oxytocin, which can promote feelings of trust, love and bonding.

With trust comes power! Nasal sprays containing oxytocin have been marketed as 'trust sprays'. These are being advertised as having commercial value as they may contribute to feelings of trust in potential clients and customers. The development of oxytocin nasal sprays also provides an opportunity for researchers to investigate the potential use of this hormone in the development of treatments for specific autism spectrum disorders (ASD) and in treatments to increase empathy and learn successful face recognition.

5.5 Exercises: Understanding and inquiring

To answer questions online and to receive **immediate feedback** and **sample responses** for every question, go to your learnON title at www.jacplus.com.au. *Note:* Question numbers may vary slightly.

Remember

1. Outline the differences between a zygote, an embryo and a fetus.
2. List the following in the correct order: birth, fertilisation, ovulation, growth, implantation.
3. Identify the part of the female reproductive system in which the fetus develops.
4. Explain why sexual intercourse doesn't always result in fertilisation and pregnancy.
5. Describe the difference between:
   (a) fertilisation and implantation
   (b) fraternal and identical twins.

Using data

6. Construct a graph, using information in this section, to show the changes in length of the embryo from 2 to 8 weeks, and the length and weight of the fetus from 8 to 40 weeks.

Investigate

7. Research and report on one of the following: endometriosis, prolapse of the uterus, cervical cancer, hysterectomy, ectopic pregnancy.
8. Research and report on one of the following antenatal tests: ultrasound scanning, amniocentesis, chorionic villus sampling.
9. (a) Investigate the commercial availability and uses of the hormone oxytocin.
   (b) In your team, construct a PMI chart based on your findings.
   (c) Discuss the ethics and issues regarding the use of oxytocin in conditions not involving childbirth and breastfeeding.
   (d) If you were on an ethics committee or governing body, what regulations would you suggest be considered concerning the availability and use of synthetic versions of hormones (such as oxytocin)?
   (e) Organise your team's discussion and findings into a format that enables it to be shared with others.

Think and create

10. Make scale models of the fetus at each age shown in the table on page 210.
11. Correctly match each key event with its day of occurrence in the menstrual cycle.

Day	Key event	Day	Key event
14	Implantation	16	Fertilisation
15	First cell division	23	Ovulation

12. Suggest why menstruation must stop during pregnancy. What would happen if this were not the case?
13. Construct storyboards for the following:
    (a) how you have changed between birth and ages two, four, six, eight and ten, and your current age
    (b) the 'life of a sperm' or the 'life of an egg'
    (c) how you could tell a Year 3 primary student 'the facts of life'.
14. Read through the information in the 'Week by week' article below and on the following page. See section 5.13 for more information on storyboards and Gantt charts.
    (a) Mind map what you consider to be the key points.
    (b) Construct storyboards to show:
       (i) the changes experienced by the mother
       (ii) the baby's development.
    (c) Construct a Gantt chart to sequence your key points.

Week by week pregnancy

Week 0–4

You... One of your eggs is fertilised by a sperm, resulting in the formation of a zygote. At this stage conception has occurred — although you may not be aware of it.

And your baby… The zygote divides to form a collection of cells known as a blastocyst. About 3–7 days later, the developing blastocyst moves down the fallopian tube to embed into the lining of the uterus. At this stage, the blastocyst is about 0.23 mm long.

Week 5

You… may experience tiredness, breast tenderness and need to urinate more often.

And your baby… is only about 1.25 mm long. During this week differentiation into ectoderm, mesoderm and endoderm occurs.

Week 6

You… may be feeling nauseous (and possibly vomiting), have sore or tender breasts, fatigue, constipation and need to urinate more frequently.

And your baby… is now called an embryo and has a beating heart, although it's still reliant on your blood supply. By the end of this week the neural tube will have closed, later to develop into a brain. Although only between 2–4 mm long, development can be affected by alcohol, nicotine and other chemicals.

Week 7

You… Although there may not be any external signs of your pregnancy, quite a lot is happening internally! You may feel quite tired and nauseous in the mornings (morning sickness).

And your baby… During this week, your baby grows significantly from 4–5 mm to 11–13 mm. The brain begins to form and develop, the heart begins to form valves, digestive systems begin to take shape and limbs are visible.

A baby's arms and legs start to form very early on. Soon, her limbs will lengthen, fingers and toes will develop and she'll be able to move them.

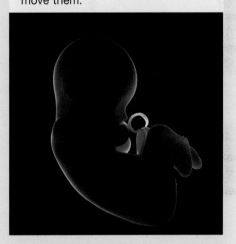

Week 8

You… Your clothes may become tighter around your torso as your uterus is increasing in size.

And your baby… is now 14–20 mm long and a face is beginning to take shape. A nose tip, nasal passage, ear openings, fingers and toes are visible. Lungs are developing and a skeleton is starting to form.

Week 9

You… Your clothes are getting even tighter as your waist continues to thicken. Your uterus has grown from the size of your fist to that of a grapefruit.

And your baby… About 22–30 mm in length, even though eyes are still forming, now eyelids and tiny external ears are forming and are visible.

Week 10

You… Changes in your hormone levels may result in mood changes and feeling more emotional.

And your baby… 31–42 mm in length, all organs are present and most of the major structures have formed.

Week 11

You… Due to the baby's rapid growth, you may feel very tired. You may also feel your uterus above your pubic bone.

And your baby… is about 44–60 mm in length with a head that takes up almost half of this length! This is a period of rapid growth; eyes are formed, face completed, and bones and ribs appear.

Week 12

You… may feel better as morning sickness improves. Your breasts may be getting larger and there is more thickening of your waist.

And your baby… is around 60 mm, nails and teeth are beginning to form, nervous system continues to develop, digestive system and external genital organs are growing. Around this time, a heartbeat can be detected.

 RESOURCES — ONLINE ONLY

Complete this digital doc: Worksheet 5.5: Inside the womb (doc-18734)

5.6 To breed or not to breed

Science as a human endeavour

5.6.1 Preventing pregnancies

Conception involves the production of a zygote and its implantation into the wall of the uterus. Techniques that prevent this happening are called **contraception**.

Contraceptives are the devices or substances used to prevent unplanned pregnancies. There are two main types of contraceptives: those that prevent fertilisation taking place and those that prevent the fertilised ovum from implanting in the uterus.

The 'cafeteria approach' to contraceptives enables people to weigh all the risks, benefits, advantages and disadvantages, and to find out which type best suits them.

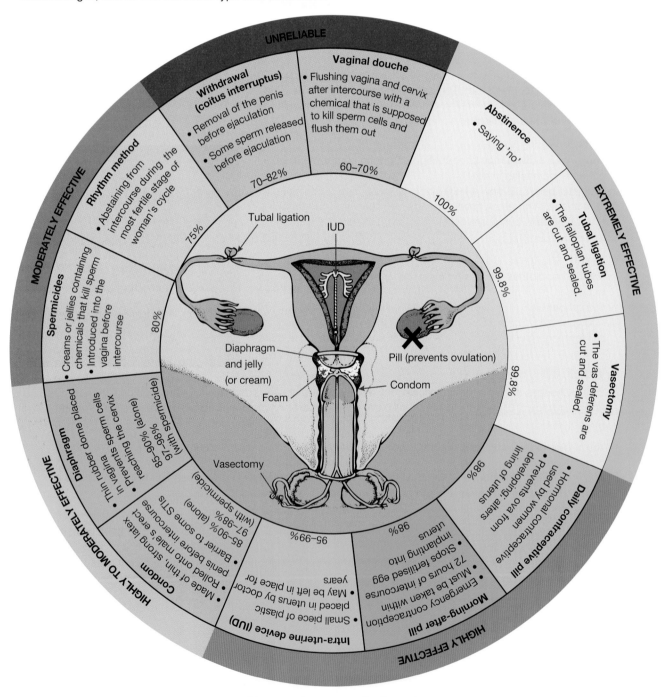

The World Health Organization estimates that, worldwide, half of all pregnancies are unplanned. The information that follows gives you the facts about a variety of contraceptives. Quite often, television shows and magazines introduce sexual activity to young adults without giving them the full story. This can deliver a distorted message. But remember, the most effective method of contraception of all is to not have sexual intercourse!

5.6.2 New and improved products

Throughout history, people have tried to find methods that would enable them to have sex but not make babies — for example, people have tried swallowing tadpoles in spring, using lemons as a 'diaphragm' and using pig intestines as condoms. Because there is still no such thing as the perfect contraceptive, many new products are being invented and tested.

For women, some of these include **transdermal patches** that stick onto the skin and release hormones, **vaginal pills** that dissolve into spermicide when inserted into the vagina before intercourse, and the **Filshie clip**, a type of fallopian tube clamp. In some countries, even a female condom is available.

New products for men include **testosterone injections** or implants to reduce sperm levels, **anti-fertility vaccines** that regulate sperm and testosterone production, **sperm duct plugs** that inject liquid plastic into the vas deferens, **chemical sterilisation**, and **gossypol**, a chemical that reduces sperm production.

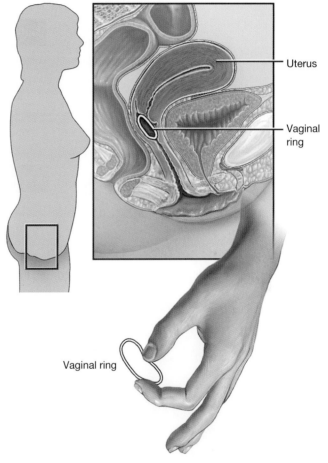

This vaginal ring can function for up to three weeks, secreting hormones similar to those in contraceptive pills.

Uterus

Vaginal ring

Vaginal ring

5.6.3 In for the long haul

There are various long-acting contraceptive methods available. Once 'introduced', these require no further action by the user for a long time. A disadvantage of most is that they require medical intervention for insertion and removal. Examples include:

- *Depo injections*: Also known as Depo–Provera, this is a hormone injected into the user's buttock muscles that prevents ovulation for about three months.
- *implants*: A contraceptive implant (about the size of a matchstick) called Implanon is inserted under the skin of the inner, upper arm and prevents ovulation for about three years.
- *hormone releasing intra–uterine devices (IUD)*: A Mirena is a T-shaped plastic device that releases hormones, acting directly on the lining of the uterus to make it thin and unreceptive to implantation of the fertilised egg. It also changes the fallopian tube lining, the mucus produced by the cervix and can stop ovulation in some women. It provides continuous contraceptive protection for about five years.

'Mapping' contraceptives — are some types of contraceptives better than others?

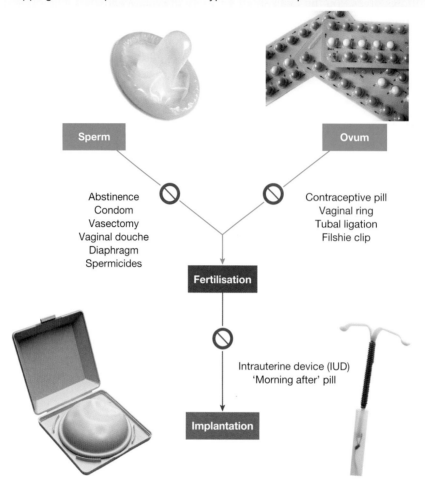

Will the future include contraceptive drugs that disable sperm or inhibit sperm production?

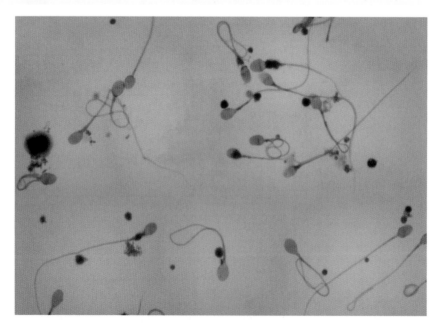

5.6 Exercises: Understanding and inquiring

To answer questions online and to receive **immediate feedback** and **sample responses** for every question, go to your learnON title at www.jacplus.com.au. *Note:* Question numbers may vary slightly.

Remember

1. Identify which methods of contraception:
   (a) prevent the sperm from reaching the egg
   (b) prevent the release of the egg.
2. Construct a table that has six columns, with headings for: the type of contraceptive, a summary of how it works, who uses it (male or female), suggested advantages, suggested disadvantages, and a prediction of how many pregnancies may occur if 100 sexually active, fertile couples were to use it.

Think

3. After a vasectomy:
   (a) does a male still produce sperm
   (b) what does the ejaculate contain?
4. After a tubal ligation, does a woman:
   (a) still ovulate
   (b) menstruate?
5. Suggest why some contraceptives may be more effective than others.

Investigate and discuss

6. Research a method of contraception and present your findings to your team or class as a poster, PowerPoint presentation or concept map.
7. Find out about four of the following types of contraceptives and then present your findings with your team as a poster, PowerPoint presentation or concept map.
   • Combined oral contraceptive pill
   • Progesterone–only pill (or the mini pill)
   • Depo–Provera (injectable contraceptive)
   • Hormonal implants
   • Morning–after pill
   • Today sponge
   • Lng–Levonova IUD
   • Vaginal ring
   • Female condoms
8. Find out the advantages and disadvantages of each type of contraceptive listed above. Present your findings as a matrix. Compare and discuss your findings with others in the class.

Create

9. Design a 'future' contraceptive, using your imagination and knowledge of reproductive systems. Decide how you would scientifically test the safety, effectiveness and popularity of your contraceptive. Produce a brochure that promotes your invention.

Some IUDs secrete hormones that act directly on the uterus to make it thin and unreceptive to implantation of the fertilised egg.

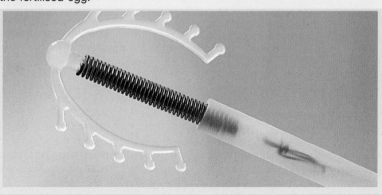

5.7 Reproductive technologies

5.7.1 Infertility

Making babies is not always easy and simple. Not everyone can make their own babies.

The term **infertility** describes the inability to conceive or carry a pregnancy to a live birth. About 20 per cent of all couples are infertile. One of the commonest causes of infertility is the inability of either the male or the female to produce gametes. Such a person is **sterile**.

Some of the other reasons couples may not be able to have children are listed in the table below. Reproductive technologies have been developed to help people overcome some of these problems.

Frozen egg technology

Fallopian tube

Needle

Follicles

Needle placed into follicle (sac containing egg) to take out egg

Ripened eggs placed with antifreeze

Ripe egg with antifreeze put into glass straw

Slices of ovary containing unripened eggs with antifreeze in tube

Years later, eggs thawed and IVF methods used to create embryos and implant them into women.

Straws and tubes stored in liquid nitrogen at −196°C

Eggs reduced in temperature in automatic freezer to −150°C

Reasons couples may not be able to have children

Type of problem	Definition/reason
Gametes	Sperm or ova are not produced in sufficient quantity or quality.
Impotence	Some men cannot maintain an erection during sexual intercourse.
Blockage or damage	Some women may have blockages in their reproductive system (e.g. fallopian tubes), preventing fertilisation.
Miscarriage	The zygote or embryo is not maintained until the full term of the pregnancy.

5.7.2 Artificial insemination (AI)

This technique involves injection of sperm into the woman's uterus close to the time of ovulation. The sperm may be collected from her partner, or from another male if her partner is sterile. Artificial insemination is also used in agriculture in the production of prime farm animals, and in the breeding programs for endangered species.

5.7.3 In-vitro fertilisation (IVF)

In IVF, the sperm and the egg are fertilised outside the female's body. The fertilised egg is incubated until it develops into an embryo, which is then introduced into the female's uterus.

In this technology, eggs are surgically removed using a needle, laparoscope and forceps. A laparoscope has the lens of a microscope. To improve the chances of success, the woman is often treated with drugs that cause super-ovulation, resulting in several ova maturing at the same time (instead of one, as is usually the case). It is possible to freeze any fertilised eggs that are not used so that they may be available at a future time. This could enable 'twins' to be born years apart. Some women may use **donor eggs** (ova from other women).

The removed egg is then fertilised outside the mother's body in a small glass tube or dish. The sperm used is treated to remove its outer protein coat (an event that usually occurs in the female's reproductive tract). The fertilised egg is incubated in the laboratory until it is at the two- or four-cell stage. A four-cell embryo is obtained about 35–46 hours after fertilisation.

The embryo is then placed into the woman's uterus for implantation. Babies born using this technique are often referred to as **test-tube babies**. In 1980, Australia's first IVF baby, a girl, was born in Melbourne; the first frozen embryo baby was also born in Melbourne, in 1984.

Some women are unable to maintain the growing embryo inside their uterus. A woman may, for example, be born without a uterus. The development of IVF technology has also opened up the field of **surrogacy**. In this situation, eggs are surgically removed from one woman and fertilised using IVF techniques. After this they are placed into another woman who undergoes the pregnancy.

Although Louise Brown's conception using IVF in 1978 was considered a miracle of science at the time, more than 80 000 Australians have since started life in a dish. Louise is now in her thirties and is a mother herself, with her children being conceived naturally.

Hormone injections are used to increase ovulation. Collected eggs may be incubated and mixed with sperm to allow the possibility of fertilisation and formation of an embryo. The embryo may be inserted into the uterus. Surplus eggs or embryos may be frozen.

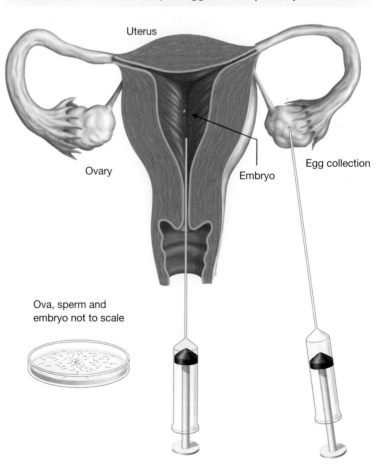

Uterus

Ovary

Embryo

Egg collection

Ova, sperm and embryo not to scale

Needle penetrating the follicle containing the ovum and surrounding fluid

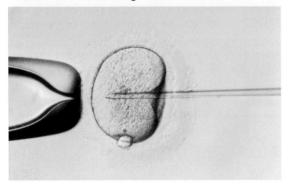

5.7.4 Testing the unborn child

There are a variety of technologies that can be used to test genetic composition and development. Some of these can be performed when the new life is in its very early stages. **Pre-implantation genetic diagnosis** (PGD) may be used to diagnose and exclude genetic abnormalities in embryos before potential implantation.

Other techniques, such as **ultrasound, amniocentesis** and **chorionic villus sampling** can be used at later stages of development. These techniques enable the gender and a variety of abnormalities to be identified.

A cell can be removed from the developing embryo to be tested for genetic abnormalities.

Ultrasound involves the use of sound waves to produce images of an unborn child inside the mother's body.

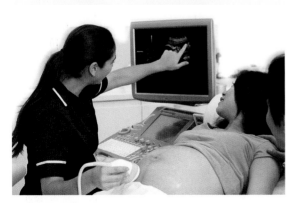

During amniocentesis, a fine needle is inserted into the amniotic sac of the fetus at around 14–16 weeks of the pregnancy and a small amount of fluid is drawn out to be tested.

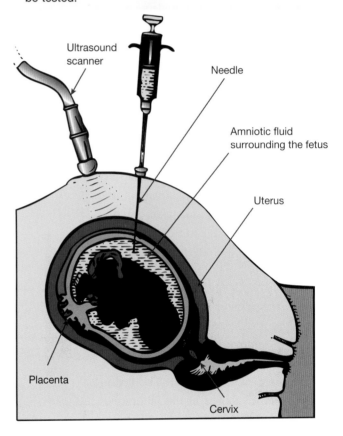

In chorionic villus sampling, cells from the developing placenta are removed for testing at around 10–12 weeks of pregnancy.

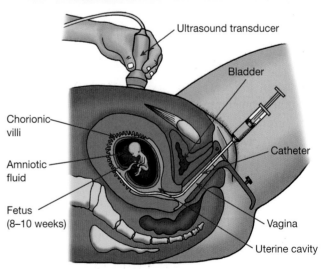

HOW ABOUT THAT!

Cadence Minge, a University of Adelaide researcher, was a winner of South Australia's Young Investigator of the Year award in 2007. Her research provided scientific evidence that high-fat diets could cause infertility in obese women. Her investigation involved using mouse eggs and showed that diets high in fat caused damage to eggs stored in the ovaries. She also found that a particular protein called PPAR-gamma could reverse the effects, but warned that it should not be considered a 'quick fix' for infertile women.

Cadence Minge (Winner of South Australia's Young Investigator of the Year award in 2007)

HOW ABOUT THAT!

There are many different types of reproductive technologies. While assisted reproductive technologies (ART) can be used to treat infertility, other types (such as contraceptives) may be used to reduce fertility. Other types of reproductive technologies may be used to determine the likelihood of developing a particular genetic disease or to increase the number of offspring with particular features.

What's the difference between a GIFT and a ZIFT? Which of the reproductive technologies in this figure increase fertility and which decrease it? What's cryopreservation and why bother with it?

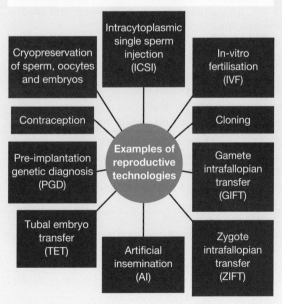

5.7 Exercises: Understanding and inquiring

To answer questions online and to receive **immediate feedback** and **sample responses** for every question, go to your learnON title at www.jacplus.com.au. *Note:* Question numbers may vary slightly.

Remember

1. Discuss the techniques that would help a couple reproduce if:
   (a) the male was infertile
   (b) the male was impotent
   (c) the female had blocked fallopian tubes
   (d) the female had a history of miscarriages.
2. Distinguish between:
   (a) artificial insemination and in-vitro fertilisation
   (b) ultrasound and amniocentesis.
3. Outline, in point form, the steps involved in IVF.
4. What are 'test-tube' babies? Is this an adequate name for them? Explain.

Think, investigate and discuss

5. In groups of four or more, discuss each issue statement in the table below.
   (a) Write a list of people's 'gut reactions' or immediate responses to each statement.
   (b) Make a list of arguments for, and a list of arguments against, each statement.
   (c) Suggest what factors influenced your opinions on these issues.
   (d) Did the opinions differ between members of your group? Suggest reasons why.
   (e) Report your findings back to the class, or organise a debate.
   (f) Write a summary paragraph about the class's overall response to each statement.

Technique	Issue statement	Your opinion (Explain your response with arguments for and against the issue.)
AI	• Sperm should be used only from males with a high IQ, blue eyes and red hair. • All women should be artificially inseminated with sperm selected by their parents.	
IVF	• The IVF program is too expensive and should be abandoned. • IVF technology should be used to build a superior race.	
Donor gametes; surrogacy	• Donors and surrogates should be anonymous and have no rights over the offspring produced. • Sperm should be collected from all males at the age of 18 and only this is to be used for fathering children.	
Frozen embryos	• These embryos should be available to other couples if they are not used within six months. • These embryos should be developed so that they provide a supply of blood and organs for transplants.	
Ultrasound; amniocentesis	• These tests should be made compulsory for all women. Any abnormalities should result in immediate removal of the fetus. • These techniques should be used to select the gender of the child.	

6. Who should decide who is entitled to access these technologies? Discuss this with your team and report back to the class.
7. What are the risks linked to reproductive technologies?
8. (a) Find out about the South Australian Young Investigator of the Year award and outline examples of scientific research that winners have been involved in.
   (b) Find out more and report on one of these research areas.

9. (a) Find out similarities and differences between the fields of obstetrics and gynaecology.
   (b) Select a topic that interests you in the area of reproductive technologies, and investigate and report on research in that field.
10. Research and report on the scientific contributions of two leading reproductive technology pioneers:
    (a) Carl Wood (IVF pioneer, assisted reproductive technology)
    (b) Alan Trounson (IVF and stem cell pioneer).
11. Developments in reproductive technology rely on scientific knowledge from different areas. Find examples that provide evidence for this claim.

learn **on** RESOURCES — ONLINE ONLY

Complete this digital doc: Worksheet 5.6: IVF — discussing the issues (doc-18727)

5.8 Reproduction issues

Science as a human endeavour

5.8.1 Issues and ethics

Research in the area of reproductive technology has brought rapid change. There is a big difference between what is scientifically possible and what is socially acceptable.

This section provides examples of situations where the use of reproductive technology has raised ethical, social, legal or economic issues. Each snippet is a summary of a news story that was either published in a newspaper or was presented on the news.

5.8.2 Surrogacy

Surrogacy is when a woman carries and delivers a baby for a childless couple. In some cases sperm from the father is used to artificially inseminate the surrogate mother. In other cases an embryo from the childless couple is produced by IVF and then transferred into the womb of the surrogate. In commercial surrogacy the surrogate is paid, sometimes large amounts of money, to carry the child. Altruistic surrogacy does not involve the exchange of money. A friend or sister might carry a child for a couple for example, and not expect payment in return. In 2008, altruistic surrogacy was legal in all states except Queensland where women could be fined or sent to jail for acting as a surrogate. A 2010 Act made altruistic surrogacy legal in Queensland.

5.8.3 Baby farms

Indian clinics specialising in surrogacy are offering rich childless couples from around the world the opportunity to use a poorer Indian woman as a surrogate. The women are paid to act as surrogates. They live together while they are pregnant, and they receive regular health checks and have all their meals prepared. For American couples (another country where commercial surrogacy is legal) the cost of an Indian surrogate is much less than an American surrogate.

Dr Kakoli Ghosh Dastidar works in a clinic in India where local women are paid to act as surrogates for other couples.

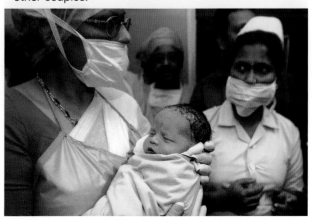

5.8.4 Freezing eggs

The option of freezing some eggs is increasingly offered to women for a variety of reasons. Treatment for certain types of cancer can make women infertile or damage their eggs. Having some eggs collected and frozen before starting cancer treatment would give these women a chance to have children after they recover from the cancer. For women who have not met the right partner or do not yet feel ready to be mothers by the time they reach their late thirties, egg freezing might offer the possibility of extending their reproductive years. The chance of producing a baby from frozen eggs is not very high at this stage. In 2007 the American Society for Reproductive Medicine calculated that for every 100 frozen and thawed eggs only 2 to 4 pregnancies would result. Eggs are more difficult to freeze than semen or embryos because they contain more water and ice crystals can form inside the egg.

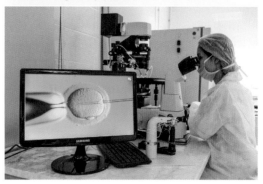

A human egg (below) shown next to the tip of a thin glass pipette

5.8.5 Frozen sperm

A UK woman confirmed that she was pregnant by her dead husband's sperm. Her husband died from meningitis in 1995. Some sperm samples were collected and frozen. She was told that she could not use the sperm samples for artificial insemination because her husband had not given his consent in writing before he died. She went to court and won the right to use the semen samples.

Semen can be collected from a recently deceased man and frozen for later use, but is it ethical to do so?

5.8.6 Designer babies

Professor Julian Savulescu, an ethicist from Oxford University, argues that parents should be able to use genetic testing combined with IVF to choose the genetic characteristics of their children. Currently it is possible to test embryos created by IVF to find out which carry certain disease-causing genes. Embryos found to carry the genes are not implanted. Professor Savulescu argues the technique should be further developed to allow parents to select genes for anything ranging from hair colour to intelligence or sporting ability.

Should parents have the right to select certain characteristics in their children?

5.8.7 IVF mistake

A Victorian couple sued doctors at an IVF clinic. The couple decided to use IVF to conceive their child because they wanted to avoid giving birth to a child with haemophilia. The mother knew she was carrying a gene for haemophilia. Haemophilia is a disease where blood does not clot properly. A person who has severe haemophilia will usually require a transfusion of a special component of blood any time they have even a minor injury such as a cut or bruise. If a woman is a carrier for haemophilia, she does not have haemophilia herself. If she has a daughter and the father does not have haemophilia, the daughter will not have haemophilia either. If the same couple have a son, however, there is a 50 per cent chance that he will have haemophilia.

The couple used IVF because they wanted the doctors to test the embryos to find out whether they were boys or girls before transferring them to the mother's womb. The doctors made a mistake and transferred a male embryo. The couple gave birth to a son who has severe haemophilia.

The couple sued the doctors who carried out the IVF treatment. They argued that the unexpected arrival of a boy caused them shock and anxiety. They also wanted to be compensated for the cost of medical treatment for their son as well as the pay they lost as a result of not being able to go to work when their son needed treatment.

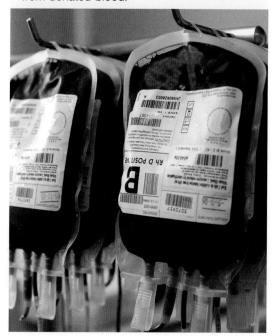

Treatment for haemophilia usually includes regular transfusions of a product obtained from donated blood.

5.8 Exercises: Understanding and inquiring

To answer questions online and to receive **immediate feedback** and **sample responses** for every question, go to your learnON title at www.jacplus.com.au. *Note:* Question numbers may vary slightly.

Remember

1. Distinguish between the following terms:
   (a) surrogacy, artificial insemination and IVF
   (b) commercial surrogacy and altruistic surrogacy.
2. Outline some situations where women may consider having their eggs frozen.
3. Explain what the chance is of producing a baby from a frozen egg. Why is the success rate so low?

Think

4. Discuss whether there should be an age limit for IVF treatment. Should this age limit apply to the mother only or to both parents? Justify your answer.
5. Create a PMI chart about the following statement: A woman should be allowed to use her dead husband's sperm to conceive a child.
6. Using IVF and genetic testing it is currently possible for parents to choose certain characteristics in their children. The technology can be used to screen out certain genetic diseases and to select the sex of the child. In the future it may be possible to select a much greater number of characteristics.
   (a) Discuss whether this particular technology is harmful or beneficial.
   (b) Should parents be allowed to select any characteristics for which there is a test available or should there be restrictions on the characteristics that parents can select? Justify your answer.
   (c) IVF and genetic testing are expensive procedures, so they may not be available to poorer couples. Explain how this could have an impact on society.

Investigate

7. Use EBSCO or another database to locate other news stories about reproductive technology. Summarise the key points in each article.

5.9 Stem cells: A source of issues

Science as a human endeavour

5.9.1 What are stem cells?

Have you heard or read about the issues regarding stem cells? What's the stem of the trouble? What are stem cells and why are people arguing about them?

Stem cells are unspecialised cells that can reproduce themselves indefinitely. They have the ability to differentiate into many different and specialised cell types. Stem cells in a fertilised egg or zygote are **totipotent** — they have the ability to differentiate into any type of cell. The source of the stem cell determines the number of different types of cells that it can differentiate into.

The ability to differentiate into specific cell types makes stem cells invaluable in the treatment and possible cure of a variety of diseases. For example, they may be used to replace faulty, diseased or dead cells. The versatility of stem cells is what makes them very important.

Stem cells can be divided into categories on the basis of their ability to produce different cell types.

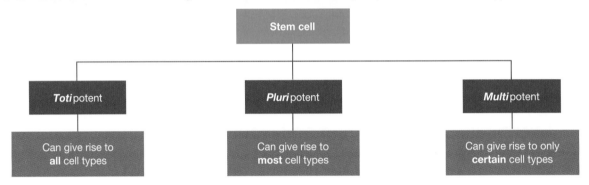

5.9.2 What are the sources of stem cells?

Embryonic stem cells can be obtained from the inner cell mass of a blastocyst. Blastocyst is the term used to describe the mass of cells formed at an early stage (5–7 days) of an embryo's development. Embryonic stem cells are **pluripotent** and can give rise to most cell types; for example, blood cells, skin cells, nerve cells and liver cells.

Stem cells can be described in terms of their source.

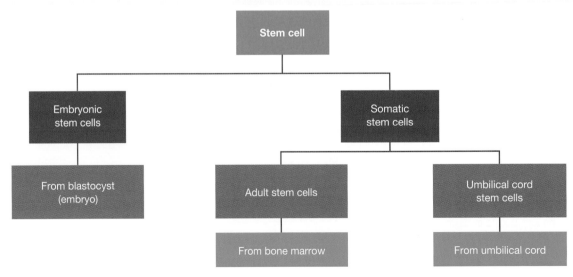

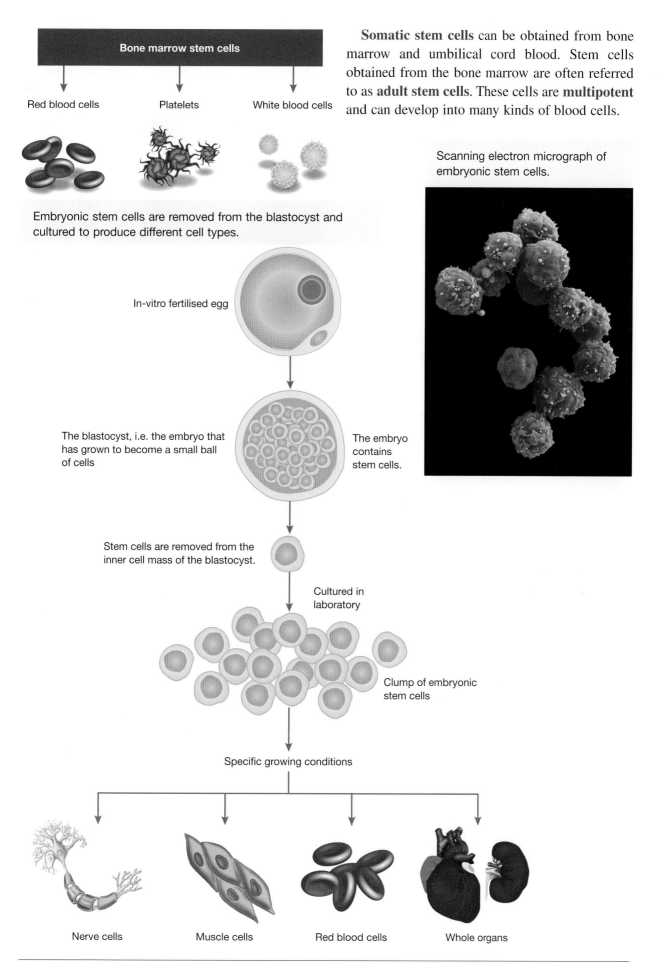

Bone marrow stem cells

Red blood cells

Platelets

White blood cells

Somatic stem cells can be obtained from bone marrow and umbilical cord blood. Stem cells obtained from the bone marrow are often referred to as **adult stem cells**. These cells are **multipotent** and can develop into many kinds of blood cells.

Scanning electron micrograph of embryonic stem cells.

Embryonic stem cells are removed from the blastocyst and cultured to produce different cell types.

In-vitro fertilised egg

The blastocyst, i.e. the embryo that has grown to become a small ball of cells

The embryo contains stem cells.

Stem cells are removed from the inner cell mass of the blastocyst.

Cultured in laboratory

Clump of embryonic stem cells

Specific growing conditions

Nerve cells

Muscle cells

Red blood cells

Whole organs

The umbilical cord is the cord that connects the unborn baby to the placenta. This is how the baby gets nutrients and oxygen while it is still inside its mother's body. This cord contains stem cells that can develop into only a few types of cells, such as blood cells and cells useful in fighting disease. **Umbilical cord stem cells** can be taken from this cord when the baby is born.

5.9.3 Can stem cells be made to order?

While the information in your genetic instructions tells your cells which types of cells they should become, scientists have also been able to modify the 'future' of some types of cells. By controlling the conditions in which embryonic stem cells are grown, scientists can either keep them unspecialised or encourage them to differentiate into a specific type of cell. This provides opportunities to grow replacement nerve cells for people who have damaged or diseased nerves. Imagine being able to cure paralysis or spinal cord injury. In the future, stem cells may also be used to treat and cure Alzheimer's disease, motor neurone disease, Parkinson's disease, diabetes and arthritis.

5.9.4 So what's the problem?

The source of embryonic stem cells raises many issues. Embryonic stem cells can be taken from spare human embryos

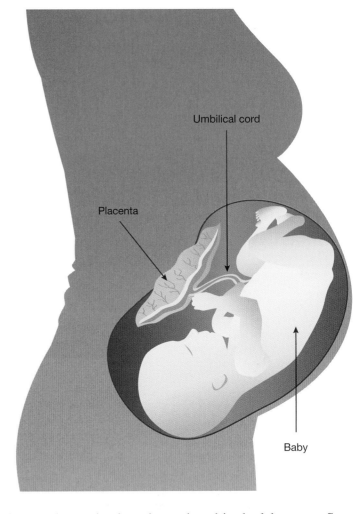

that are left over from fertility treatments or from embryos that have been cloned in the laboratory. Some argue that this artificial creation of an embryo solely for the purpose of obtaining stem cells is unethical.

An embryo is the result of a sperm fertilising an egg. If this happens outside a woman's body, it is called in-vitro fertilisation.

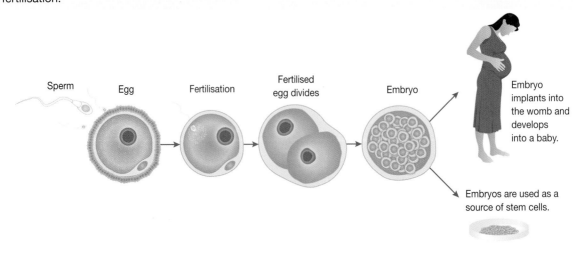

There has also been concern about the fate of the embryo. In the process of obtaining stem cells, the embryo is destroyed.

Some parents have decided to have another child for the sole purpose of being able to provide stem cells for a child who is ill or has a disease. In this case, the blood from the umbilical cord or placenta is used as the source. Some suggest that this is not the 'right' reason to have a child and that children should not be considered to be a 'factory' for spare parts for their siblings.

HOW ABOUT THAT!

Professor Alan Trounson is an Australian scientist who has spent a great part of his working life perfecting the technique for creating embryos outside the human body. He was part of the team that produced the first test-tube baby in Australia in 1980. He has also done a lot of work on embryonic stem cells. In 2000, his team showed that it was possible to produce nerve cells from embryonic stem cells.

He was recently appointed as the president of a Californian institute that specialises in stem cell research. It is the best-funded facility of its kind in the world, so Trounson will have the best technologies at his disposal to move stem cell research forwards.

Alan Trounson, an Australian scientist who is one of the world's top stem cell researchers

5.9 Exercises: Understanding and inquiring

To answer questions online and to receive **immediate feedback** and **sample responses** for every question, go to your learnON title at www.jacplus.com.au. *Note:* Question numbers may vary slightly.

Remember

1. What are stem cells?
2. Distinguish between the terms 'totipotent', 'pluripotent' and 'multipotent'.
3. Outline the importance of stem cells.
4. List sources of stem cells.
5. Outline issues regarding stem cell research.
6. Describe a scientific contribution made by the Australian scientist Professor Alan Trounson.

Investigate, share and discuss

7. Investigate some of the following questions.
    (a) Which inherited genetic diseases are potentially treatable with stem cells?
    (b) How many different kinds of adult stem cells exist and in which tissues can they be found?
    (c) Why have adult stem cells remained undifferentiated?
    (d) What are the factors that stimulate adult stem cells to move to sites of injury or damage?
8. In a team, discuss the following questions to suggest a variety of perspectives.
    (a) Is it morally acceptable to produce and/or use living human embryos to obtain stem cells?
    (b) Each stem cell line comes from a single embryo. A single cell line allows hundreds of researchers to work on stem cells. Suggest and discuss the advantages and disadvantages of this.
    (c) If the use of human multipotent stem cells provides the ability to heal humans without having to kill another, how can this technology be bad?
    (d) Parents of a child with a genetic disease plan a sibling whose cells can be used to help the diseased child. Is it wrong for them to have another child for this reason?
9. Find out how stem cell research is regulated in Australia and in one other country. What are the similarities and differences of the regulations? Discuss the implications of this with your team mates.

10. Investigate aspects of stem cell research and put together an argument for or against the research and its applications. Find a class member with the opposing view and present your key points to each other. Ask questions to probe any statements that you do not understand or would like to clarify. Construct a PMI to summarise your discussion.
11. Investigate and report on research at the Australian Stem Cell Centre.

5.10 Comparing reproductive strategies

5.10.1 Big families

Reproduction can be a risky business — but when the stakes are high it can be worth it! Some animals have some pretty tricky ways of reproducing …

Many organisms produce more eggs than can survive. Imagine what would happen if the 2000 eggs laid by a female house fly all survived! Environmental factors and predators kill many offspring before they get a chance to develop to the stage at which they themselves can reproduce. Sea urchins, for example, discharge millions of gametes into the sea at one time. The coordinated timing of this release increases the chances of fertilisation occurring. However, most of the young sea urchins die. These deaths are caused by many factors, such as competition for food and resources, and predation by other animals. If this reduction in the numbers of sea urchins did not occur, they would soon over-populate the oceans. A high juvenile death rate is also quite common in many other organisms.

5.10.2 Dad's having a baby

Seahorses are very unusual fish, especially when it comes to making babies! It is the female that inserts part of her body (an ovipositor) into the male. She pumps eggs into a pouch at the front of his body and he then fertilises them with his sperm. Labour can sometimes take two days. Dad gives birth to 50–100 little seahorses, squeezing them out one at a time. No wonder he's called a big-bellied seahorse.

There are some amazing stories to tell about other types of seahorses. The male *Photocorynus* seahorse never grows larger than 10 cm and leads a parasitic life in which he is permanently attached to the female, hanging on by his mouth! This is useful to the female because it means that she doesn't have to search dark ocean depths to find a mate when her eggs are ready for fertilisation.

5.10.3 Guess who's coming to dinner?

In some fish species in which the male is in charge of protecting a clutch of eggs, it is not unusual for him to indulge in eating some of his own offspring. Honey, I ate the kids!

This trend also appears in some spider groups. The male Australian red-back spider, for example, is usually

Male *Hippocampus abdominalis* seahorses try to get females to select them to carry eggs by inflating their pouches into a white balloon.

eaten by his sexual partner while mating with her. He is even considerate enough to position his body directly in front of her jaws after he has inserted his coil-shaped sexual organ into her. Male red-backs have a short lifespan; locating a female is extremely competitive and often the tip of their sexual organ breaks off during sex!

Recent studies have found that males that are consumed increase their chances of fertilising the female's eggs. By being eaten, they distract the female so that they may mate for longer. It was found that males that were eaten were able to mate for 25 minutes compared with 11 minutes for those that escaped. Hence, the eaten males had twice the chance of fertilising the eggs with their sperm. So, although being eaten for dinner seems like a high price to pay for sex, it does have some long-term rewards.

5.10.4 Did you know that …

- Some reptiles and rodents actually 'cement' up the female's genitalia by using some of the semen, which sets into a hard plug, not allowing other sperm to get in.
- Male starworms are 'live-in lovers', spending their entire lives within the female's vagina. Her eggs are fertilised by these parasitic males (which live off her vaginal fluids) as soon as they are released.
- Some butterflies have eyes on their genitals to help guide the hooks and claspers of the male to the appropriate nooks and crannies in the female during copulation.
- The Australian gastric brooding frog (now thought to be extinct) swallowed its externally fertilised eggs and then developed them in its stomach. A special chemical produced by the eggs stopped them from being digested. More than 25 baby frogs would crawl out of the female's stomach and into her mouth.

Some male damselflies have a penis with a special hook on the end. He uses it to remove other sperm left inside his mate by previous lovers before he makes his own deposit.

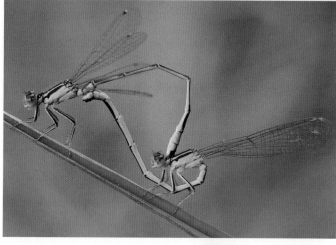

5.10.5 Sending out signals

Using smell

Chemicals called **pheromones** can play an important role in communications between members of the same species. This type of communication makes it very easy for animals to locate a mate, even in sparsely populated areas.

When a female dog is about to ovulate, she comes 'on heat'. During this time she releases a pheromone into her urine to notify male dogs that she is ready for mating. Likewise, female moths use scented chemicals that sexually attract male moths from as far away as 8 kilometres.

5.10.6 Using light

Fireflies can make part of their body glow different colours A chemical reaction produces a bright yellow, green or blue colour, which is used to help males and females find each other so that they can mate. Not all females, however, have reproduction on their minds. Females of a particular type of firefly have a different activity in mind. They flash their glowing abdomens on and off in a particular pattern, usually suggestive of a mating invitation. Sadly, instead of a romantic rendezvous, the males become a tasty meal.

5.10.7 Using sound

Whales may become separated by long distances, so in order to reproduce it is important that they can communicate. The male humpback whale sings a song during the mating season to advertise his sexual availability to females.

Birds also use their songs to attract potential mates. Frogs and crickets may not sound so melodic, but they have their own way of making it known that they are available for sex. Male crickets make their chirping song by rubbing their forewings together. Often they build their own version of a stereo amplifier by digging an underground nest with a twin-horned tunnel entrance. By sitting at the junction of the horns they can beam out their message loud and clear for all to hear.

5.10.8 Tammar trends

Researchers are studying the reproductive biology of the Tammar wallaby, which may help us to understand more about ourselves.

A baby Tammar wallaby is born about 26 days after conception. At birth, it weighs only 400 mg, is about the size of the end of your little finger, and is blind and hairless. After leaving the birth canal, it crawls up into its mother's pouch and attaches itself to one of her teats. At this stage, its external sex organs have not yet developed; researchers already know that these develop in stages quite different from those in many other mammals.

A newborn Tammar wallaby sucking on its mother's teat

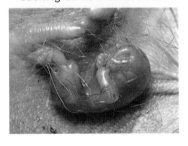

After suckling for about five months, it emerges from the pouch as a young joey. Although a joey can continue to suckle for up to a year, the mother can suckle another wallaby at a different stage of development at the same time. She does this by simultaneously producing two different types of milk. Research on how she does this could help us to improve milk production in farmed animals and our own human nutrition.

Will Tammar wallabies provide clues to our future reproductive technologies?

The mother Tammar wallaby can suspend the development of a fertilised egg until its older brother or sister has left the pouch, or until environmental conditions are more suitable. Finding out how she achieves this may help us develop new fertility and development technologies for other mammals, including humans.

5.10.9 Get a look at that!

Did you know that ancestral reptiles were the first vertebrates to have a penis, and that snails contain both male and female reproductive organs? While there is considerable diversity in the organisation of reproductive systems in organisms, there are also patterns and similarities. Although reproductive organs may appear structurally different, they often perform similar specialised functions that enable their species to survive and reproduce. In the figure of the snail at right, can you identify similarities to our human reproductive systems? If so, what are they?

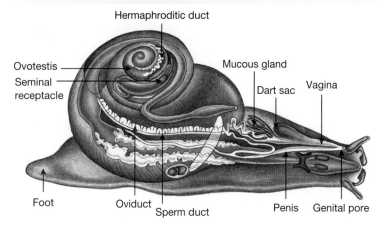

Snails are **hermaphrodites** because they have both egg-producing and sperm-producing organs.

5.10 Exercises: Understanding and inquiring

To answer questions online and to receive **immediate feedback** and **sample responses** for every question, go to your learnON title at www.jacplus.com.au. *Note:* Question numbers may vary slightly.

Remember

1. Explain why a male red-back spider has a difficult life.
2. Describe why male seahorses are unusual fish.
3. Why didn't the babies in the Australian gastric brooding frog get digested in their mother's stomach?
4. Describe one way in which the following males may increase the chances of their sperm fertilising a female gamete:
   (a) some butterflies
   (b) starworms
   (c) damselflies
   (d) some reptiles and rodents.
5. What is the name of the group of chemicals that can play an important role in communications between members of the same species?
6. Suggest three ways in which smell is important to reproduction.
7. Describe what it means when a dog is 'on heat'.
8. How do fireflies advertise their 'sexual availability'?
9. Identify which animals use sounds as a key invitation for a sexual interlude.
10. Describe what the Tammar wallaby looks like when it is born.
11. Outline some ways that Tammar wallaby research may assist studies in reproductive biology.

Think and discuss

12. Suggest why reproduction is worth the 'risks' that may be involved.
13. Explain why internal fertilisation is generally more efficient than external fertilisation.
14. Describe three ways in which animals may increase their chance of successful reproduction by having specialised reproductive structures or techniques.

Investigate

15. Find out more about the reproductive systems and mating behaviour of two different animals. Draw up a summary table to describe how they are similar and how they are different.
16. Research the reproductive system of an animal of your choice. Describe how it reproduces and draw a diagram of its reproductive parts. Present your information in a poster.
17. Research some other methods that plants and animals use to increase their chances of producing offspring. Report your findings to the class.

18. Find out more about Tammar wallaby research and present your findings in a PowerPoint presentation, mind map, newspaper article or poster.
19. What is a hermaphrodite? Suggest possible advantages and disadvantages of this condition.
20. Find out more about the evolution of reproductive structures in vertebrates, and report to others on current scientific research in this field.
21. Research the reproductive systems of at least three different animals and then suggest modifications to the design of reproductive systems that could improve their efficiency.
22. (a) Investigate the reproductive systems of the following animals so that you can complete the table.

Features of	Mammal	Fish	Turtle	Snail	Insect
Gametes					
Male reproductive structures					
Female reproductive structures					
Fertilisation					

(b) Comment on the (i) similarities and (ii) differences between animals recorded in your table.
(c) Which animals are most similar? Which animals are most different? Suggest possible reasons for this pattern.
(d) Identify a reproductive research question that could be investigated for each animal.
(e) Select one of the animals and build a model of its reproductive system.
(f) Suggest criteria that could be used to distinguish between the reproductive systems of the animals in the table.
(g) Explain, using examples, how the structure of a reproductive cell, organ or structure is well suited to its function.

Create

23. Write a story, play or poem about the life cycle (from gamete production to death) of an animal of your choice.
24. Make up a crossword with some amazing reproductive stories discovered from your own research.
25. Design a new breed of organisms. Make your 'organisms' out of plasticine or bread dough. Create a booklet that describes their lifestyle, how they find their mates and how they reproduce.

5.11 The sex life of plants

5.11.1 Flowers

Like animals, many plants can reproduce sexually. Flowering plants (angiosperms) have their reproductive structures located in their flowers.

Flowers make up the sexy bits of plants. The **petals** and **nectaries** are often used to lure insects and other animals to assist in the delivery of 'sperm' or **pollen**. Flowers are designed to increase the chances of pollen grains making contact with the sticky **stigma**.

5.11.2 Pollination

Pollination describes the way in which pollen grains reach the stigma. Plants may pollinate themselves (**self-pollination**). More often, however, they obtain the pollen from the flower of a different plant of the same species (**cross-pollination**). Cross-pollination increases the variation among the offspring and gives them a better chance of survival. The pollen grains may be transferred to other flowers by wind, insects or other animals.

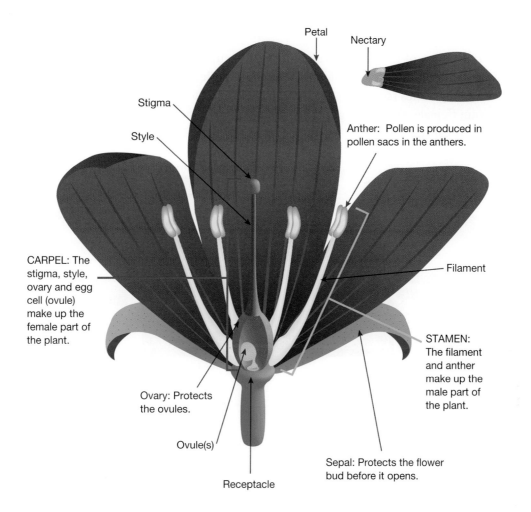

Petal

Nectary

Stigma

Style

Anther: Pollen is produced in pollen sacs in the anthers.

CARPEL: The stigma, style, ovary and egg cell (ovule) make up the female part of the plant.

Filament

STAMEN: The filament and anther make up the male part of the plant.

Ovary: Protects the ovules.

Ovule(s)

Sepal: Protects the flower bud before it opens.

Receptacle

Insect-pollinated flowers usually have attractive, brightly coloured petals and nectaries. The pollen grains themselves may be in a shape that makes them become easily attached to the insect.

Wind-pollinated flowers are usually less conspicuous and have no large scented petals or nectar. Their shape enables small, light pollen grains to be shaken from the plant and carried away with even the slightest gust of wind. The **anthers** hang outside the flower and the feathery stigmas spread out to catch airborne pollen grains.

The difference between self-pollination and cross-pollination

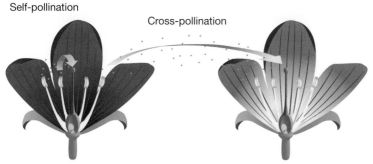

Self-pollination

Cross-pollination

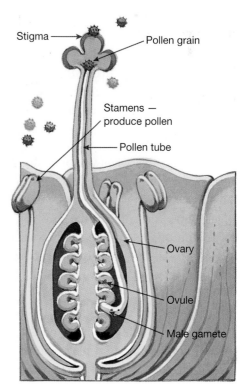

Stigma

Pollen grain

Stamens — produce pollen

Pollen tube

Ovary

Ovule

Male gamete

5.11.3 Fertilisation

As in animals, only a few of the **pollen grains** produced actually fertilise an egg cell. After pollen grains are on the stigma of a flower, a long hollow tube called a **pollen tube** is formed. This pollen tube grows down the **style**. Male gametes (sex cells) travel down these tubes to the **ovules** inside the **ovary**, where they fuse with the ovum (female gamete or egg). This joining of male and female gametes is called **fertilisation**. The fertilised egg is called a zygote.

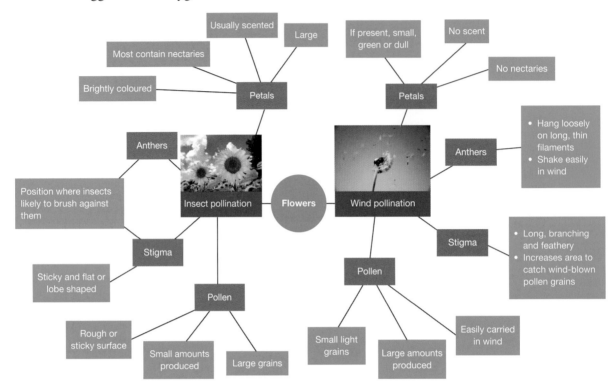

5.11.4 Plant babies

Once the flower has done its job and the egg cell has been fertilised by the pollen nucleus, another sequence of events takes place. Inside the ovule, the fertilised egg, or zygote, divides into a little ball of cells that becomes an embryo. Special tissue called **endosperm** surrounds the embryo and supplies it with food. The ovule becomes the **seed**, and tissue forms around it to provide a protective **seed coat**.

Seeds and fruit

Are you aware that when you bite into an apple, cherry or orange you are actually eating the enlarged ovary of the plant? Did you know that these swollen ovaries contain the plant's 'babies' in their embryonic form? The plants are using you as a way of distributing their 'young' out into the world.

During the formation of the seed, the ovary expands and turns into a **fruit**.

The fruit of some plants can be sweet, which makes them attractive to animals, including humans, as a source of food. The animals that eat the fruit aid the plant by dispersing the seeds over a much wider area than the plant could achieve by itself.

5.11.5 Seed dispersal

One of the main jobs that fruits do is to help **disperse** or spread the seeds. Plants disperse their seeds in a variety of ways: dispersal may involve animals, including birds (such as in tomatoes, grapes and apples); water (such as in coconuts); or wind (such as in grasses and dandelions). Some plants can disperse their seeds by themselves. For example, the fruits of some plants in the pea family (legumes) split open suddenly when they are ripe and dry, throwing the seeds long distances.

Wind dispersal (left) and self dispersal (right)

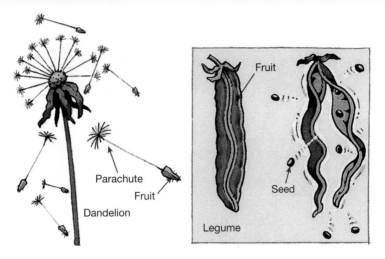

The life cycle of a flowering plant

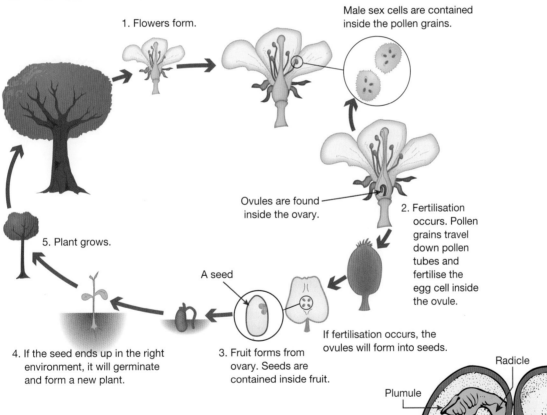

1. Flowers form.

Male sex cells are contained inside the pollen grains.

Ovules are found inside the ovary.

2. Fertilisation occurs. Pollen grains travel down pollen tubes and fertilise the egg cell inside the ovule.

5. Plant grows.

A seed

If fertilisation occurs, the ovules will form into seeds.

4. If the seed ends up in the right environment, it will germinate and form a new plant.

3. Fruit forms from ovary. Seeds are contained inside fruit.

5.11.6 Seeds and germination

The embryo, inside the seed, is made up of three different parts: the baby shoot (**plumule**), the baby root (**radicle**) and one or two thick, wing-like **cotyledons**.

When the conditions are right, the seed bursts open and a new plant grows out. This process is called germination. When **germination** is complete, the embryo has become a young plant or **seedling**.

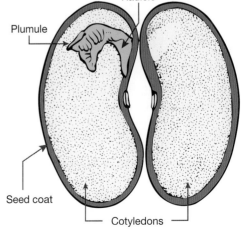

Germination of a broad bean

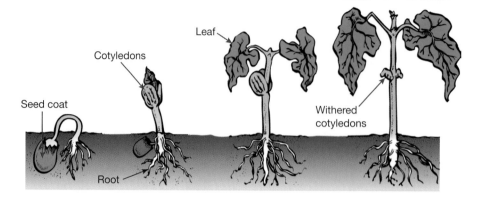

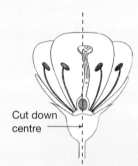

INVESTIGATION 5.3

What's in a flower?

AIM: To identify the parts of a flower and relate their structure to their function

Materials:
flowers
sharp knife or razor blade
cutting board
hand lens
tweezers

Cut down centre

Method and results

1. Draw a diagram of your flower. Locate, count and label the petals and sepals.
2. Identify and label the male and female parts you can see.
   - Place the flower on the cutting board and hold it with the tweezers.
   - Carefully cut the flower in half down the middle (a vertical cross-section).
   - Use the hand lens to look at the ovary and eggs.
3. Draw the cross-section and label the female parts inside the flower.

Discuss and explain

4. Identify ways in which the flowers you observed were (a) similar and (b) different.
5. Suggest reasons for (a) similarities between the flowers and (b) differences between the flowers.
6. (a) Predict which parts of the flower (i) become seeds and (ii) may grow into fruit.
   (b) Justify your predictions.
   (c) Check references to see if your predictions were accurate and comment on your findings.
7. Describe possible relationships between the parts of the flower in your diagram.
8. Describe how the various structures of the flower that you have observed assist the plant in reproduction.
9. Suggest how the investigation could be improved.

INVESTIGATION 5.4

Investigating features of flowers

AIM: To identify a feature of a flowering plant and investigate its relationship to reproduction

> **CAUTION**
>
> Be responsible in your fieldwork and handle the plant parts very gently and carefully. Do not pick, break, tread, trample or climb the plants. Remember that you are dealing with living things.

Materials:
5 pieces of blank A4 paper
pencil
flowering plants growing in local environment

Method and results

1. Identify a research question that relates to either the structure or a feature of a flower that may increase its chances or effectiveness of pollination.
   - Find five plants, each with different types of flower.
2. Using a separate page for each plant: at the top of the page:
   - record your name and the date
   - record the plant's name, or, if unknown, record it as 'specimen A, B, C' etc.
   - give a general description of the location in which the plant is found
3. Divide the rest of your A4 sheet into three sections:
   (i) half-page sketch of a flower
      - Try to show the parts listed in the table at the bottom of page 240 and label them.
      - Count or estimate how many stamens, stigma, petals and sepals are present.
   (ii) quarter-page sketch of a leaf — include any veins that you see.
   (iii) quarter-page sketch of the plant's overall appearance.
4. Record the structure or feature of your flower identified in your research question from part 1.

Discuss and explain

5. In regards to your chosen floral structure or feature, identify ways in which the flowers you observed were (a) similar and (b) different.
6. Suggest reasons for (a) similarities between the flowers and (b) differences between the flowers.
7. Research your observed plants using databases and the internet. Construct a table, field guide, cluster map or multimedia format to summarise your findings on the following:
   (a) possible identification
   (b) labelled sketch or image of flower and fruit
   (c) type of pollination and type of seed dispersal
   (d) an interesting fact.
8. Based on your observations and your research:
   (a) suggest how your chosen floral structure or feature may influence the effectiveness of the pollination of the plant to which it belongs
   (b) construct a relevant hypothesis that may be investigated.

Evaluation

9. Identify strengths and limitations of this investigation and suggest possible improvements.

5.11 Exercises: Understanding and inquiring

To answer questions online and to receive **immediate feedback** and **sample responses** for every question, go to your learnON title at www.jacplus.com.au. *Note:* Question numbers may vary slightly.

Remember

1. Match the words in the left-hand column (below) with those in the right-hand column.

sepal	sperm
petal	sugar
pollen	leaflet
nectary	colour
ovule	egg cell

2. Copy and complete the table on next page for the flower parts shown in the flower figure on page 235.

Name of flower part	Function	Male, female or neither

3. Describe the relationship between:
   (a) stigma and stamen
   (b) ovule and seed
   (c) ovary and fruit
   (d) pollen and anthers.
4. Distinguish between the following terms.
   (a) Self-pollination and cross-pollination
   (b) Pollination and fertilisation
   (c) Plumule and radicle
   (d) Germination and fertilisation
5. Rearrange the following terms to construct a flowchart that shows the correct sequence for flowering plant reproduction: fertilisation, seed dispersal, germination, pollination.
6. Construct a mind or concept map to summarise the functions of different parts of a flower.

Think

7. Suggest why some orchid flowers closely resemble female wasps.
8. Use storyboards, cartoons or timelines to summarise how plants reproduce.
9. Use a bubble or mind map to show some foods that are seeds or products of seeds.

Investigate and create

10. Find and research examples of wind-pollinated and insect-pollinated plants. Construct models that show what you have found out about their structures.
11. What does pollen have to do with hay fever? Make a model to show the relationship.
12. What are the conditions needed by most plants for germination?
13. Find four examples of different ways that the seeds of plants can be dispersed. Construct a story or play that includes these examples.
14. Find out more about the life cycle of a plant of your choice. Report your findings as models, puppets or a poster.
15. Construct a Gantt chart or storyboard that includes seed dispersal, pollination, fertilisation, germination and development into a seedling.
16. Write a poem about the sex life of plants. Include as many of the bold-typed words in this section as possible.

Investigate

17. Is there a relationship between the colour of a flower and the strength of its scent? Design and then carry out an investigation to determine whether the colour of the flower influences how strong the scent is.
18. Do some insects prefer some types of flower colour? Research or devise your own investigation. Share your conclusions with others in your class.
19. Find out more about the seed dispersal of five different types of plants and report your findings in a visual map.
20. Use information in this section and other resources to relate structural features of the following parts of the reproductive system of a flowering plant to their functions.

Part of system	Structural features	Function
Stigma		
Style		
Petal		
Ovary		
Pollen		
Pollen tube		
Seed		
Nectary		

21. The figure below shows stages in the formation of a pear (fruit) from the flower. Use the internet to find details of this process for three other fruits. Present your findings in labelled diagrams.

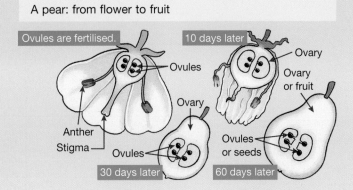

A pear: from flower to fruit

5.12 Multiplying by dividing

5.12.1 Asexual reproduction

Imagine looking exactly like your parent — and all of the rest of your family!

Not all organisms reproduce by sexual methods. In some types of organisms, a single parent produces one or more genetically identical offspring. This is called **asexual reproduction**. Binary fission, spore formation, budding and vegetative propagation are examples of this type of reproduction.

Unlike sexual reproduction, asexual reproduction does not require the fusion of sex cells. It also does not require sex cells from another organism. Because all the genetic information comes from a single individual, all offspring of asexual reproduction are identical to each other — and to their parent.

Individuals that have identical genetic information to each other are called **clones**. As well as occurring in nature, technology has also used **cloning** to produce genetically identical organisms.

5.12.2 Binary fission — let's split

Some unicellular organisms reproduce by **binary fission**. In this type of asexual reproduction, when an organism has grown to a certain size, it divides into two. Prior to this division, the genetic material in the cell is replicated. The cytoplasm then divides, producing two cells with identical genetic information.

Binary fission can occur in both prokaryotes (such as bacteria) and eukaryotes (such as *Amoeba*, *Euglena* and *Paramecium*). While the same term is used, the actual processes involved for these different types of organisms are different. In eukaryotes, a type of cell division called mitosis is involved, whereas in prokaryotes it is not. The process in prokaryotes is less complex and faster. For example, one bacterial cell could produce about 16 million offspring in eight hours. Some types of bacteria can also produce more than two cells per division. This is called **multiple fission**. Multiple fission is very efficient and allows for an even greater increase in numbers within a short time frame.

Unicellular organisms reproduce simply by dividing into two cells. This is called binary fission.

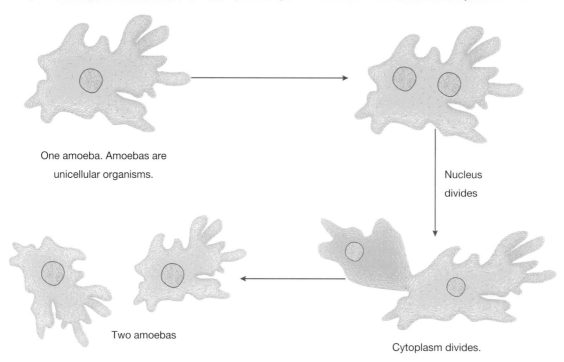

One amoeba. Amoebas are
unicellular organisms.

Nucleus
divides

Two amoebas

Cytoplasm divides.

5.12.3 Budding offspring!

Imagine your offspring beginning as a simple swelling on your side and then developing its own mouth and features. When its development is complete, it merely detaches itself and independently continues its own life. This is the sequence of events that happens in yeasts and also in freshwater hydra. The initial swelling is called a bud and hence this process is often called **budding**.

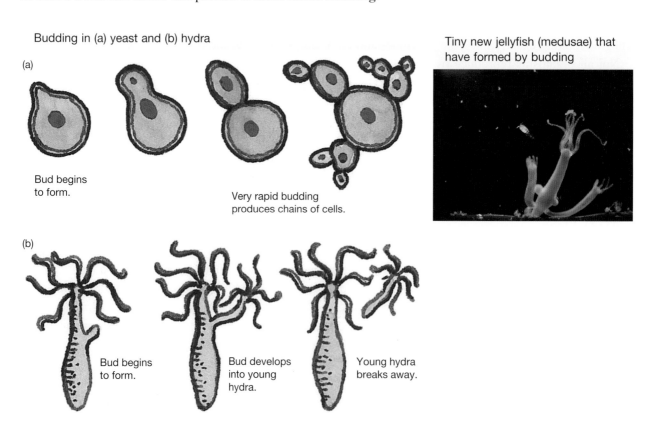

Budding in (a) yeast and (b) hydra

(a)

Bud begins
to form.

Very rapid budding
produces chains of cells.

Tiny new jellyfish (medusae) that
have formed by budding

(b)

Bud begins
to form.

Bud develops
into young
hydra.

Young hydra
breaks away.

5.12.4 Spores on the wind

Some fungi (such as mushrooms, and bread and fruit mould) have spores that, when released, may develop into offspring identical to the parent fungi. These spores are merely a group of unspecialised body cells, combined with a source of nutrients and packaged in a resistant coat. They can provide an effective means of dispersing future generations, and may also overcome adverse conditions by waiting until conditions are favourable before they begin to grow.

5.12.5 Taking a short cut

In **vegetative propagation**, the non-sexual parts of the plant are used to develop new individuals of the same type. Examples include bulbs (e.g. daffodils), stem tubers (e.g. potatoes), runners (e.g. native violets) and cuttings (e.g. roses).

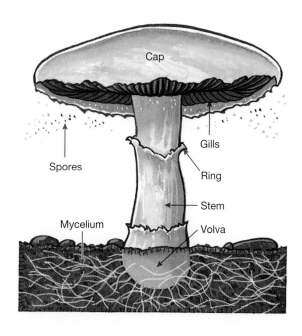

Examples of vegetative propagation: (a) runners, (b) cuttings and (c) tubers

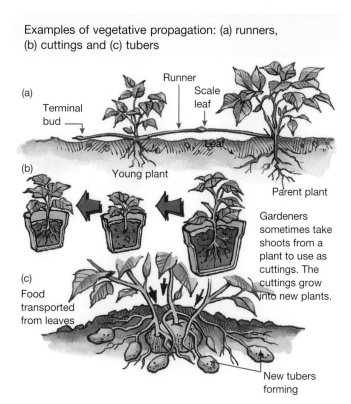

Gardeners sometimes take shoots from a plant to use as cuttings. The cuttings grow into new plants.

5.12.6 Fragmentation and regeneration

Flatworms and starfish are animals with some strange reproductive abilities. Fragmentation is commonly observed in flatworms. During this type of reproduction, the parent flatworm breaks into several pieces and, over time, each piece develops into a new adult flatworm. Regeneration is a similar type of reproduction that can be seen in starfish. While some starfish can regenerate replacement new limbs, others, such as the *Linckia* starfish, can regenerate completely new organisms from a severed arm.

Tiny new starfish growing at the end of a discarded *Linckia* starfish arm

5.12.7 Parthenogenesis — girls only?

In some animals, the females produce eggs, but these develop into embryos without fertilisation taking place. The scientific name for the development of new individuals from an unfertilised egg is **parthenogenesis**. Worker bees, for example, develop from unfertilised eggs laid by the queen bee.

Some gecko lizard groups are parthenogenetic and form all-female families. An example is Bynoe's gecko (*Heteronotia binoei*), which is found only in Australia. A population of these geckos would contain only females. Births that result without any meeting between eggs and sperm are often referred to as **virgin births**.

INVESTIGATION 5.5

Asexual reproduction
AIM: To observe asexual reproduction in plants

Materials:
large onion
potato
grass runner
leaf–stem cutting from geranium or impatiens. Note: a leaf–stem cutting is a piece
of the plant's stem that is cut just below a joint or growing point and has at least
three leaves.
leaf from an African violet, jade plant or snake plant
rooting medium (this can be purchased from a nursery)

An onion with its base in water

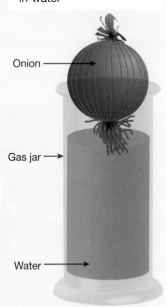

Method

- Fill a gas jar almost to the top with water and place the onion in the mouth of the gas jar so that its base is sitting in the water as shown in the diagram.

A leaf–stem cutting

- Leave the potato in a dark cupboard.
- Remove the lower leaves from the leaf–stem cutting. Quarter fill a beaker or glass jar with water and place the cutting in the water.
- Place some rooting medium in a pot. Add water to the rooting medium until it feels moist. Cut a 3 cm section from the leaf of the African violet, jade or snake plant. Stand the piece of leaf upright in the rooting medium.
- Cut a piece of the grass runner. Ensure the section you have cut has at least one growing point. Press the piece of grass runner into the rooting medium (laying it flat on the surface).
- Leave all the plant parts undisturbed for two weeks. You may need to top up the water over that time.

Results

1. Copy and complete the table below. You may need to dig the leaf–stem cutting and the runner from the rooting medium and wash them to see what has happened to them.

Plant part	Description after two weeks	Diagram
Onion		
Potato		
Leaf–stem cutting		
Leaf		
Runner		

Discuss and explain

2. In your own words, summarise your observations for each of the plant parts.
3. Based on your observations, what conclusions can you make?
4. Explain why each of the examples in the table above are forms of asexual reproduction.
5. What are the advantages of growing plants using one of the techniques described above rather than growing them from seeds?
6. Suggest improvements to the design of the investigation.

5.12 Exercises: Understanding and inquiring

To answer questions online and to receive **immediate feedback** and **sample responses** for every question, go to your learnON title at www.jacplus.com.au. *Note:* Question numbers may vary slightly.

Remember

1. State what is meant by the term 'asexual reproduction'.
2. List four examples of asexual reproduction.
3. Describe what is meant by the term 'clone'.
4. Identify one way in which asexual reproduction differs from sexual reproduction.
5. Explain why the offspring produced by asexual reproduction are all identical.
6. Match (a) binary fission, (b) fragmentation and (c) budding with the correct description of the reproductive process.
    (i) Offspring starts out as a growth on the parent.
    (ii) Single-celled parent grows to a certain size, genetic material replicates, then cytoplasm divides the cell in two.
    (iii) Parts of the parent break into pieces and each piece develops into a separate organism.
7. List three types of vegetative propagation and examples of plants associated with them.
8. Describe what is meant by the term 'parthenogenesis', including two examples in your response.

Think and discuss

9. Are you a clone? Explain.
10. Sexual reproduction results in variation among the offspring, whereas asexual reproduction does not. Discuss and record advantages and disadvantages for each type of reproduction.
11. Suggest why many insects, which would usually reproduce sexually, use parthenogenesis to produce offspring in favourable conditions.

Investigate

12. (a) Place a carrot top on moist cottonwool until leaves appear, then transfer the plant to a plastic pot containing moist potting mix. Record what happens.
    (b) Try this with a variety of other vegetables. Summarise your findings.
13. Find out more about parthenogenesis and virgin births.
14. Investigate and report on the impact of plant cloning techniques in agriculture in one of the following areas: horticulture, fruit production, vineyards.
15. Investigate the use of cloning technology in Australia. Organise a debate on any relevant issues.

learn on RESOURCES — ONLINE ONLY

Complete this digital doc: Worksheet 5.11: Asexual reproduction (doc-18735)

5.13 Storyboards and Gantt charts

5.13 Storyboards and Gantt charts

1. Decide how many scenes you need in your story. Often, 6–8 is a good number. Divide your page into this number of equal sections.
2. Consider which will be the three main events in your story and draw them roughly in the first, middle and last sections of your page.
3. Brainstorm the scenes that fit between these. Select the most appropriate and add them as intermediate scenes.
4. Mentally stand back and examine your story outline; make any desired changes to enhance its dramatic impact.

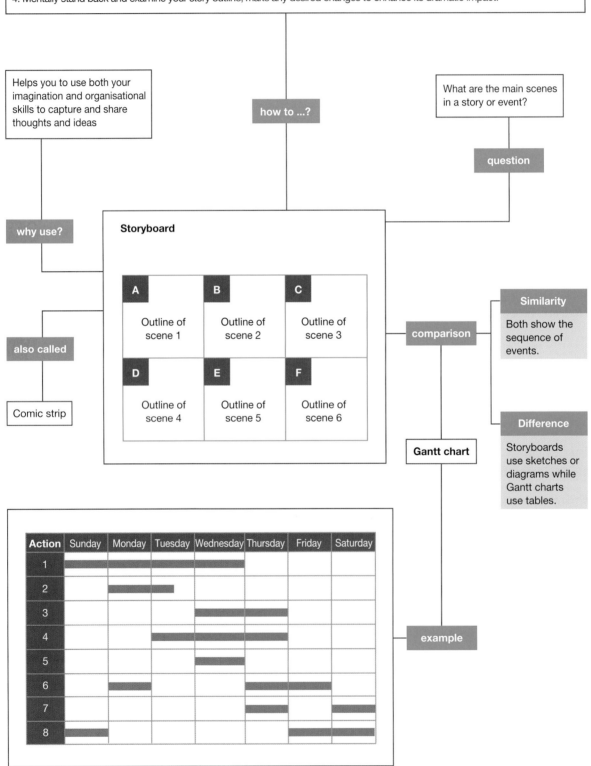

Helps you to use both your imagination and organisational skills to capture and share thoughts and ideas

how to ...?

What are the main scenes in a story or event?

question

why use?

Storyboard

A	B	C
Outline of scene 1	Outline of scene 2	Outline of scene 3
D	E	F
Outline of scene 4	Outline of scene 5	Outline of scene 6

also called

Comic strip

comparison

Similarity

Both show the sequence of events.

Difference

Storyboards use sketches or diagrams while Gantt charts use tables.

Gantt chart

Action	Sunday	Monday	Tuesday	Wednesday	Thursday	Friday	Saturday
1							
2							
3							
4							
5							
6							
7							
8							

example

5.14 Review

5.14.1 Study checklist

Asexual reproduction

- contrast sexual and asexual reproduction
- outline the role of cell division in asexual reproduction
- describe some types of asexual reproduction

Sexual reproduction

- describe fertilisation
- contrast internal and external fertilisation
- outline the life cycle of flowering plants
- compare the reproductive systems of three different animals
- contrast mammalian and plant sexual reproduction

Human reproduction

- recall the name and function of the organs of the male and female human reproductive system
- relate the structure of the organs of the male and female reproductive system to their function
- explain the role of the placenta
- outline the role of hormones in reproduction
- describe the process of fertilisation
- outline some changes that occur to the zygote between the time of fertilisation and implantation
- outline some changes that occur to a fetus as it develops inside the womb
- describe the birth process in humans
- describe some birth control techniques

Life cycle of flowering plants

- recall the names and functions of the parts of flowering plants
- distinguish between pollination and fertilisation in flowering plants
- describe how fruit and seeds are formed from flowers
- outline ways in which seeds can be dispersed

Reproductive technologies

- evaluate the benefits and disadvantages of a number of reproductive technologies
- identify some causes of infertility
- assess the impact of reproductive technologies
- describe fertility treatments including artificial insemination and in-vitro fertilisation

---

Individual pathways

ACTIVITY 5.1	ACTIVITY 5.2	ACTIVITY 5.3
Growth and reproduction	Investigating growth and reproduction	Analysing growth and reproduction
doc-6087	doc-6088	doc-6089

learnon ONLINE ONLY

5.14 Review 1: Looking back

To answer questions online and to receive **immediate feedback** and **sample responses** for every question, go to your learnON title at www.jacplus.com.au. *Note:* Question numbers may vary slightly.

1. Identify the parts labelled A–T in the diagrams below. Write one function of at least two parts in each diagram.

The male reproductive system

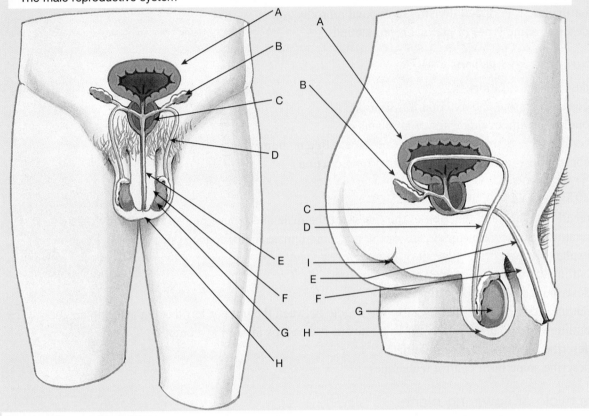

The female reproductive system

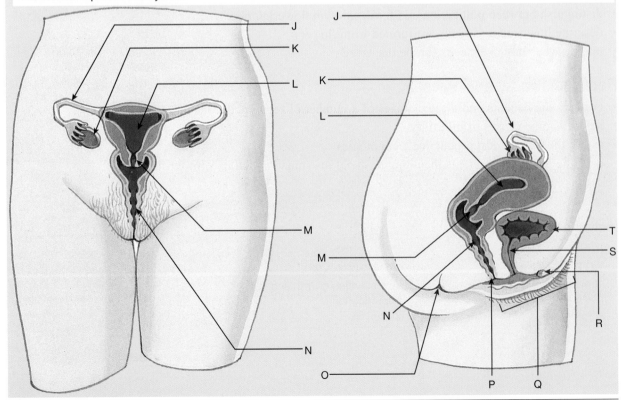

2. Design a calendar of the menstrual cycle and then outline the events that occur at each stage on your calendar.
3. Unscramble the following types of asexual reproduction.

taevvegeti gatponproai   gatienreoner

narybi sfionis   sheneipartognes

4. Summarise the disadvantages and advantages of sexual and asexual reproduction.
5. Invent, design and make your own creature. Describe its courting and mating behaviour, and give details about the way it reproduces.
6. A paramecium is a single-celled organism that reproduces asexually.
   (a) Make a list of the advantages and disadvantages of reproducing this way.
   (b) Compare your list with that of your team. Discuss any differences.
   (c) Find out more about paramecia and, as a team, write and perform a paramecium puppet play about their lives.
7. Match the contraceptives below with the way they prevent conception and their effectiveness.

Contraceptive	How it prevents conception	Effectiveness
Condom with spermicide	Prevents ova from developing	Extremely effective
Diaphragm without spermicide	The fallopian tubes or vas deferens are cut and sealed	Unreliable
Daily contraceptive pill	Keeps sperm and semen from entering the woman's vagina after ejaculation	Highly effective
Surgical: vasectomy and tubal ligation	Removal of male's penis from the vagina before ejaculation	Highly effective
Coitus interruptus	Prevents sperm cells from reaching the cervix	Moderately effective

8. Label the parts A–G in the diagram below.

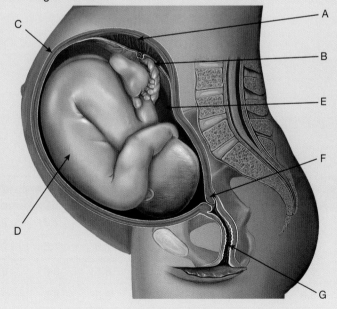

9. (a) Using information from this chapter, make at least 20 'reproduction game cards' with a question on one side and the answer on the other side.
  (b) Write a list of any questions that you still have about reproduction.
  (c) Research at least three of these questions and summarise your findings on 'reproduction game cards' with the question on one side and the answer on the other.
  (d) Design and construct a game board with at least one diagram idea from this chapter.
  (e) Design a game that uses the cards and the board that you have made. Create a game rules book so that others will know how to play. Create any other materials that you need for your game.
  (f) Play each other's games.
10. Construct a working model that simulates some aspect of this chapter.
11. It has been said that we are currently in the midst of a biotechnological revolution with new technologies offering us many more reproductive options. Is this true for all parts of the world? Hold a discussion about the global impact of reproductive technologies.
12. Construct a table naming the organs of the human male and female reproductive systems. For each organ, describe its structure and function.
13. Suggest how scientific knowledge about the life cycles of plants and animals can be used to develop regulations about importation of foodstuffs into Australia. Suggest reasons for these regulations.
14. Suggest how knowledge of the life cycle of a particular plant or animal may influence the practices of an agriculturalist.
15. On the basis of what you have learned in this section of your studies, suggest responses to the following questions.
  (a) How can there be weeds in the garden if I didn't plant them there?
  (b) Why don't twins always look the same?
  (c) Why doesn't a caged bird lay eggs that can hatch into baby birds?
16. Write down in your workbook which letter in the following diagram corresponds to each of these terms: ovules, sepals, filament, style, stigma, ovary, anther, petals, stamen, carpel.

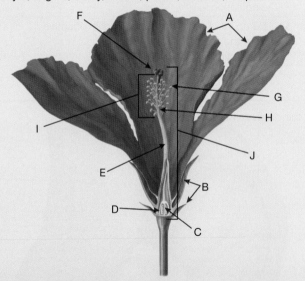

Heights (cm) of seedlings

Position	Day 1	Day 2	Day 3	Day 4	Day 5	Day 6	Day 7	Day 8	Day 9	Day 10
Fridge	5.0	5.5	6.0	6.2	6.6	7.0	7.3	7.5	7.7	8.0
Garage	5.0	5.6	6.2	6.6	7.0	7.3	7.6	7.9	8.4	8.8
Windowsill	5.0	6.0	6.7	7.5	8.0	8.5	9.0	9.6	10.2	10.6
Desk	5.0	5.8	6.3	7.0	7.5	8.0	8.5	9.1	9.6	10.0

17. Charlotte wanted to find out if temperature affects the growth of plants. She bought four seedlings. She put one seedling in the fridge and one in her garage (which has no windows so is dark and cooler than her house). She put the third seedling on the windowsill (in full sun) and the fourth seedling on her desk (out of the sun but in daylight). Charlotte measured the height of each seedling every day for 10 days. Her results are shown in the table on previous page.
    (a) Write an aim for Charlotte's experiment.
    (b) Suggest three improvements to Charlotte's experiment.
    (c) Graph Charlotte's results.
    (d) Write a conclusion for this experiment.

Complete the following activities to produce a learning and thinking journal for this chapter.

18. Draw a diagram of an insect-pollinated flower and use descriptive labels to show what each part does.

19. (a) Use a table to show differences between the sizes, shapes and structures of a fly during each stage of its life cycle.
    (b) Construct a graph to show the differences in length during the adult, egg, larval and pupal stages of the life cycle.
    (c) Suggest possible survival advantages for the differences throughout the life cycle.

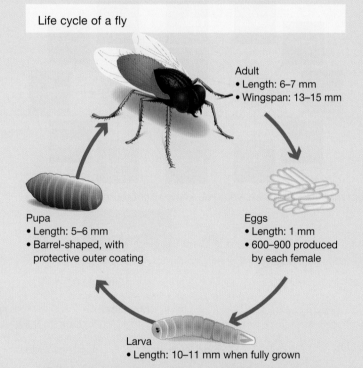

Life cycle of a fly

Adult
• Length: 6–7 mm
• Wingspan: 13–15 mm

Eggs
• Length: 1 mm
• 600–900 produced by each female

Pupa
• Length: 5–6 mm
• Barrel-shaped, with protective outer coating

Larva
• Length: 10–11 mm when fully grown

20. Use a matrix to compare flowers that are wind pollinated with those that are insect pollinated.

Topic	Feature A	Feature B	Feature C	Feature D	Feature E
1	✓		✓	✓	✓
2		✓			✓
3		✓		✓	✓
4			✓	✓	✓

21. Construct a Venn diagram or double bubble map to show the similarities and differences between:
    (a) structures involved in plant and animal reproduction
    (b) fertilisation in plant and animal reproduction
    (c) fusion of gametes in plants and animal reproduction

(d) embryo development in plants and animals

(e) 'birth' in plants and animals.

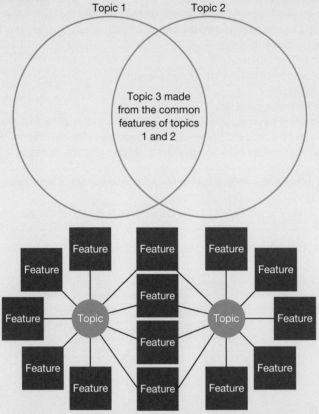

22. Construct a concept map to summarise what you know about reproduction.

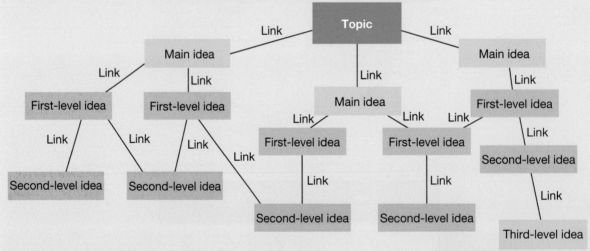

23. Increased knowledge and understanding of reproductive processes have led to the development of new reproductive technologies. Construct a PMI for issues associated with one of these technologies.

24. Use a flowchart to show an example of a life cycle of a flowering plant. Include pollination, fertilisation, development, seed dispersal and germination.

25. Use the figure below to help you construct a summary of the differences between sexual and asexual reproduction. What other features can you add?

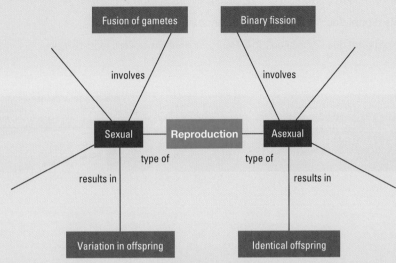

26. Use a tree map to show two sides of a discussion about plant reproduction and animal reproduction.

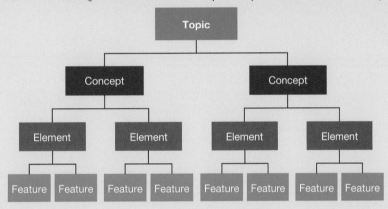

27. Make up (and perform) a song or poem to summarise something that you have learned in this chapter.
28. Label the parts of the plant in the diagram using the following terms: stigma, male gamete, pollen grain, pollen tube, stamen, ovary, ovule.

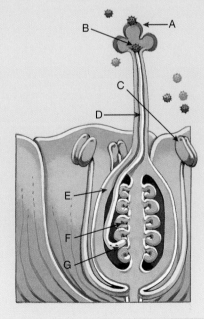

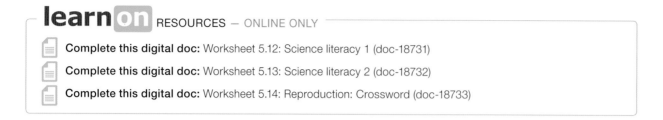

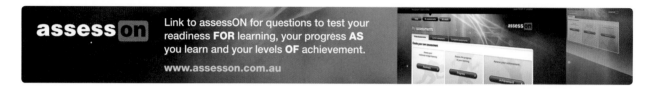

TOPIC 6
States of matter

6.1 Overview

Almost all substances on Earth can be grouped as solids, liquids or gases. By comparing the properties of solids, liquids and gases, you can begin to answer questions such as 'What are substances made of?' This question has fascinated people for thousands of years, and scientists are still looking for more answers.

6.1.1 Think about particles

- Why does ice melt?
- What is dry ice and why doesn't it melt?
- Why do car windows fog up in winter?
- What are clouds made of?
- What is the difference between hail and snow?
- Why are there small gaps in railway lines?

Numerous **videos** and **interactivities** are embedded just where you need them, at the point of learning, in your learnON title at www.jacplus.com.au. They will help you to learn the content and concepts covered in this topic.

6.1.2 Your quest

Bathroom science

1. Why does the mirror fog up in the bathroom after someone has had a hot shower?
2. On really hot days, you may have a cold shower to cool down. Does the bathroom mirror fog up when you do this?
3. Some showers have shower curtains rather than glass shower screens. When people have warm showers, the curtain tends to move in towards the person and stick to them. Give possible explanations for why this happens.

What is water vapour — a gas, a liquid, or both?

4. When you have a hot shower, the bathroom fills with water vapour. Is this water vapour a gas or a liquid or both? Explain your reasoning.
5. How hot does water have to be before it can burn you?
6. Does water vapour always rise?
7. Are water vapour and steam the same thing?
8. Can you see water vapour or steam?

6.1.3 Ranking substances

1. Working in small groups, rank the following substances in order from most solid-like to most liquid-like to most gas-like.
   - a brick
   - jelly
   - sugar
   - Vegemite®
   - orange cordial
   - steam
   - plasticine
   - tomato sauce
   - air
   - green slime
2. Compare your rankings with those of other groups. Comment on any differences between the rankings.
3. Which substances were most difficult to classify as solid, liquid or gas? Explain why they were difficult to classify.
4. Draw a three-column table, like the one below, and separate the substances into three categories: solid, liquid or gas.

Green slime — is it solid or liquid? How do you know?

Solid	Liquid	Gas

6.2 States of matter

6.2.1 Solids, liquids and gases

Every substance in the universe is made up of matter that can exist in a number of different forms called states. Almost all matter on Earth exists in three different states: **solid**, **liquid** and **gas**. These states of matter have very different **properties**. That is, they are different in the way they behave and appear.

Solids

Solids such as ice have a very definite shape that cannot easily be changed. They take up a fixed amount of space and are generally not able to be compressed.

Most solids cannot be poured, but there are some, such as salt, sand and sugar, that can be.

Liquids

Water is a liquid and its shape changes to that of the container in which it is kept. Like solids, liquids take up a fixed amount of space.

Iodine diffusing in a fume cupboard

INVESTIGATION 6.1

Comparing solids, liquids and gases

AIM: To compare the properties of solids, liquids and gases

Materials:

ice cube	plastic syringe	spatula
balloon	beaker of water	

Method and results

- Pick up an ice cube and place it on the bench. Using a spatula, try to squash it or compress it to make it smaller.
- Take the beaker of water and draw a small amount up into the syringe. Place your finger over the opening at the end of the syringe and press down on the plunger.
- Partially inflate a balloon with air and hold the opening tightly closed. Try to squeeze the balloon.
- Release your hold on the opening of the balloon.
1. Copy the table below and use your observations to complete it.

Properties of solids, liquids and gases

Substance	State of substance	Can the shape be changed easily?	Does it take up space?	Can it be compressed?
Ice	Solid			
Water	Liquid			
Air	Gas			

Discuss and explain

2. Where did the air in the balloon go when you released the opening?

If a liquid is poured into a glass, it will take up the shape of the glass. If you continue to pour, it will eventually overflow onto the bench or floor.

Gases

Gases spread out and will not stay in a container unless it has a lid. Gases move around, taking up all of the available space. This movement is called **diffusion**. In the illustration at right, iodine gas is being formed and is spreading, or diffusing, throughout the gas jar.

Gases, unlike solids and liquids, can be compressed, making them take up less space. An inflated balloon can be compressed by squeezing it.

The purple iodine gas diffuses, taking up all of the available space. What will happen to the gas if the lid is removed?

6.2.2 Measuring matter

The amount of matter in a substance, whether solid, liquid or gas, is called **mass**. The most commonly used unit of mass is the kilogram (kg), which is equal to 1000 grams (g). Mass is measured with an electronic scale or beam balance.

The amount of space taken up by a substance is called its **volume**. The volume of solids is usually measured in cubic metres (m^3) or cubic centimetres (cm^3). The volume of fluids is measured in millilitres (mL). One millilitre occupies the same volume as $1 cm^3$. A **fluid** is a substance that can flow. All liquids and gases are fluids.

$$1 \text{ mL} = 1 \text{ cm}^3$$
$$1 \text{ L} = 1000 \text{ cm}^3$$
$$1000 \text{ L} = 1 \text{ m}^3$$

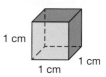

This cube has a volume of $1 cm^3$ and can hold 1 mL of a fluid.

1 cm
1 cm
1 cm

INVESTIGATION 6.2

Measuring the volume of an irregular-shaped solid
AIM: To measure the volume of an irregular-shaped solid

Materials:
100 mL beaker
100 mL measuring cylinder
stone or pebble that will fit into the measuring cylinder

Method and results
- Half-fill (approximately) a 100 mL beaker with water.
- Carefully pour the water into the measuring cylinder.
- Carefully place the pebble into the measuring cylinder. Take care not to spill any water.

Discuss and explain
1. Read and record the volume of water in the measuring cylinder using the technique shown in the diagram at right.
2. Read and record the new volume.
3. What was the volume of the solid in mL?
4. What was the volume of the solid in cm^3?
5. Suggest another way of measuring the volume of the solid object.

Reading the volume of a liquid in a measuring cylinder. The curved upper surface is called the meniscus. Your eye should be level with the flat part in the centre of the meniscus.

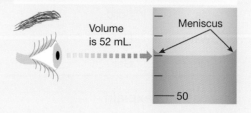

Volume is 52 mL.

Meniscus

50

6.2 Exercises: Understanding and inquiring

To answer questions online and to receive **immediate feedback** and **sample responses** for every question, go to your learnON title at www.jacplus.com.au. *Note:* Question numbers may vary slightly.

Remember

1. List as many as you can remember of the solids, liquids and gases you came into contact with before leaving for school today. Organise them into a table under three headings, 'Solids', 'Liquids' and 'Gases', or into a cluster, mind or concept map.
2. (a) Write down three properties that most solids have in common.
   (b) Would liquids have the same three properties? If not, what differences might be expected?
3. Which properties of gases are different from those of liquids?
4. What is the unit used for measuring small volumes such as for liquid medicines? How could you measure such a volume?

Think

5. Both steel and chalk are solids. What properties of steel make it more useful than chalk for building bridges?
6. Are plasticine and playdough solids or liquids? Explain why.
7. What is diffusion? Give two examples of this occurring around your house.
8. Is it possible for a solid to behave like a fluid? Explain your answer.
9. At the petrol station, the safety sign asks for the car engine to be switched off before you fill the petrol tank. Why is this necessary?

Investigate

10. There is a fourth state of matter known as plasma, which is not very common on Earth. Research and report on:
    (a) how plasma is different from solids, liquids and gases
    (b) where plasma can be found
    (c) how plasma can be used on Earth.
11. Different liquids pour or flow in different ways. Test this by pouring honey, shampoo, cooking oil and water from one container to another. Time how long they take to pour. Make sure it is a fair test. Record the results in a table and write a conclusion based on your observations and results.

6.3 Changing states

6.3.1 Changing states

Water is the only substance on Earth that exists naturally in three different states at normal temperatures. It is in the oceans, in the polar ice and in the air as water vapour. Water is constantly moving and changing states. You can observe water changing states in the kitchen. To change the state of any substance, including water, it must be heated or cooled, or the pressure changed.

Melting point and boiling point

The state of matter of any substance depends on its temperature. The temperature at which a substance changes from a solid into a liquid (melts) is called its **melting point**. A liquid changes into a solid

Unfortunately, the ice sculpture in the photograph won't last for very long. Even as the sculptor works, it is melting as heat moves into it from the warmer air around it.

(freezes) at the same temperature. Water has a melting point of 0°C, so to melt ice it has to be heated to a temperature of 0°C. To freeze water it has to be cooled to a temperature of 0°C.

The **boiling point** is the temperature at which a substance boils. At this temperature, the substance changes from liquid into gas (evaporates) quickly. At the same temperature, a gas changes into a liquid (condenses). The boiling point of water is 100°C. The melting and boiling points of some common substances are shown in the table at the bottom of this page.

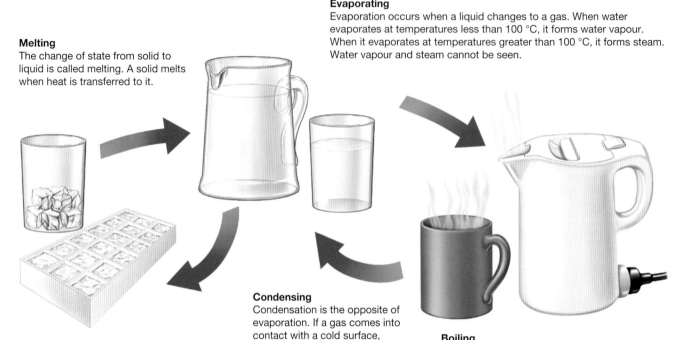

Evaporating
Evaporation occurs when a liquid changes to a gas. When water evaporates at temperatures less than 100 °C, it forms water vapour. When it evaporates at temperatures greater than 100 °C, it forms steam. Water vapour and steam cannot be seen.

Melting
The change of state from solid to liquid is called melting. A solid melts when heat is transferred to it.

Condensing
Condensation is the opposite of evaporation. If a gas comes into contact with a cold surface, it can turn into a liquid.

Freezing
The change of state from a liquid to a solid is called freezing. A liquid turns into a solid when heat is transferred away from it.

Boiling
During boiling, the change from liquid to gas (evaporation) happens quickly. The change is so fast that bubbles form in the liquid as the gas rises through it and escapes. During boiling, the entire substance is heated. A liquid remains at its boiling point until it has all turned into a gas.

HOW ABOUT THAT!

Melting and boiling points change with the height above sea level. This is because the air gets thinner as you move away from the Earth's surface. If you were climbing Mount Everest and made a cup of coffee near its peak, you would find that the water boiled at about 70 °C instead of 100 °C.

At a concert or special event, you may have seen a thick 'smoke' used for effect. This smoke is produced when solid carbon dioxide, called 'dry ice', changes state from a solid directly to a gas. This very unusual change of state is called **sublimation**. The 'smoke' is actually tiny droplets of water that condense from the air as the cold dry ice sublimes. Dry ice sublimes at a temperature of −78.5 °C. Iodine also sublimes. Diamonds sublime at a temperature of 3550 °C.

Melting and boiling points of some common substances at sea level

Substance	Melting point (°C)	Boiling point (°C)
Water	0	100
Table salt	804	1413
Iron	1535	2750
Aluminium	660	1800
Oxygen	−218	−183
Nitrogen	−210	−196

6.3 Exercises: Understanding and inquiring

To answer questions online and to receive **immediate feedback** and **sample responses** for every question, go to your learnON title at www.jacplus.com.au. *Note:* Question numbers may vary slightly.

Remember

1. (a) Copy and complete the diagram below, labelling the changes of state.
   (b) Use a labelled arrow to add 'sublimation' to your diagram.

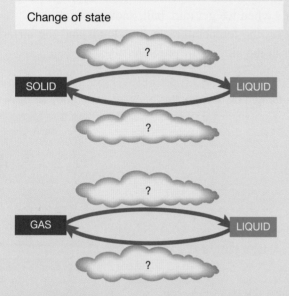

2. What is the name given to the change of state from liquid water to steam? What happens to make this occur?
3. What happens to liquid water when it is cooled below 0 °C? Has heat moved into or out of the liquid?
4. When water evaporates it can change state from liquid to a gas in the form of either steam or water vapour. Explain the difference between steam and water vapour.

Analyse

Use the table from bottom left of previous page to answer these questions.

5. At what temperature would you expect table salt to melt? At what temperature would it freeze?
6. Would you expect aluminium to be found as a solid, liquid or gas at:
   (a) 200 °C
   (b) 680 °C
   (c) 1900 °C?
7. Which substance — oxygen or nitrogen — would freeze first if the temperature were gradually lowered?

Think

8. Explain the difference between evaporation and boiling.
9. Why is dry ice useful to produce a 'smoke' effect? What other uses are there for dry ice?

Investigate

10. Dry the outside of a very cold can of soft drink or carton of milk and allow it to stand on a table or bench for about ten minutes. (Don't forget to put it back in the fridge afterwards.)
    (a) What change occurred on the outside of the can?
    (b) Where did the water come from?
    (c) What change of state has occurred?

---

learn on RESOURCES — ONLINE ONLY

Complete this digital doc: Worksheet 6.1: Changing the boiling point of water (doc-18738)

6.4 The state of the weather

6.4.1 Water and the weather

Rain, hail, snow and sleet are all types of **precipitation**. Precipitation is falling water, whether in solid or liquid form. All precipitation occurs because energy from the sun melts ice and causes liquid water to evaporate to become water vapour in the atmosphere. When the temperature in the atmosphere gets low enough, the water vapour condenses or freezes. That's when we get rain, hail, snow or sleet.

The type of precipitation we get depends mostly on the temperature in the clouds and the air around them. It also depends on the amount of water vapour in the air and air pressure.

Rain

Rain forms when water vapour condenses in cold air, forming tiny droplets of water. These droplets are so small that they are kept up by moving air, forming clouds.

As the droplets join together they become too heavy to remain in the air. They fall to the ground as rain. When air currents are low, very tiny drops of rain may fall as a fine mist known as drizzle.

Clouds are formed by tiny droplets of water, kept up by air currents.

Hail

If drops of rain freeze, they may form hailstones. Air currents within clouds move raindrops from the bottom of the cloud upwards to the top of the cloud. The top of the cloud is much colder than the bottom and the rising raindrops freeze very quickly. The frozen raindrops fall back towards the bottom of the cloud. If the air currents are strong enough, the frozen raindrops rise again, adding a new layer of ice. They fall again, then rise again to form another layer of ice. This can happen over and over again, each time adding a new layer of ice. When the ice has built up many layers, it gets heavy enough to fall to the ground as a hailstone. Hailstones can be extremely large and cause extensive damage.

In summer, warm rising air helps to keep the hailstones in the clouds for longer, forming even more layers than usual. These hailstones can reach masses of over one kilogram before they fall.

Snow

Snow consists of crystals of ice that have frozen slowly in clouds. Many different shapes and patterns can be found in snowflakes. The shape and size depend on how cold the cloud is, its height and the amount of water vapour it holds. Crystals of ice form when clouds have temperatures below −20 °C. The crystals join together and fall. As they fall, they become wet with moisture but then refreeze as snowflakes.

If the air between the cloud and the ground is colder than 0 °C, the snowflakes fall as very powdery, dry snow. If the air is warmer, the ice crystals melt and fall as rain or sleet.

Sleet

Sleet is snow that is melting or raindrops that are not completely frozen. Sleet forms when the air between the clouds and the ground is warm enough to melt ice.

6.4.2 Predicting the weather

The scientists who predict, or forecast, the weather are **meteorologists**. Meteorology is the study of the atmosphere and includes the observation, explanation and prediction of weather and climate. Numerous observations of temperature, precipitation, wind speed, air pressure, humidity and more are needed to make weather forecasts. Humidity is a measure of the amount of water vapour in the air.

Before the first weather balloon was launched in 1882, observations with instruments such as thermometers, barometers and rain gauges could be made only on land or ships. Not long after the invention of the first 'flying machine' in 1903, weather instruments were attached to the wings of planes, allowing them to be taken higher in the atmosphere.

As new technology becomes available, the number and quality of observations improve. Improved weather balloons, together with radar, satellite images and computer modelling, allow meteorologists to make predictions further ahead and more accurately than ever before.

Snowflakes form many different shapes and patterns but always have six 'sides'.

A meteorologist releases a weather balloon in Antarctica.

6.4 Exercises: Understanding and inquiring

To answer questions online and to receive **immediate feedback** and **sample responses** for every question, go to your learnON title at www.jacplus.com.au. *Note:* Question numbers may vary slightly.

Remember

1. What are clouds made of?
2. Using words or a labelled diagram, explain how hailstones are formed.

3. How can hailstones get as large as the one in the photograph on page 262?
4. Explain the difference between snow and sleet.
5. What is meteorology concerned with?
6. What is humidity a measure of?

Think

7. Suggest why extra-large hailstones are more common in summer than winter.
8. Ski resort operators suffer a shortage of snow in some years. What conditions would they look for to predict coming snowfalls?

Investigate

9. Make a list of leisure activities that rely on predictions about the weather.
10. In which occupations do each of the following types of weather prevent activity?
    (a) Extreme heat
    (b) Heavy rain
    (c) Thunderstorms
11. Record the predictions of the maximum temperature of your nearest capital city made in a 7-day forecast. For each day of the 7-day period, also record the maximum temperature predicted on the day before. These forecasts can be found online on the Bureau of Meteorology website (www.bom.gov.au), on the TV news or in daily newspapers.

    Then record the actual maximum temperature for each day as reported on the evening news or www.bom.gov.au. Use a table like the one below to record your data.

Daily maximum temperatures (°C)							
	Day 1	**Day 2**	**Day 3**	**Day 4**	**Day 5**	**Day 6**	**Day 7**
Prediction in 7-day forecast							
Forecast the day before							
Actual maximum temperature							

(a) How does the accuracy of the 7-day forecast compare with the accuracy of the previous day's forecast?
(b) State your opinion about the accuracy of the forecast made on the day before.
(c) Apart from temperature, what other aspects of the weather forecast are reported in newspapers and on the TV news?
12. Find out what relative humidity is and with which instrument it is measured.
13. Research and report on what a hydrologist does.
14. Find out the difference between weather and climate.

learn on RESOURCES — ONLINE ONLY

▦ **Watch this eLesson:** Understanding a weather forecast (eles-0161)

6.5 Matter and energy: The particle model

6.5.1 The particle method

How do you explain why ice has properties that are different from those of water or steam? Scientists use a model to explain the different properties of solids, liquids and gases. This model is called the **particle model**.

According to the particle model:

- all substances are made up of tiny particles
- the particles are attracted towards other surrounding particles
- the particles are always moving
- the hotter the substance is, the faster the particles move.

A particle model for different states

Liquid

Gas   Solid

6.5.2 Particles in a solid

In solids the particles are very close together, so they cannot be compressed. The attraction between neighbouring particles in a solid is usually strong. Because there are such strong bonds between the particles, solids usually have a fixed shape and a constant volume. The particles in solids cannot move freely; instead they vibrate in a fixed position.

6.5.3 Particles in a liquid

In liquids the particles are held together by attraction, but the bonds between them are not as strong as those in solids. The weak particle attraction allows the particles to roll over each other, but they can't 'escape'. For this reason, liquids have a fixed volume but the rolling motion of the particles allows them to take up the shape of their container. As in solids, the particles in liquids are still very close together; liquids cannot be compressed into smaller spaces.

> **WHAT DOES IT MEAN?**
>
> The word *particle* comes from the Latin word *particula*, meaning 'part'.

6.5.4 Particles in a gas

The particles in a gas have much more energy than those in solids or liquids, and they are in constant motion. The attraction between the particles in a gas is so weak that they are able to move freely in all directions. They spread out to take up any space that is available. This means that gases have no fixed shape or volume. Because of the large spaces between particles, gases can be compressed.

Particles in a liquid and a gas

Spreading out

The spreading of one substance through another is called diffusion. This can happen only when the particles of one substance can spread through the particles of another substance. Diffusion is possible in liquids and gases because the particles move around. You would expect diffusion to happen faster in gases than in liquids because the particles move faster. Particles in a solid vibrate in a fixed position, so diffusion can't occur.

The particle model and balloons

The particle model can be used to explain what happens to a balloon when you inflate it. Particles of air inside the balloon constantly move in all directions. They collide with each other and with the inside wall of the balloon. But the wall is not rigid. It can stretch as more particles are added. The balloon **expands** until it can't stretch any more. When you let some of the air out of the balloon, fewer particles collide with the inside wall of the balloon. It gets smaller or **contracts**.

6.5.5 Heating solids, liquids and gases

When a substance is heated, the particles gain energy, move faster, become further apart and take up more space and substance expands as the temperature increases.

The tyres on a moving car get quite hot. This makes the air inside expand and may even cause a blowout in extreme circumstances. Heating usually causes gases to expand much more than solids or liquids. Gases expand easily because the particles are spread out and not attracted to each other strongly. Solids, liquids and gases contract when they are cooled because the particles lose energy, slow down, need less space to move in and become more strongly attracted to each other.

Hot air balloons rise when the air inside them expands. The particles in the heated air move faster and take up more space. This makes each cubic centimetre of air inside the balloon lighter than each cubic centimetre of air outside the balloon, so the air inside the balloon rises, taking the balloon with it.

Architects and engineers allow for expansion and contraction of materials when designing bridges and buildings. Bridges have gaps at each end of large sections so that in hot weather, when the metal and concrete expand, they will not buckle. Railway lines also have gaps to allow for expansion.

The particles of two gases spread through each other until they are evenly mixed.

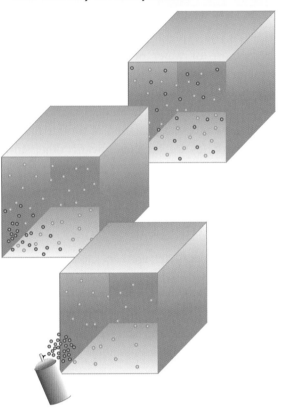

These hot air balloons rise when the air inside them expands. How do they get back down to the ground?

The volume of a substance changes when it is heated or cooled.

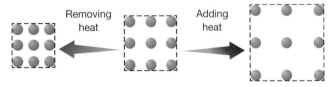

Removing heat   Adding heat

Contraction
• Particles move more slowly.
• Distance between particles gets smaller.
• The attraction between the particles increases.

Expansion
• Particles move faster.
• Distance between particles increases.
• The attraction between the particles decreases.

Electrical wires are hung from poles loosely so that when the weather cools, they will not become too tight and break as they contract. The amount by which a structure will expand or contract depends on the material it is made from; so when choosing a material, it is important to find out how much that material will expand or contract. The table on page 271 shows how much some commonly used materials expand when the temperature increases by 10°C.

INVESTIGATION 6.3

Investigating diffusion

AIM: To compare diffusion of liquids and gases

Materials:
250 mL beaker
water
food colouring
eye-dropper
fragrant spray

Method and results

- Place a drop of food colouring into a beaker of water and record your observations for several minutes, making sure the beaker is not moved.
- Release some of the fragrant spray in one corner of the classroom. Move away and observe by smell.
1. Draw a diagram to show the movement of the food colouring through the water.

Discuss and explain

2. Explain how the fragrant spray moved through the air.
3. This investigation shows diffusion in a liquid (water) and in a gas (air).
   (a) In which state does diffusion occur faster?
   (b) Why do you think this is so?

INVESTIGATION 6.4

Explaining gases

AIM: To observe and explain the expansion and contraction of a gas

Materials:
balloon   *piece of string*
ruler   *small conical flask*
2 large beakers   *ice cubes*
hot and cold water

Method and results

1. Copy the table at end of this investigation 6.4 into your workbook.
   - Inflate the balloon to its maximum size. Then deflate it. This makes it easier to stretch.
   - Inflate the balloon again, to a size slightly larger than an orange. Fit the neck of the balloon over the conical flask to seal it.
   - Wrap the string once around the widest part of the balloon to find its circumference. With a ruler, measure the length of the string that encircled the balloon.
2. Record the measurement in your table.
   - Half-fill one of the beakers with ice cubes and a small amount of cold water.
   - Place the conical flask in the ice-water beaker and observe the balloon. After a few minutes, use the string to measure the circumference of the balloon again.
3. Record your measurement in your table.
   - Put some hot water into the second beaker. Take the conical flask from the ice-water and place it into the hot water.
4. Leave for a few minutes, then measure and record the balloon's circumference.

Discuss and explain

5. Was any air added to or removed from the balloon after it was placed over the conical flask?
6. After being in ice-water and hot water, were there any changes in the size of the balloon?
7. Using the particle model, try to explain what might have made the balloon contract and expand.
8. What quantity was varied or changed in this experiment? What things were kept the same?

Effect of temperature on air

Temperature of surroundings	Circumference of balloon (cm)
Room temperature	
Cold (ice-water)	
Hot (hot water)	

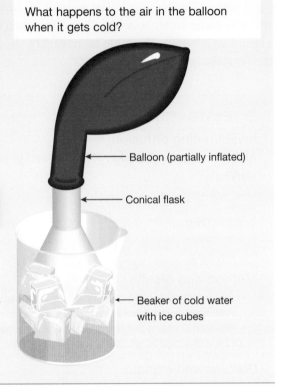

What happens to the air in the balloon when it gets cold?

Balloon (partially inflated)

Conical flask

Beaker of cold water with ice cubes

An exception to the model

According to the particle model, the spaces between the particles in a liquid get smaller as the liquid is cooled, and the particles are closest once the liquid has become a solid. However, water is one of the few substances that do not behave exactly as the particle model predicts.

While the temperature of water is cooled from 100°C to 4°C, the particles behave as expected, with the spaces between them growing smaller. As water temperature drops below 4°C, however, something strange happens — the spaces between the particles start to get larger again. By the time water freezes at 0°C, the particles are further apart than they were at 4°C! In general, the volume taken up by water particles increases by nearly 10 per cent when it becomes ice; you may have noticed this if you have ever put a full bottle of water in the freezer!

6.5.6 Thermometers

Bulb thermometers, like the one pictured at top left on the next page, use the expansion of liquids when they are heated to measure temperature. Most bulb thermometers consist of thin tubes and a bulb that contains a liquid. As the temperature rises, the liquid expands, moving up the tube, which is sealed at the top.

The two most commonly used liquids in thermometers are mercury and alcohol. Mercury has a low freezing point (−39°C) and a high boiling point (357°C). Alcohol, however, is much more useful in very cold conditions because it does not freeze until the temperature drops to −117°C. On the other hand, alcohol boils at 79°C, so it cannot be used for measuring higher temperatures.

The temperature of the human body ranges between 34°C and 42°C; it is normally about 37°C. A clinical thermometer is designed to measure this range.

Oops! The reason why you shouldn't put a bottle full of water in the freezer.

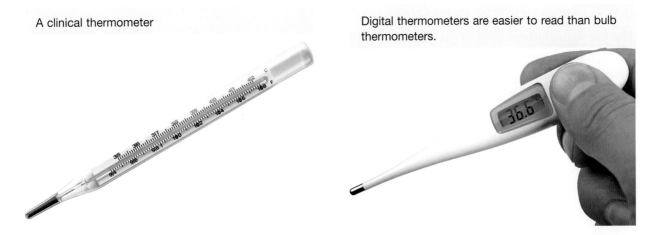

A clinical thermometer

Digital thermometers are easier to read than bulb thermometers.

Look at the photograph above left. The tube narrows near the bulb. Once the mercury has expanded, this narrowing prevents the mercury contracting and moving back into the bulb before the temperature can be read. Once a reading has been taken, the mercury has to be shaken back into the bulb before the thermometer can be reused.

Bulb thermometers are gradually being replaced by digital thermometers, which don't rely on expansion and contraction of mercury or any other liquid. Digital thermometers contain a thermostat, which is a sealed solid, embedded inside. The thermostat's resistance to electric current depends on temperature. A tiny computer measures the thermostat's resistance and calculates the temperature, which is displayed on a small screen.

INVESTIGATION 6.5

Expansion of solids

AIM: To observe and explain the expansion of a solid

Materials:
metal ball and ring set
Bunsen burner and heatproof mat
tongs

A ball and ring set

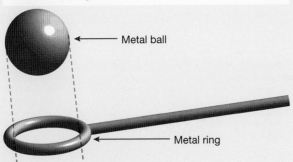

Metal ball

Metal ring

Method and results

- Try to put the ball through the ring.
- Use the Bunsen burner to heat the ring and use tongs to try to put the ball through it. Take care not to touch the hot metal.
- Let the ring cool and try to put the ball through the ring again.

Discuss and explain

1. What happened to change the size of the ring?
2. Use the particle model to explain the change that took place in the ring.

6.5.7 Gases under pressure

The fire extinguishers used to put out electrical fires are filled with carbon dioxide gas. Carbon dioxide can be used in this way only because huge amounts of it can be compressed, or squeezed, into a container. Gases can be compressed because there is a lot of space between the particles. Gases compressed into cylinders are used for barbecues, scuba diving, natural gas in cars and aerosol cans.

1. Gases, including carbon dioxide, have lots of space between their particles.
2. Carbon dioxide is compressed into a cylinder. The particles are squashed closer together.
3. The carbon dioxide particles are now under increased pressure. This means that the particles in the gas collide frequently with the walls of the cylinder and push outwards. The particles are trying to escape, but are held in by the container.
4. When the nozzle is opened, the pressure forces the carbon dioxide gas out very quickly through the opening.
5. The particles of gas quickly spread out over the fire. The gas smothers the fire, stopping oxygen from the air getting to it. Fires cannot burn without oxygen, so the fire goes out.

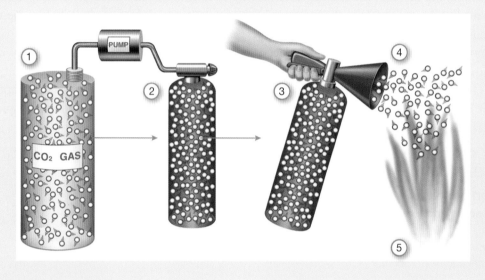

INVESTIGATION 6.6

Expansion of liquids

AIM: To observe and explain the expansion of a liquid

Materials:
500 mL conical flask   *narrow glass tube*
tripod and gauze mat   *food colouring*
eye-dropper   *marking pen*
rubber stopper with one hole to fit the tube
Bunsen burner, heatproof mat and matches

Method and results

- Use an eye-dropper to place two or three drops of food colouring in the flask, then fill it with water right to the top.
- Place the stopper in the flask with the glass tube fitted. Some coloured water should rise into the tube. Mark the level of the liquid in the tube with the marking pen.

- Place the flask on the tripod and gauze mat, light the Bunsen burner and gently heat the liquid. Observe what happens to the liquid level in the tube.
- After about five minutes of heating, turn off the Bunsen burner and watch what happens to the liquid level in the tube.
1. What happens to the level of the liquid while it is being heated?
2. What happens to the level of the liquid while it is cooling down?

Discuss and explain

3. Use the particle model to explain why liquids expand.

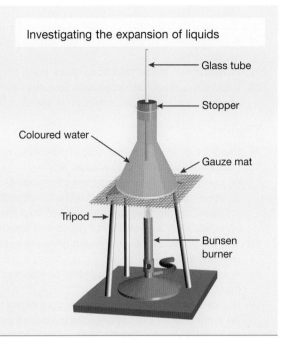

Investigating the expansion of liquids

Glass tube
Stopper
Coloured water
Gauze mat
Tripod
Bunsen burner

6.5 Exercises: Understanding and inquiring

To answer questions online and to receive **immediate feedback** and **sample responses** for every question, go to your learnON title at www.jacplus.com.au. *Note:* Question numbers may vary slightly.

Remember

1. Explain why a model is needed to explain the properties and behaviour of different states of matter.
2. List the four main ideas of the particle model.
3. What is diffusion?
4. Explain why solids generally expand when they are heated.
5. The following statements are incorrect. Rewrite them correctly.
   (a) Heating a liquid might make its particles stick closer together.
   (b) Solids have a definite shape because their particles are free to move around.
   (c) You can compress a gas because its particles are close together.
6. When a substance is heated, its temperature increases. What other change might be observed?
7. (a) Describe what change you expect to see when hot metal objects are cooling.
   (b) Why does this happen? Explain, using the particle model.
8. List two examples of structures that contain gaps to prevent them buckling in hot weather.
9. Give one reason why overhead electric power lines are not hung tightly.
10. What happens to the particles in carbon dioxide gas when they are compressed into a fire extinguisher?

Analyse

Expansion of materials

Substance	Expansion (mm) of 100 m length when temperature increases by 10 °C
Steel	11
Platinum	9
Concrete	11
Glass — soda	9
Glass — Pyrex	3
Lead	29
Tin	21
Aluminium	23

Use the table on the previous page to answer the following questions.

11. If a steel rod of 10 metres in length was heated so that its temperature rose by 10 °C, how long would the rod become?
12. Why is Pyrex, rather than soda glass, used in cooking glassware such as casserole dishes and saucepans?
13. Concrete is often reinforced with steel bars or mesh to make it stronger. Why is steel a better choice than another metal, such as aluminium or lead?

Think

14. Describe an everyday example of diffusion.
15. Use the particle model to explain why:
    (a) perfume can be smelled from a few metres away
    (b) steam can be compressed while ice cannot
    (c) water vapour takes up more space than the same amount of liquid
    (d) solids do not mix well, but gases and liquids mix easily in most cases.
16. Use the particle model to predict what will happen to the length and width of a solid substance if it is heated (without melting).
17. A jar with the lid jammed on tightly can be hard to open. If hot water is run over the lid, it becomes easier to open. Why?
18. Hot air balloons have a gas heater connected to them.
    (a) What happens to the particles inside the balloon when the heater is turned on?
    (b) Explain why the balloon rises.
19. Under what conditions might you use an alcohol thermometer rather than a mercury thermometer?
20. List the advantages of digital thermometers over mercury bulb thermometers for measuring human body temperature.
21. Use the particle model to explain what keeps car or bicycle tyres in the right shape when they are pumped up to a high air pressure.

Investigate

22. The mercury thermometer was invented by a German named Gabriel Fahrenheit (1686–1736). A different set of markings is used to scale Fahrenheit thermometers. At what temperatures does water boil and freeze on this scale?
23. Why do icebergs float in Arctic and Antarctic waters? Do you think much of the iceberg is under the water, or is it mostly above? How could you test your hypothesis? Design a suitable experiment.
24. Use the internet to investigate the safe storage of gas cylinders. Make a list of requirements and state the reason for each of them.

---

learn on RESOURCES — ONLINE ONLY

Watch this eLesson: Under pressure (eles-0058)

Complete this digital doc: Worksheet 6.2: Fire! Fire! (doc-18739)

Complete this digital doc: Worksheet 6.3: Expansion of liquids (doc-18740)

Complete this digital doc: Worksheet 6.4: Particles in our lives (doc-18741)

6.6 Energy in, energy out

6.6.1 Energy and change of state

A change of state involves the heating or cooling of matter. As a substance is heated, energy is transferred to it. When a substance cools, energy moves away from it to another substance or to the environment. The change in energy causes the particles in the substance to move at different speeds.

An increase in the energy of the particles of a substance results in an increase in the temperature of the substance. A decrease in the energy of particles results in a decrease in the temperature of the substance.

The flowchart below shows what happens to the particles that make up a substance when it changes from a solid state into a gas state. When a gas is cooled, the direction of the flowchart can be reversed as the substance changes from a gas state into a solid state.

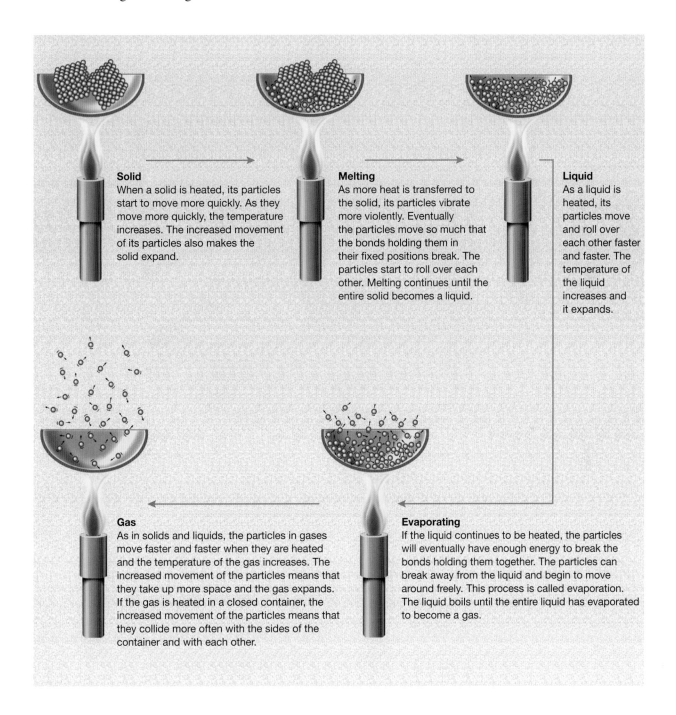

Solid
When a solid is heated, its particles start to move more quickly. As they move more quickly, the temperature increases. The increased movement of its particles also makes the solid expand.

Melting
As more heat is transferred to the solid, its particles vibrate more violently. Eventually the particles move so much that the bonds holding them in their fixed positions break. The particles start to roll over each other. Melting continues until the entire solid becomes a liquid.

Liquid
As a liquid is heated, its particles move and roll over each other faster and faster. The temperature of the liquid increases and it expands.

Gas
As in solids and liquids, the particles in gases move faster and faster when they are heated and the temperature of the gas increases. The increased movement of the particles means that they take up more space and the gas expands. If the gas is heated in a closed container, the increased movement of the particles means that they collide more often with the sides of the container and with each other.

Evaporating
If the liquid continues to be heated, the particles will eventually have enough energy to break the bonds holding them together. The particles can break away from the liquid and begin to move around freely. This process is called evaporation. The liquid boils until the entire liquid has evaporated to become a gas.

6.6.2 Foggy mirrors

Have you noticed how the mirror in the bathroom 'fogs up' after a hot shower? The 'fog' is actually formed by invisible water vapour in the air cooling down when it contacts the cold glass. It condenses to become water.

Fog in the air

Some of the energy of the particles in the water vapour is transferred away from the vapour to the air. The transfer of energy leaves the water vapour with less energy — so much less energy that its particles slow down. The transfer of energy away from the water vapour means it cools down and turns into tiny droplets of water. These tiny droplets form clouds. This process is called condensation.

Invisible gas

Water vapour forms when particles in the hot water gain enough energy to escape from each other and become a gas. You can't see water vapour. The particles in the water vapour move around freely. They have more energy than the particles in the liquid water.

Fog on the mirror

The energy from some of the particles in the water vapour is transferred to the cold mirror. This causes the water vapour to condense on the mirror.

6.6 Exercises: Understanding and inquiring

To answer questions online and to receive **immediate feedback** and **sample responses** for every question, go to your learnON title at www.jacplus.com.au. *Note:* Question numbers may vary slightly.

Remember

1. Describe the motion of the particles in a liquid.
2. When a substance changes state from a solid into a liquid:
    (a) what happens to the bonds between the particles?
    (b) how does the motion of the particles change?
3. Describe what happens to the movement of particles as a substance changes its state from a gas to a liquid.

Think

4. Describe two changes in the properties of a substance when its particles move faster.
5. In movies, you sometimes see a mirror being held up to the mouth and nose of someone who is unconscious to check whether they are breathing. Explain why this would work.
6. Construct a flowchart like the one on the previous page to show how a gas changes state to become a liquid and then a solid. Include the names and descriptions of the two changes of state that take place.
7. For each of the following changes of state of a substance, identify whether it involves adding energy to the particles or transferring energy away from the particles.
    (a) Melting   (b) Condensation   (c) Boiling
    (d) Freezing   (e) Sublimation   (f) Evaporation

6.7 Concept maps and mind maps

6.7.1 Concept maps and mind maps

1. On small pieces of paper, write down all the ideas you can think of about a particular topic. You could also use software or an app to construct concept and mind maps.
2. Select the most important ideas and arrange them under your topic. Link these main ideas to your topic and write the relationship along the link.
3. Choose ideas related to your main ideas and arrange them in order of importance under your main ideas, adding links and relationships.
4. When you have placed all of your ideas, try to find links between the branches and write in the relationships

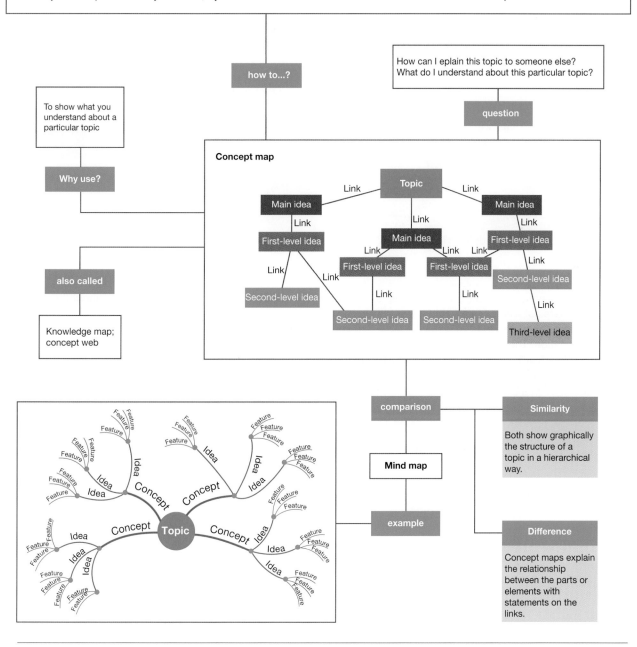

6.7 Exercises: Understanding and inquiring

To answer questions online and to receive **immediate feedback** and **sample responses** for every question, go to your learnON title at www.jacplus.com.au. *Note:* Question numbers may vary slightly.

Think and create

1. The concept map below represents some of our knowledge about the states of matter. This concept map is just one way of representing ideas about matter and how they are linked. However, all but one of the keywords in the boxes are missing.

   Copy the concept map and complete it by writing suitable keywords in the boxes.

Select the keywords from the list below. One keyword is used three times.

fill space	liquid	sliding
fixed shape	particles	solid
free	pour	vibrating
gas		

A concept map shows links between important ideas.

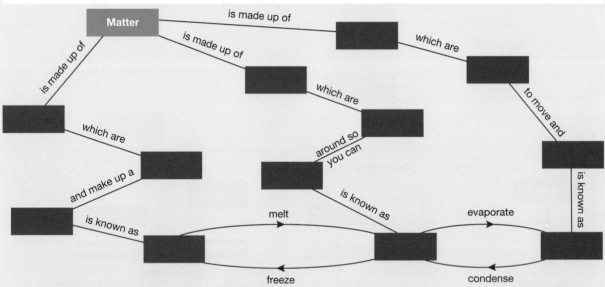

2. Using sticky notes, create a concept map similar to the one above to represent your knowledge about the three states of water. Start with the word 'water' at the top.
3. Create a concept map of your own to represent your knowledge of how water in the atmosphere affects precipitation. The diagram below shows one way of starting your concept map. You'll need to write in suitable link words yourself. (*Hint:* Start by writing down all of the important words or terms related to precipitation that you can think of and use as many of them as you can in your concept map.)

A concept map of precipitation could begin like this.

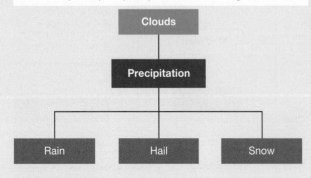

4. A mind map is similar to a concept map, but the topic is placed in the centre instead of at the top. There are no other boxes — just branches on which the keywords and terms are listed. Complete the mind map below to represent your knowledge of the states of matter.

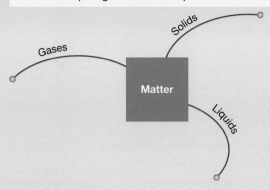

A mind map begins with the topic in the centre.

5. The states of matter can be represented by a concept map or a mind map. Which map did you find easier to construct? Explain why.

6.8 Review

6.8.1 Study checklist

States of matter

- describe the properties and physical behaviour of solids, liquids and gases
- define fluid as a substance that flows
- measure the volume of solids and liquids
- explain how mass is measured

Changes of state

- describe the changes in the physical properties of substances during melting, freezing, evaporation, boiling, condensation and sublimation
- relate changes of state to heating and cooling
- define melting point and boiling point

The particle model of matter

- list the four major assumptions of the particle model of matter
- describe the arrangement and movement of particles in solids, liquids and gases
- describe the diffusion of gases and why it occurs in terms of the particle model
- use the particle model to explain why solids, liquids and gases expand when they are heated
- describe the behaviour of gases under pressure in terms of the particle model
- link the energy of particles to heating and cooling
- link changes in state to the flow of energy into or out of a substance and the subsequent changes in the behaviour of the particles of the substance

Science as a human endeavour

- describe the role of meteorologists in observing, explaining and predicting the weather
- relate weather events such as rain, hail and snow to changes of state
- outline the implications of expansion and contraction of materials for engineers and architects

6.8 Review 1: Looking back

To answer questions online and to receive **immediate feedback** and **sample responses** for every question, go to your learnON title at www.jacplus.com.au. *Note:* Question numbers may vary slightly.

1. Use the particle model to explain why steam takes up more space than liquid water.
2. In which state are the forces of attraction between the particles likely to be greatest?
3. List all of the changes of state that take place in the water cycle.
4. In which state — solid, liquid or gas — do the particles have:
   (a) the most energy
   (b) the least energy?
5. Explain why perfume or aftershave lotion evaporates more quickly than water.
6. Copy and complete the table below to summarise the properties of solids, liquids and gases. Use a tick to indicate which properties each state *usually* has.

Property	Solid	Liquid	Gas
Has a definite shape that is difficult to change			
Takes up a fixed amount of space			
Can be poured			
Takes up all of the space available			
Can be compressed			
Is made of particles that are strongly attracted to each other and can't move past each other			
Is made of particles that are not held together by attraction			

7. Copy and label the three diagrams below to show which represents solids, liquids and gases. Make an improvement to each diagram so that it describes the particle model more fully.

Which states are represented by these diagrams?

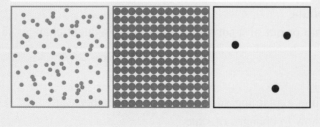

_____ _____ _____

8. Snow and hail are water in a solid state. Describe the difference between snow and hail, and explain how each of them is formed.

9. Name the process that is taking place in the following diagrams and explain why it occurs only in liquids and gases.

10. Which of the diagrams below (A, B or C) best represents the particles of a solid after expanding?

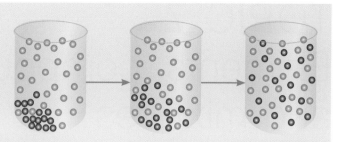

Original solid

A

B

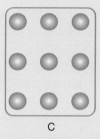

C

11. Explain how mercury and alcohol thermometers are able to provide a measure of temperature.

12. Copy and complete the diagram below, labelling the missing state and changes of state.

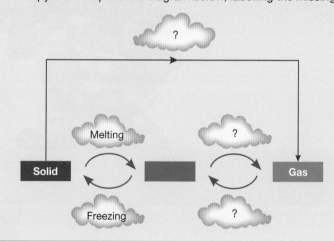

TOPIC 7
Elements, compounds and mixtures

7.1 Overview

There are millions of different substances in the world. Some, like water, occur naturally. Others, like paper and plastic, are made in factories. Some substances, like sugar and blood, are made by living things. All substances have one important thing in common: they are all made of the tiny building blocks of matter that we call atoms.

7.1.1 Think about substances **assess⊙n**

- How did plumbers get their name?
- Which metal can drive you crazy?
- Can water be split?
- Can you breathe nitrogen gas?
- What is most of an atom made up of?
- Just what is 'plastic' made from?
- Which precious gem is made from the same substance as charcoal and soot?

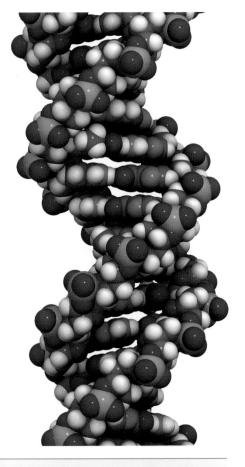

LEARNING SEQUENCE

Numerous **videos** and **interactivities** are embedded just where you need them, at the point of learning, in your learnON title at www.jacplus.com.au. They will help you to learn the concepts covered in this topic.

7.1.2 Your quest
How small are the bits that matter?

INVESTIGATION 7.1

AIM: To investigate division of matter

Materials:
a strip of paper cut from A4 paper (about 30 cm long)
pair of scissors
ruler

Method
- Construct a table like the one below and record the length of the strip of paper.
- Cut the strip of paper in half. Put one half aside. Measure the length of the other half.
- Cut the measured half in half again. Again, put one half aside, and measure and record the length of the other half.
- Before you go any further, predict how many times you will be able to cut the strip in half.

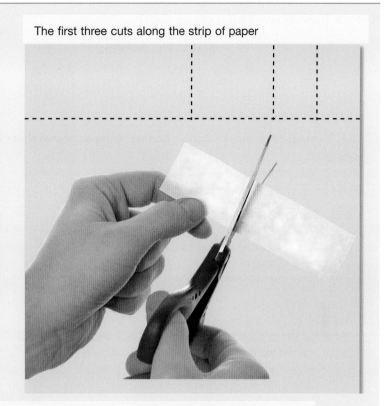

The first three cuts along the strip of paper

How small are the bits?

Number of cuts	Length of strip (approximate)
0	30 cm
1	15 cm
2	7.5 cm (easy?)
3	
4	
5	
6	
7	
8	1 mm (you're doing well to get this far!)
9	
10	
12	
14	
18	1 micron (1 millionth of a metre, one thousandth of a millimetre)
22	
26	
31	The size of a single atom

- Continue this process until you can no longer cut the strip in half.

Discuss and explain
1. How many cuts were you able to make? Was it more or less than your prediction?
2. Estimate the number of cuts you would need to make before the strip would be too small to see.

7.1.3 What's inside?

How do you know what's inside a substance when you can't actually see inside it and it is so small that you can't see it even with the most powerful microscope? It seems impossible — but it can be done!

INVESTIGATION 7.2

AIM: To experience the difficulties of describing an object that cannot be seen

Materials:
sealed box 1, containing a mystery object
sealed box 2, which is empty

Method and results
- Together with a partner, and without opening the boxes, work out what is inside box 1.
- Ask your teacher for any equipment that you think might be helpful. But remember that you are not allowed to open the boxes!
1. Write down any information you can find out about the object.

Discuss and explain
2. What do you think is in box 1?
3. What are the reasons for your decision about what is in the box?
4. What was the reason for having box 2 in this activity?

7.2 It's elementary

Science as a human endeavour

7.2.1 The alchemists

About 1000 years ago, when kings and queens lived in castles and were defended by knights in shining armour, there lived the **alchemists**.

They chanted secret spells while they mixed magic potions in their flasks and melted metals in their furnaces. They tried to change ordinary metals into gold. They also tried to find a potion that would make humans live forever. They studied the movements of the stars and claimed to be able to see into the future. Kings and queens took the advice of the alchemists very seriously.

The alchemists never found the secrets they were looking for, but they did discover many things about substances around us. During the same period people who worked with materials also helped us to understand many everyday substances. Blacksmiths worked with metals to make stronger and lighter swords and armour, fabric dyers learned how to colour cloth, and potters decorated their work with glazes from the Earth. Without the knowledge passed down by these people, the world as we know it would be very different! Twelve important substances were discovered during these ancient times: gold, iron, silver, sulfur, carbon, lead, mercury, tin, arsenic, bismuth, antimony and copper. Alchemists discovered five of these.

7.2.2 Real science

In about the seventeenth century, people stopped thinking about magic and instead carried out **investigations** based on careful **observations**. These new seekers of knowledge were called **scientists**. They discovered that the 12 substances could not be broken down into other substances. Scientists investigated many common everyday substances as well, including salt, air, rocks, water and even urine! They discovered that nearly everything

around us could be broken down into other substances. They gave the name 'element' to the substances that could not be broken down into other substances. Between 1557 and 1925, another 76 **elements** were discovered. We now know that 92 elements exist naturally. In recent years scientists working in laboratories have been able to make at least another 24 artificial elements.

learn on RESOURCES — ONLINE ONLY

Watch this eLesson: Lavoisier and hydrogen (eles-1772)

HOW ABOUT THAT!

In days gone by, substances containing the element mercury were used to make hats. In those days it was not known that mercury is a very poisonous element. Poisoning by mercury can affect your nervous system and your mind. This sometimes happened to hat makers who were exposed to mercury for a long time; hence the expression 'mad as a hatter'!

7.2.3 Warning! Danger!

Many elements are safe to handle. However, there are also many that are not. For example, the elements sodium, potassium and mercury need special care and handling. Sodium and potassium are soft metals that can be cut with a knife. They both get very hot if they come into contact with water. They are stored under oil so that water in the atmosphere cannot reach them.

7.2.4 Elements are rare

Most of the substances around you are made up of two or more elements. You will not be able to find many of the 92 naturally occurring elements in their pure form. It is possible, however, to examine many of the elements in the school laboratory.

INVESTIGATION 7.3

Checking out appearances

AIM: To examine and describe the properties of a selection of elements

Materials:

samples of chemical elements (e.g. carbon, sulfur, copper, iron, aluminium, silicon)

Method and results

1. Copy the table below into your workbook or obtain a copy from your teacher.
   - Carefully examine each of the elements in the set (look for colour, appearance, hardness).
2. Complete the table by filling in the description and discussing with other students the substances that might include the element. One example is completed for you.

Elements	State	Description	In which substances might the element be present?
Hydrogen	Gas	Clear, colourless, explosive	Acids, water

7.2 Exercises: Understanding and inquiring

To answer questions online and to receive **immediate feedback** and **sample responses** for every question, go to your learnON title at www.jacplus.com.au. *Note:* Question numbers may vary slightly.

Remember

1. What problems were the alchemists of medieval times trying to solve?
2. What is an element?
3. Which types of substances did blacksmiths help us to understand?
4. How did the scientists differ from the alchemists?
5. Why do sodium and potassium need to be stored under oil?
6. State one harmful effect of mercury on humans.

Think

7. Is water an element? Give a reason for your answer.
8. Give one reason for displaying chemical safety symbols at the entrances of many buildings.

Investigate

9. Many years ago, balloons were filled with hydrogen so that they could float high in the sky. However, hydrogen is no longer used in balloons because it explodes too easily. At fairs, carnivals and in florists' shops, you can often buy colourful gas-filled balloons that fly high into the sky if you let them go. These balloons are filled with an element called helium. Find out who discovered helium, where it was discovered and when.
10. The element mercury was known to ancient people and was very important to the alchemists. Find out all you can about this liquid metal. What does its name mean? Where is it found? What has it been used for in the past? What is it used for now? What is the safety procedure if mercury is spilt?

learn on RESOURCES — ONLINE ONLY

📄 **Complete this digital doc:** Worksheet 7.1: How big is an atom? (doc-18744)

7.3 Elements: The inside story

7.3.1 Atoms and elements

About 2500 years ago a teacher named Democritus lived in ancient Greece. He walked around the gardens with his students, talking about all sorts of ideas.

Democritus suggested that everything in the world was made up of tiny particles so small that they couldn't be seen. He called these particles *atomos*, which means 'unable to be divided'. Other thinkers at the time disagreed with Democritus. It took about 2400 years for evidence of the existence of these **atoms** (as we now call them) to be found.

We now know that each element is made of its own particular kind of atom. Gold contains only gold atoms, oxygen contains only oxygen atoms, carbon contains only carbon atoms and so on. But what is it that makes atoms different from one another? To answer this question we need to know a little bit more about the atom.

For scientists, the atom was like the mystery box on page 282. Even though the atom couldn't be seen, scientists did experiments over many years and they thought carefully about the information they gathered.

Finding evidence for the existence of atoms was not possible until Galileo wrote about the need for controlled experiments and the importance of accurate observations and mathematical analysis in the 16th century. Galileo's 'scientific method', along with the development of more accurate weighing machines, was used by John Dalton in 1803 to show that matter was made up of atoms. He proposed that atoms could not be divided into smaller particles and that atoms of different elements had different masses.

For the next 100 years, scientists thought the atom was a solid sphere, but discoveries including radioactivity and electric current, and new technology such as the vacuum tube and Geiger counters, allowed scientists to 'peek' inside.

In 1911, New Zealander Sir Ernest Rutherford used some of the new discoveries and inventions to prove that atoms were not solid particles.

He fired extremely tiny particles at a very thin sheet of gold. Most of the particles went straight through. Only sometimes did they bounce off as if they had hit something solid. He concluded that the tiny particles could be getting through only if most of each atom consisted of empty space.

Niels Bohr proposed the next model of the atom. He suggested that the electrons changed their orbits around the positively charged nucleus and formed electron 'clouds'.

In 1932 James Chadwick found another type of particle in the nucleus of the atom — the neutron.

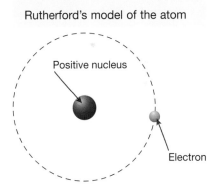

Rutherford's model of the atom

Positive nucleus

Electron

Bohr's model of the atom

Orbital electrons negatively charged

Nucleus, containing positively charged protons

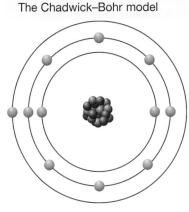

The Chadwick–Bohr model

7.3.2 Inside the atom

It is now understood that all atoms are made up of small particles.

The amount of negative charge carried by each electron is the same as the amount of positive charge carried by each proton. In an atom, the number of protons is equal to the number of electrons, so there is no overall electric charge.

Moving very rapidly around the nucleus are **electrons**. Electrons are much smaller in size and weight than both protons and neutrons. Each electron carries a negative electric charge.

The particles in the centre of an atom are called **protons** and **neutrons**. Together they form the **nucleus**. Each proton carries a positive electric charge. Neutrons have no electric charge.

Helium atoms are lighter than all others except hydrogen atoms. This blimp is filled with helium.

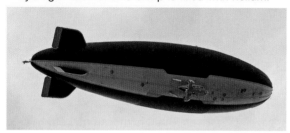

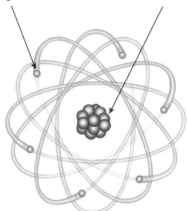

7.3.3 Atomic numbers

The number of protons in an atom is called its **atomic number**. Each element has a different atomic number. The blimp above is filled with helium, which has an atomic number of 2. Helium atoms are lighter than all others except hydrogen atoms. All carbon atoms have six protons inside the nucleus, so the atomic number of carbon is 6. For each proton in the carbon atom it also has one electron, meaning a carbon atom has six electrons. Carbon atoms can have 6, 7 or 8 neutrons in their nuclei. The lightest element is hydrogen, which has one proton in each atom and an atomic number of 1. The heaviest natural element is uranium with 92 protons in each atom.

INVESTIGATION 7.4

Modelling an atom

AIM: To model an atom and observe what makes up most of an atom

Materials:

1 hula hoop	*1 straw*
rice grains	*cotton thread*
table-tennis ball	*sticky tape*
broom and dustpan	

Method and results

- Set up the equipment as shown in the diagram.
- From across the room, but within target distance, use the straw as a peashooter to fire rice grains at the table-tennis ball.
- Count how many grains go right through and how many hit the table-tennis ball. (*Note:* Hits to the cotton thread do not count!)
- Use the broom and dustpan to clean up the mess you've left on the floor.
1. Construct a bar graph to display your results.
2. Construct a bar graph to show the class results.

CAUTION

Ensure that the rice grains are not fired towards any person.

3. Which part of the atom does the table-tennis ball represent in this model?
4. What does the hula hoop represent?

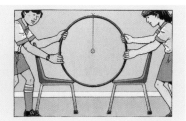

7.3.4 What's in a name?

As the early scientists discovered more and more elements, it became more important that they all agreed on what to call them. Each element was given a name and a **chemical symbol**.

The chemical symbols of most elements are very easy to understand. The symbol sometimes starts with the capital letter that is the first letter of the element's name. For some elements that is the complete symbol. For example:

$$O = oxygen, C = carbon,$$
$$N = nitrogen, H = hydrogen.$$

When there is more than one element starting with the same capital letter, a small letter is also used. For example:

$$Cl = chlorine, Ca = calcium,$$
$$Cr = chromium, Cu = copper.$$

If an element has a symbol that doesn't match its modern name, that's because the symbol is taken from the original Greek or Latin name.

For example:

Na = sodium (*natrium*)
Pb = lead (*plumbum*)
Hg = mercury (*hydro argyros*)
Ag = silver (*argentum*)
K = potassium (*kalium*)
Fe = iron (*ferrum*).

The names and symbols of some of the elements have some interesting origins.

- Einsteinium (Es) is named after the famous scientist Albert Einstein.
- Polonium (Po) was discovered by another famous scientist, Marie Curie. She named polonium after Poland, the country of her birth.
- Helium (He) was first discovered in the sun. It is named after Helios, the Greek god of the sun.
- Sodium (Na) was first called by the Latin name *natrium*.
- Lead (Pb) also used to have a Latin name, *plumbum*. That's where the word 'plumber' comes from. The ancient Romans, who spoke Latin, used lead metal to make their water pipes.

7.3 Exercises: Understanding and inquiring

To answer questions online and to receive **immediate feedback** and **sample responses** for every question, go to your learnON title at www.jacplus.com.au. *Note:* Question numbers may vary slightly.

Remember

1. What idea did Democritus have around 2500 years ago about what substances were made up of?
2. How many types of natural atoms are there?
3. Name the three parts of an atom and state the location of each part.

4. What does the atomic number of an element tell you?
5. Explain why electrons don't fly off their atoms.
6. In what way is an atom of carbon different from an atom of uranium?
7. What makes up most of every atom?
8. List the symbols of each of the following elements: hydrogen, carbon, oxygen, nitrogen, iron, calcium, copper, lead, mercury.

Think

9. List some of the discoveries and inventions that scientists have used to learn more about the atom since the time of Democritus.
10. What type of electric charge does the nucleus of every atom have?
11. What is the atomic number of uranium?
12. To which element does the atom illustrated on page 286 belong?
13. Why do carbon atoms have six electrons?
14. Draw a diagram of an atom that has three protons and one neutron in its nucleus. How many electrons will it have?
15. Why is it important for scientists around the world to agree on the names and chemical symbols of the elements?

Create

16. Construct a model of an atom with three protons, three neutrons and three electrons. Use any appropriate materials. For example, you could use plastic balls, papier-mâché and wire, or perhaps a bowl of jelly with lollies in it.

Investigate

17. Find out the names, atomic numbers and uses of the elements with the following symbols. Construct a table in which to display your findings.
    Sn Au Cu N Ne Sr Ca
18. Research and report on what nanotechnology is and what connection it has with atoms.

 RESOURCES — ONLINE ONLY

 Try out this interactivity: Periodic table (int-0229)

7.4 Compounding the situation

7.4.1 Elements, compounds and mixtures

There are millions and millions of different substances in the world. They include the paper of this book, the ink in the print, the air in the room, the glass in the windows, the wool of your jumper, the cotton and polyester in your shirt or dress, the wood of your desk, the paint on the walls, the plastic of your pen, the hair on your head, the water in the taps and the metal of the chair legs. The list could go on and on.

All substances can be placed into one of three groups: elements, compounds or mixtures.

- Elements are substances that contain only one type of atom. Very few substances exist as elements. Most substances around us are either compounds or mixtures.
- **Compounds** are usually very different from the elements of which they are made. In compounds, the atoms of one element are **bonded** very tightly to the atoms of another element or elements. The elements that make up a compound are completely different substances from the compound. For example, pure salt (sodium chloride) is a compound made up of the elements sodium (a silvery metal)

and chlorine (a green, poisonous gas). A compound always contains the same relative amounts of each element. For example, the compound carbon dioxide is always made up of two atoms of oxygen for each atom of carbon. Its chemical formula is therefore CO_2. The compound sodium chloride always has one sodium atom for each chlorine atom and its formula is simple, $NaCl$.

- **Mixtures** can be made up of two or more elements, two or more compounds or a combination of elements and compounds. Unlike compounds, the parts of a mixture are not always in the same proportion. Sea water is the most common mixture on the Earth's surface, but the percentage of salt is not always the same. It can also include a variety of other elements and compounds in different quantities. Coffee is a mixture that can contain different relative amounts of water, milk, coffee beans and sugar. Brass is a mixture of metals that can have different relative amounts of copper and zinc. There can be no unique chemical formula for mixtures.

The substances that make up mixtures can usually be easily separated from each other. When the parts of a mixture are separated, no new substances are formed. For example, fizzy soft drink contains water, gas, sugar and flavours. If you shake the soft drink, the gas bubbles separate from the water and go into the air. You still have the water in the bottle and the gas in the air; they are just not mixed together any more. The parts of the mixture can be separated quite easily. The gas escapes when the lid of the container is opened, and the water can be separated by evaporation, leaving behind sugar and some other substances.

When the atoms of different elements bond together, a compound is formed. When heated together, the elements iron and sulfur form a new compound called iron sulfide. Iron sulfide has the formula FeS. Every compound has a formula comprising the symbols of the elements that make it up. Unlike mixtures, the elements within a compound cannot easily be separated from each other.

Elements can be separated from compounds in several ways. These include:
- passing electricity through a compound
- burning the compound
- mixing the compound with other chemicals.

Each of these methods involves a chemical reaction in which completely different substances are formed.

WHAT DOES IT MEAN?
The word *compound* comes from the Latin word *componere*, meaning 'to put together'.

Some common substances

Substance	Type	Composed of:	Scientific name
Gold	Element	Gold	Gold
Diamond	Element	Carbon	Carbon
Water	Compound	Hydrogen and oxygen	Dihydrogen oxide
Pure salt	Compound	Sodium and chlorine	Sodium chloride
Brass	Mixture	Copper and zinc	Brass
Soft drink	Mixture	Water, sugar, carbon dioxide and other compounds	
Sea water	Mixture	Water, sodium chloride and other compounds	

7.4.2 Splitting water

We are surrounded by water. It is in our taps, in our bodies, in the rivers, in the sea, in the air and it comes down as rain. We wash in it, cook with it and drink it. We cannot live without water. Water is not an element — it can be broken down into simpler substances. The illustration at right shows an apparatus called a Hofmann voltameter. Water is placed in the voltameter, which is connected to a battery. The electricity splits the water into the elements of which it is made: **hydrogen** and **oxygen**.

Hydrogen and oxygen are both elements. They are both gases, and they look the same; they have no colour and no smell. Oxygen is necessary for substances to burn — even hydrogen does not burn without it. Hydrogen is a much less dense gas than oxygen. This means that a balloon filled with hydrogen will float up very high, but one filled with oxygen will not.

The element hydrogen is present in almost all acids. By placing a piece of metal in an acid, the hydrogen is forced out. The hydrogen can be collected and tested with a flame.

The element oxygen is present in water, air, rocks and even hair bleach. Oxygen is the gas that all living things need to stay alive. It is also necessary for all substances to burn. When hydrogen gas is burned, it combines with oxygen in the air to form water. This releases a lot of energy. If large amounts of hydrogen and oxygen are used, enough energy can be released to lift a space rocket.

Water is split in a Hofmann voltameter. The clear gas in the left tube is hydrogen. The gas in the right tube is oxygen. What do you notice about the amounts of hydrogen and oxygen that are produced?

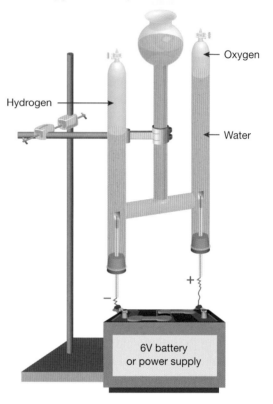

INVESTIGATION 7.5

Making a compound from its elements
AIM: To use a chemical reaction to make a compound from its elements

Materials:
4–5 cm strip of clean, shiny magnesium ribbon. It can be coiled to fit in the crucible.
crucible with lid
pipeclay triangle, tongs and safety glasses
Bunsen burner, heatproof mat and matches

Method and results
- Examine the piece of magnesium and note its appearance before placing it in the crucible and covering it with the lid.
- Put the crucible on the pipeclay triangle as shown in the diagram on next page.
- Heat the crucible with a strong blue flame, monitoring the reaction by occasionally lifting the lid a little with tongs.
- When all the magnesium ribbon has been changed, turn off the flame and leave the crucible on the tripod to cool.
1. Describe the substance in the crucible.

Discuss and explain
2. Is magnesium an element or a compound? Give a reason for your decision.
3. Magnesium is one of the reactants in this experiment. What is the other reactant?

---

4. Is the substance remaining in the crucible an element or a compound? What is its name?
5. What is the evidence that a new substance has been made?
6. Apart from observing whether the reaction is complete, give another reason for lifting the lid of the crucible a little with tongs during the burning.

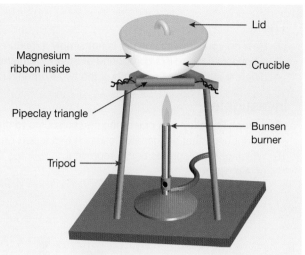

Lid

Magnesium ribbon inside

Crucible

Pipeclay triangle

Bunsen burner

Tripod

INVESTIGATION 7.6

Let's collect an element

AIM: To observe a chemical reaction between a metal and an acid

Materials:

safety glasses
matches
measuring cylinder

2 test tubes and test-tube rack
dilute hydrochloric acid
magnesium metal

Method and results

- Measure 10 mL of hydrochloric acid and pour it into the test tube.
- Add a piece of magnesium and place the second test tube on top of the first as shown in the diagram. Carefully observe what happens.
- After one minute, take the second test tube off the first. While it is still inverted, immediately light the gas in the second test tube with a match.

1. Describe what happened in the test tube containing the metal and the acid.
2. What does hydrogen gas look like?
3. What happened when you lit the gas?
4. Look closely at the second test tube. Describe what you see inside it.

Collect the hydrogen gas by placing the second test tube over the first.

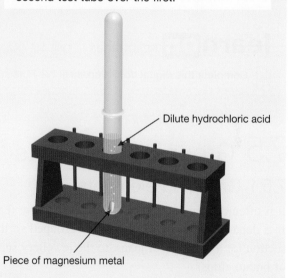

Dilute hydrochloric acid

Piece of magnesium metal

7.4 Exercises: Understanding and inquiring

To answer questions online and to receive **immediate feedback** and **sample responses** for every question, go to your learnON title at www.jacplus.com.au. *Note:* Question numbers may vary slightly.

Remember

1. How do compounds differ from elements?
2. What are the important differences between a mixture and a compound?
3. Fizzy soft drink is a mixture of several compounds. List three of the compounds and suggest how each of them could be separated from the mixture.
4. What happens when atoms are bonded together?
5. Which elements are bonded together to form table salt?
6. List three ways in which elements can be separated from their compounds.

Think

7. How do you know that water is not simply a mixture of hydrogen and oxygen?
8. Magnesium oxide is a compound of magnesium and oxygen. How do you know that it is a completely different substance from each of the two elements it is made up of?

Investigate

9. Joseph Priestley was one of the first scientists to discover the element oxygen. He also discovered many compounds that are gases. Research and report on the life of Joseph Priestley.
10. Which of the diagrams below represent:
    (a) elements
    (b) compounds
    (c) mixtures?

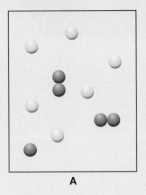 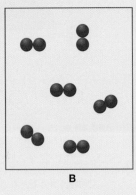 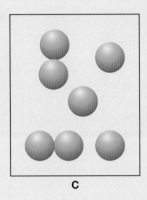 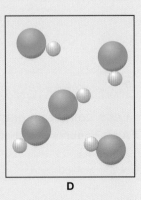

| A | B | C | D |

7.5 Grouping elements

7.5.1 Grouping elements

It is often convenient to group objects that have features in common. Shops provide a good example of this. In a department store, the goods are grouped so that you know where to buy them. You go to the clothing section for a new pair of jeans, to the jewellery section for a new watch and to the food section for a packet of potato chips.

Scientists also organise objects into groups. Biologists organise living things into groups. Animals with backbones are divided into mammals, birds, reptiles, amphibians and fish. Geologists organise rocks into groups. The elements that make up all substances can also be organised into groups.

7.5.2 Metals and non-metals

Scientists have divided the elements into two main groups: the **metals** and the **non-metals**.

Metals

The metals have several features in common:
- They are solid at room temperature, except for mercury, which is a liquid.
- They can be polished to produce a high shine or lustre.
- They are good conductors of electricity and heat.

- They can all be beaten or bent into a variety of shapes. We say they are malleable.
- They can be made into a wire. We say they are ductile.
- They usually melt at high temperatures. Mercury, which melts at –40°C, is one exception.

Non-metals

Only 22 of the elements are non-metals. At room temperature, 11 of them are gases, 10 are solid and 1 is liquid. The solid non-metals have most of the following features in common:

- They cannot be polished to give a shine like metals; they are usually dull or glassy.
- They are brittle, which means they shatter when they are hit.
- They cannot be bent into shape.
- They are usually poor conductors of electricity and heat.
- They usually melt at low temperatures.

Metalloids

Some of the elements in the non-metal group look like metals. One example is silicon. While it can be polished like a metal, silicon is a poor conductor of heat and electricity, and cannot be bent

Common examples of non-metals are sulfur, carbon and oxygen.

or made into wire. Those elements that have features of both metals and non-metals are called **metalloids**. There are eight metalloids altogether: boron, silicon, arsenic, germanium, antimony, polonium, astatine and tellurium.

Metalloids are important materials often used in electronic components of computer circuits.

Looking for similarities

AIM: To describe the characteristics of a variety of elements

Materials:
safety glasses
samples of sulfur, zinc, tin, carbon, silicon, copper
steel wool or very fine sandpaper
battery or power pack
wires with alligator clips
light globe

Method and results

1. Make a copy of the table below and use it to record your observations.
    • Rub each of the elements with the fine sandpaper and observe whether they are shiny or dull.
    • Try to bend the element.
    • Connect the circuit as shown in the diagram to determine whether electricity passes through each of the elements.

Connect your element sample into this circuit.

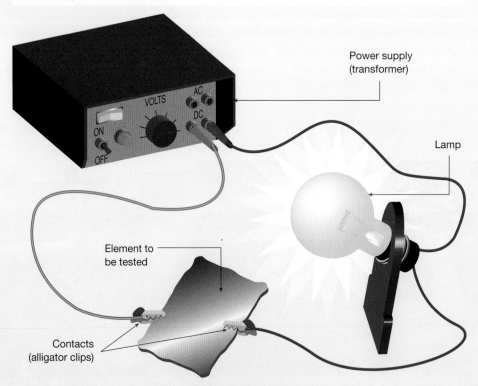

Power supply
(transformer)

VOLTS

AC

DC

ON

OFF

Lamp

Element to
be tested

Contacts
(alligator clips)

2. Which of the six elements have a shiny surface when polished?
3. Which of the six elements do not have a shiny surface when polished?
4. Which of the six elements can be bent?
5. Which of the six elements cannot be bent?
6. Which of the six elements allow electricity to pass through?
7. Which of the six elements do not conduct electricity?

Discuss and explain

8. Attempt to divide the elements into two groups based your observations. Suggest names for these groups.
9. Which of the six elements tested does not seem to fit into either of these two groups?

Characteristics of some elements

Element	Shiny or dull?	Does it bend?	Does it conduct electricity?
Sulfur			
Zinc			
Tin			
Carbon			
Silicon			
Copper			

7.5 Exercises: Understanding and inquiring

To answer questions online and to receive **immediate feedback** and **sample responses** for every question, go to your learnON title at www.jacplus.com.au. *Note:* Question numbers may vary slightly.

Remember

1. Outline four features that metals have in common.
2. Outline four features that non-metals have in common.
3. What is a metalloid? List some examples.
4. Which metal is liquid at room temperature?
5. What does 'metallic lustre' mean?

Think

6. While all metals have similar characteristics, there are also differences between them. List three ways in which metals can differ from each other.
7. Silicon is used in the 'chips' of computer circuits, but it is never used in the connecting wires of electric circuits. Why not?

Imagine

8. Imagine that you are a scientist who has discovered what appears to be a new element. It is golden in colour and very shiny. What experiments would you do to test whether it was a metal or non-metal? What results would you expect to get if it was a metal?

Investigate

9. Polonium is a metal discovered by Marie Curie. She also discovered another metal. Find out its name and the important role it played in medicine.

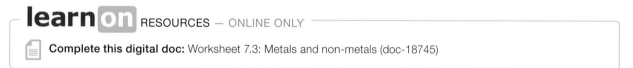

learn on RESOURCES — ONLINE ONLY

Complete this digital doc: Worksheet 7.3: Metals and non-metals (doc-18745)

7.6 Patterns, order and organisation: Chemical name tags

7.6.1 The periodic table

As more and more elements were being discovered, the early scientists began to find that some of them had things in common.

Because some elements had things in common, scientists decided to organise them into groups. It took a long time and a lot of experimenting to work out the groups. A Russian scientist called Dmitri Mendeleev finally worked out a system for grouping the elements. His system is called the **periodic table** and a modern version is used by scientists today.

Alkali metals → Group 1
Alkaline earth metals → Group 2

Key
— Atomic number
— Name
— Symbol
— Relative atomic mass

Period 1	
1 Hydrogen **H** 1.008	2 Helium **He** 4.003

Transition metals

	Group 1	Group 2	Group 3	Group 4	Group 5	Group 6	Group 7	Group 8	Group 9
Period 2	3 Lithium **Li** 6.94	4 Beryllium **Be** 9.02							
Period 3	11 Sodium **Na** 22.99	12 Magnesium **Mg** 24.31							
Period 4	19 Potassium **K** 39.10	20 Calcium **Ca** 40.08	21 Scandium **Sc** 44.96	22 Titanium **Ti** 47.87	23 Vanadium **V** 50.94	24 Chromium **Cr** 52.00	25 Manganese **Mn** 54.94	26 Iron **Fe** 55.85	27 Cobalt **Co** 58.93
Period 5	37 Rubidium **Rb** 85.47	38 Strontium **Sr** 87.62	39 Yttrium **Y** 88.91	40 Zirconium **Zr** 91.22	41 Niobium **Nb** 92.91	42 Molybdenum **Mo** 95.96	43 Technetium **Tc** 98.91	44 Ruthenium **Ru** 101.1	45 Rhodium **Rh** 102.91
Period 6	55 Caesium **Cs** 132.9	56 Barium **Ba** 137.3	57–71 Lanthanides	72 Hafnium **Hf** 178.5	73 Tantalum **Ta** 180.9	74 Tungsten **W** 183.8	75 Rhenium **Re** 186.2	76 Osmium **Os** 190.2	77 Iridium **Ir** 192.22
Period 7	87 Francium **Fr**	88 Radium **Ra**	89–103 Actinides	104 Rutherfordium **Rf**	105 Dubnium **Db**	106 Seaborgium **Sg**	107 Bohrium **Bh**	108 Hassium **Hs**	109 Meitnerium **Mt**

Lanthanides

57 Lanthanum **La** 138.91	58 Cerium **Ce** 140.122	59 Praseodymium **Pr** 140.91	60 Neodymium **Nd** 144.24	61 Promethium **Pm** (145)	62 Samarium **Sm** 150.4	63 Europium **Eu** 151.96

Actinides

89 Actinium **Ac** (227)	90 Thorium **Th** 232.04	91 Protactinium **Pa** 231.04	92 Uranium **U** 238.03	93 Neptunium **Np** 237.05	94 Plutonium **Pu** (244)	95 Americium **Am** (243)

The periodic table. Elements 1–92 all occur naturally. Those from element 93 onwards have been made in laboratories and are all radioactive. Those from element 112 onwards are not shown in this table.

7.6.2 Looking for similarities

A vertical column on the periodic table is called a **group**. Elements in the same group on the periodic table always have some features in common. Sometimes these common features are easy to observe, but some of the similarities are not so obvious. For example, neon and argon are gases that do not change when mixed with other elements except under extreme circumstances. They are said to be **inert**. These two gases are found in the last group of the periodic table along with three other inert gases. The group containing the inert gases is called the **noble gas** group.

The group number corresponds to the number of electrons in the outer shell.

The period number refers to the number of the outermost shell containing electrons.

New radioactive elements are still being produced — the most recent one at the time of publication was element 118.

Halogens **Noble gases**

Group 10	Group 11	Group 12	Group 13	Group 14	Group 15	Group 16	Group 17	Group 18
			5 Boron **B** 10.81	6 Carbon **C** 12.01	7 Nitrogen **N** 14.01	8 Oxygen **O** 16.00	9 Fluorine **F** 19.00	10 Neon **Ne** 20.18
			13 Aluminium **Al** 26.98	14 Silicon **Si** 28.09	15 Phosphorus **P** 30.97	16 Sulfur **S** 32.06	17 Chlorine **Cl** 35.45	18 Argon **Ar** 39.95
28 Nickel **Ni** 58.69	29 Copper **Cu** 63.55	30 Zinc **Zn** 65.38	31 Gallium **Ga** 69.72	32 Germanium **Ge** 72.63	33 Arsenic **As** 74.92	34 Selenium **Se** 78.96	35 Bromine **Br** 79.90	36 Krypton **Kr** 83.80
46 Palladium **Pd** 106.4	47 Silver **Ag** 107.9	48 Cadmium **Cd** 112.4	49 Indium **In** 114.8	50 Tin **Sn** 118.7	51 Antimony **Sb** 121.8	52 Tellurium **Te** 127.8	53 Iodine **I** 126.9	54 Xenon **Xe** 131.3
78 Platinum **Pt** 195.1	79 Gold **Au** 197.0	80 Mercury **Hg** 200.6	81 Thallium **Tl** 204.4	82 Lead **Pb** 207.2	83 Bismuth **Bi** 209.0	84 Polonium **Po** (209)	85 Astatine **At** (210)	86 Radon **Rn** (222)
110 Darmstadtium **Ds**	111 Roentgenium **Rg**	112 Copernicium **Cn**						

Metals ← → **Non-metals**

64 Gadolinium **Gd** 157.25	65 Terbium **Tb** 158.93	66 Dysprosium **Dy** 162.50	67 Holmium **Ho** 164.93	68 Erbium **Er** 167.26	69 Thulium **Tm** 168.93	70 Ytterbium **Yb** 173.04	71 Lutetium **Lu** 174.97

96 Curium **Cm** (247)	97 Berkelium **Bk** (247)	98 Californium **Cf** (251)	99 Einsteinium **Es** (254)	100 Fermium **Fm** (257)	101 Mendelevium **Md** (258)	102 Nobelium **No** (255)	103 Lawrencium **Lr** (256)

7.6 Exercises: Understanding and inquiring

To answer questions online and to receive **immediate feedback** and **sample responses** for every question, go to your learnON title at www.jacplus.com.au. *Note:* Question numbers may vary slightly.

Remember

1. Obtain a copy of the periodic table from your teacher. Colour in the elements that you have already seen in the laboratory.
2. Write down the symbols for the following elements: hydrogen, carbon, oxygen, nitrogen, iron, tin, calcium, sulfur, copper and krypton.
3. What is similar about all of the gases in the noble gas group of the periodic table?

Create

4. Make up a 'Guess the element' card game, finding out and using information about at least twenty of the elements.

7.7 Making molecules

7.7.1 Bonding

The naturally occurring elements are the building blocks of everything in our world. The atoms of various elements can be joined in a wide variety of ways to produce many compounds. Elements and compounds can be combined in many ways to make countless mixtures.

Atoms can join, or bond, in many different ways. In some substances, atoms are joined in groups called **molecules**. For example, in oxygen gas, oxygen atoms are joined in groups of two. In the compound carbon dioxide, one carbon and two oxygen atoms are joined in every molecule. Atoms can join to form small or large molecules of many different shapes.

Some compounds are not made up of molecules. Instead the atoms bond by lining up one after the other. Sodium bonds to chlorine, which bonds to sodium and so on. Common table salt is an example of a substance that is bonded in this way.

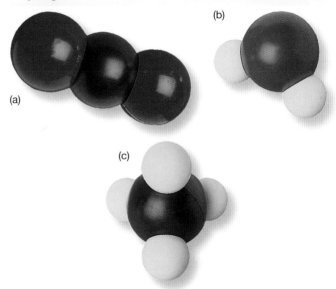

Models representing the molecules of the compounds (a) carbon dioxide, (b) water and (c) methane. The black balls represent carbon; the red, oxygen; and the white, hydrogen.

INVESTIGATION 7.8

Mix 'n' match

AIM: To model the molecules of a variety of compounds

Materials:
green, red and blue sheets of paper
scissors   pencil   ruler
1 large sheet of cartridge paper

Method and results

- Cut out 15 diamonds, each 2 cm long and 1.5 cm wide, from the green sheet of paper.
- Cut out 30 equilateral triangles, with each side 2 cm, from the red sheet of paper.
- Cut out 15 squares, with each side 2 cm, from the blue sheet of paper.
- Imagine that different types of atoms are represented by particular shapes:
  a blue square = carbon
  a green diamond = oxygen
  a red triangle = hydrogen
  and that, by placing them side by side on the sheet of paper, you are joining them.
- Place two green diamonds next to each other on the sheet. This represents the oxygen molecule, as shown in the diagram at right.
- Place one blue square on the sheet between two green diamonds. This represents the compound carbon dioxide. Label it with its name and symbol.

Cut these shapes from coloured paper.

←— 1.5 cm —→

2 cm

2 cm

2 cm

2 cm   2 cm

2 cm   2 cm

2 cm

A green diamond represents an atom of oxygen. Together, two diamonds represent a molecule of oxygen.

- Represent and label the following substances:
  - (a) water, which contains 1 oxygen and 2 hydrogen atoms
  - (b) methane (natural gas), which contains 1 carbon and 4 hydrogen atoms
  - (c) benzene (in petrol), which contains 6 carbon and 6 hydrogen atoms
  - (d) glucose (sugar), which contains 6 carbon, 12 hydrogen and 6 oxygen atoms
  - (e) hydrogen peroxide (found in hair bleach), which contains 2 oxygen atoms and 2 hydrogen atoms.

Discuss and explain

1. Which of these compounds contain only hydrogen and carbon atoms?
2. In what ways are these two substances different from each other?
3. Which of the compounds contain only oxygen and hydrogen? Do these compounds have the same characteristics?
4. Think about the appearance of the compound sugar. How does it differ in appearance from the elements from which it is made?

7.7.2 Compounds of today and tomorrow

Polymer is the name given to a compound made of molecules that are long chains of atoms. Most polymers are made up of chains containing carbon atoms. **Plastics** are synthetic polymers, while cotton and rubber are examples of natural polymers. Although scientists first developed polymers in laboratories in the 1800s, it was not until after World War II that most modern polymers were invented. Modern polymers are used in food wrapping, paint, plastic 'glass', polystyrene foam for packaging and cups, note money, cases for electronic appliances such as computers and televisions, clothing, glues, shopping bags, sports equipment and even tea bags!

WHAT DOES IT MEAN?

The word *polymer* comes from the Greek word *polymeres*, meaning 'of many ports'.

HOW ABOUT THAT!

- Nitrogen is an element. It is a clear, colourless gas made up of molecules. Each molecule is made up of a pair of atoms. Nitrogen makes up 80 per cent of the atmosphere, which means that four-fifths of each breath you take is nitrogen. Our bodies cannot use this nitrogen so we breathe it straight out again! The gases oxygen, hydrogen and chlorine also exist as molecules made up of pairs of atoms.
- Gold is the only metal element found in large amounts in its pure form, rather than bonded in compounds with other elements.

7.7 Exercises: Understanding and inquiring

To answer questions online and to receive **immediate feedback** and **sample responses** for every question, go to your learnON title at www.jacplus.com.au. *Note:* Question numbers may vary slightly.

Remember

1. What is a molecule? Name two compounds that are made up of molecules.
2. Are all compounds made up of molecules? Explain.
3. Name four elements that are made up of molecules.
4. What are polymers?

Think

5. What is the difference between an atom and a molecule?
6. Copy and complete the table below. Use the formula of each compound to work out how many elements are present and which ones they are. (The formula of a compound not only tells you which elements are present, but also indicates the ratio of atoms of the different elements. For example, in the compound NH_3 there are three hydrogen atoms for each nitrogen atom.)

Compound	Formula	Number of elements	Names of elements
Copper sulfate	$CuSO_4$	3	Copper, sulfur, oxygen
Zinc sulfide	ZnS		
Ammonia	NH_3		
Sulfuric acid	H_2SO_4		
Hydrochloric acid	HCl		
Table salt	$NaCl$		

Investigate

7. Australia has led the way in the production of polymer banknotes. Find out all you can about how these notes are made.

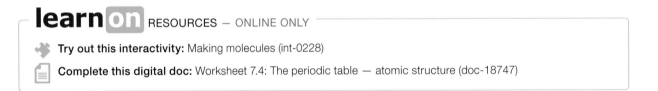

learn on RESOURCES — ONLINE ONLY

Try out this interactivity: Making molecules (int-0228)

Complete this digital doc: Worksheet 7.4: The periodic table — atomic structure (doc-18747)

7.8 Carbon: It's everywhere

7.8.1 That's carbon?

Carbon is a most amazing element. It is found naturally in three different forms. One form is diamond, another is graphite (the 'lead' in lead pencils), and the third is called amorphous carbon (coal, charcoal and soot). Diamond is the hardest substance known and is used to make drill tips and cutting tools. The three forms are different from each other because the carbon atoms are joined in different ways.

Carbon is found combined with other elements in a huge range of compounds. No other element forms as many different types of compounds as carbon. Carbon is found in everything from the skin of an elephant to paint on the walls!

7.8.2 The chemistry of life

All living things are made up of compounds including proteins, fats and carbohydrates. The main element in these compounds is carbon. Carbon is not found only in living things, but also in the air in carbon dioxide and under the sea in limestone. The carbon atoms in carbon dioxide were once carbon atoms in living things. The carbon atoms in living things will eventually become carbon atoms in the air or carbon atoms in limestone under the sea. The illustration below shows how nature constantly recycles carbon atoms.

Plants take in carbon dioxide through their leaves and, in a process known as **photosynthesis**, use the carbon dioxide and water to make sugar. Sugar is a compound made up of carbon, hydrogen and oxygen atoms. Plants use the sugar to make other substances and for energy to grow. Animals eat plants or plant-eating animals. The carbon atoms then become part of the animals' bodies.

Carbon atoms in the bodies of living things return to the air in several ways: **respiration**, **decomposition** and **burning**.

- Respiration is a process that occurs in the cells of every living thing, from a microscopic water plant to a humpback whale. Respiration releases energy and produces carbon dioxide. The carbon dioxide released by the cells in your body is taken by your blood to your lungs. The carbon dioxide that you breathe out contains carbon atoms that were once part of your body.
- Decomposition is what happens when plant or animal material breaks down, such as in a compost heap or after something is buried. Microscopic living creatures called decomposers absorb some of the substances in the dead material and release carbon dioxide to the air by respiration.
- When substances containing carbon are burned, carbon dioxide is released. Coal, natural gas and oil are all **fuels** formed from living things, and contain carbon atoms. Fuels are **combustible**; that is, they are easily ignited. When these fuels are burned in homes, cars, factories and power stations, carbon dioxide is released into the air. Bushfires also release carbon dioxide back to the air.

The flow of carbon atoms through the environment

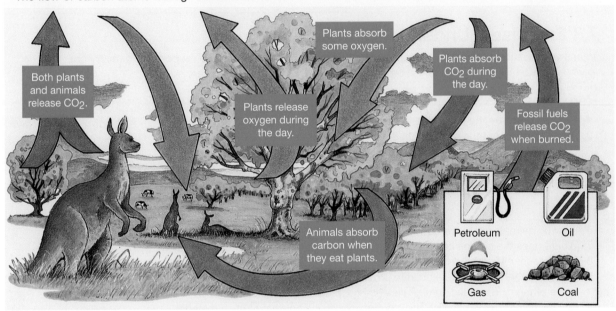

Both plants and animals release CO_2.

Plants absorb some oxygen.

Plants release oxygen during the day.

Plants absorb CO_2 during the day.

Fossil fuels release CO_2 when burned.

Animals absorb carbon when they eat plants.

Petroleum

Oil

Gas

Coal

INVESTIGATION 7.9

Looking for carbon

AIM: To test for the presence of carbon in a range of substances

Materials:
safety glasses
Bunsen burner, heatproof mat and matches
metal tongs
small samples of substances to investigate: woollen cloth, cottonwool, sugar cube, wood, bread, peanut, steel wool, glass, paper, aluminium foil

Method and results

Your task in this investigation is to find out whether the element carbon is present in some common substances. If carbon is in a substance, the substance turns black when it is burnt. Your teacher may allow you to burn some plastics in the fume hood, including artificial fabrics such as nylon and rayon.

- Hold a small piece of the substance you are going to test in the metal tongs.
- Put the substance in the blue flame of the Bunsen burner.

- When the substance catches alight, take it out of the flame and, keeping it above the heatproof mat, allow it to burn slowly. Does it turn black?
1. Draw up a table like the one below and record your observations.

Discuss and explain
2. In which of the substances tested is carbon present?
3. Can you be sure that, if the substance went black, carbon was present? Give a reason for your answer.
4. Can you be sure, if a substance didn't go black, that it didn't contain carbon? Give a reason for your answer.

Substance	Observations	Is carbon present?
Wood		
Cottonwool		

7.8 Exercises: Understanding and inquiring

To answer questions online and to receive **immediate feedback** and **sample responses** for every question, go to your learnON title at www.jacplus.com.au. *Note:* Question numbers may vary slightly.

Remember
1. List and describe the three different forms of the element carbon.
2. How do plants get the carbon that they need to make sugar?
3. Describe three ways in which carbon can return to the atmosphere.
4. Where does respiration take place?
5. Some fabrics are more combustible than others. What does this mean?

Think
6. How do animals obtain carbon?
7. Where does the carbon come from to form limestone at the bottom of the sea?
8. The amount of carbon dioxide in the Earth's atmosphere is increasing. Why is this happening?
9. Many different materials are used to provide heating. The table below shows how much carbon is in some of them. The last column indicates how much heat (in MJ) 1 kg of each material typically provides.
   (a) Draw a bar graph showing the percentage carbon content of each material.
   (b) Which is the best material for heating?
   (c) Does the table indicate any relationship between the amount of carbon in a material and the amount of heat that it provides? Explain your answer clearly.

Material	Carbon content (%)	Heat production (MJ)
Wood	11	17.9
Brown coal	73	29.5
Black coal	80	35.9
Natural graphite	90	39

Investigate
10. Fuels may be solids, liquids or gases. Search the internet or use a library to find as many examples as possible of solid, liquid and gas fuels, and complete a table like the one at right.

	State		
	Solid	Liquid	Gas
Examples			
Uses			
Advantages			
Disadvantages			

7.9 Affinity diagrams and cluster maps

7.9.1 Affinity diagrams and cluster maps

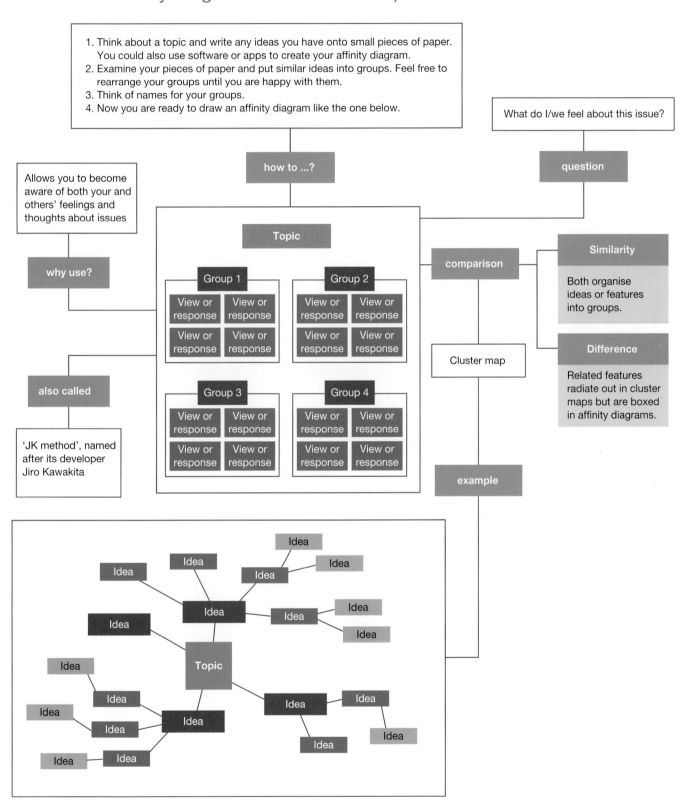

1. Think about a topic and write any ideas you have onto small pieces of paper. You could also use software or apps to create your affinity diagram.
2. Examine your pieces of paper and put similar ideas into groups. Feel free to rearrange your groups until you are happy with them.
3. Think of names for your groups.
4. Now you are ready to draw an affinity diagram like the one below.

What do I/we feel about this issue?

question

how to ...?

Allows you to become aware of both your and others' feelings and thoughts about issues

why use?

also called

'JK method', named after its developer Jiro Kawakita

Topic

Group 1
View or response | View or response
View or response | View or response

Group 2
View or response | View or response
View or response | View or response

Group 3
View or response | View or response
View or response | View or response

Group 4
View or response | View or response
View or response | View or response

comparison

Cluster map

example

Similarity

Both organise ideas or features into groups.

Difference

Related features radiate out in cluster maps but are boxed in affinity diagrams.

Idea Topic Idea (cluster map diagram)

7.9 Exercises: Understanding and inquiring

To answer questions online and to receive **immediate feedback** and **sample responses** for every question, go to your learnON title at www.jacplus.com.au. *Note:* Question numbers may vary slightly.

Think and create

1. (a) Write each of the ideas, objects or substances listed below on a small card or sticky note.

air hydrogen protons oxygen

water periodic table METALS

bond nucleus

sea water

easily separated

electrons salt carbon

CHOCOLATE THICK SHAKE

neutrons H_2O

orbit

(b) Arrange the ideas, objects or substances on the cards into four categories in an affinity diagram like the one below. You will need to work out the name of the missing category.

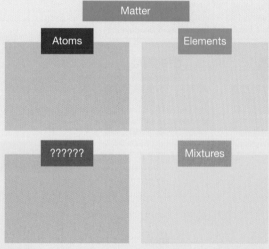

Matter

Atoms

Elements

??????

Mixtures

2. Which element is shown below?

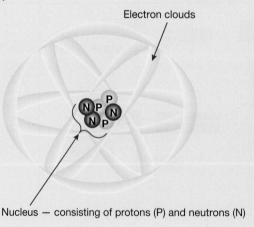

Electron clouds

Nucleus — consisting of protons (P) and neutrons (N)

3. Use the ideas, objects and substances from question 1 to create a cluster map using the four categories as the main associations. Add as many associations as you can to the diagram. Don't forget that you can sometimes make links between the different arms of your cluster map.

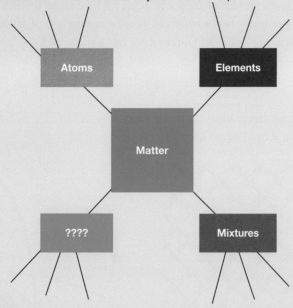

4. About 2500 years ago, when the Greek teacher Democritus suggested that all matter was made of atoms, other Greek thinkers proposed that there were four elements.
   These elements were earth, air, fire and water. All other substances were combinations of these four elements. Work in a small group to create a cluster map called 'elements' using 'earth', 'air', 'fire' and 'water' as the main associations. Add as many common substances as you can to your map.
5. Create an affinity diagram like the one below that illustrates the properties and uses of metals, fuels and fabrics. Include as many statements as you can about each group of substances.

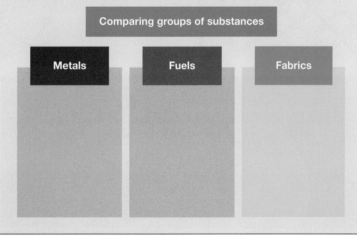

learn on

7.10 Project: Science TV

Scenario

In the media world, programs that combine entertainment and education are known as 'edutainment'. With the success of edutainment programs such as *Mythbusters* (SBS/7Mate), *Scope* (Network 10) and *The ExperiMentals* (ABC), it seems that science is attracting a bigger share of the television market than many network executives would have expected. Now, your local TV network — Channel 55 — has decided to jump on the 'science as edutainment' bandwagon and has announced that next year it will develop a program called *Science TV*.

To make *Science TV* more appealing to a younger audience, the developing executives of the program want it to be presented by a team of school students, who will do all of the introductions, explanations and experiments for each of the segments. It is important that the right team of students is found or the program will be canned after only a few episodes, so Channel 55 has announced that it is accepting online audition files from groups of students who think they have what it takes to be the *Science TV* stars.

Your task

Your group is going to put together a video submission that you could send to the Channel 55 developers to showcase how suitable you would be as the stars of *Science TV*.

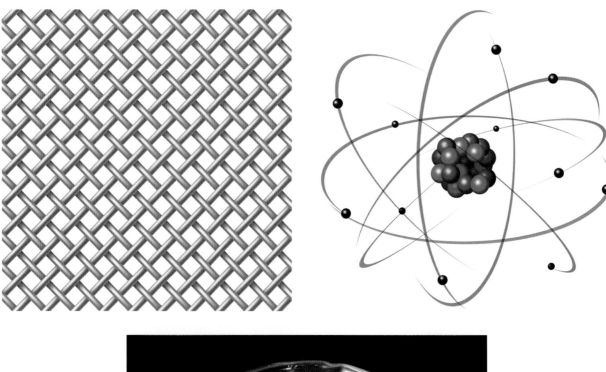

The guidelines for the video submission from the Channel 55 website are as follows:
• The video must be between four and five minutes in length.
• The target audience of *Science TV* is between 8 and 14 years old.
• At least two people must be shown on camera.

- The video must be in the form of a chemistry segment that explains one of the following:
  (a) What is the difference between a physical change and a chemical change?
  (b) What are elements, compounds and mixtures?
  (c) How would we separate a mixture of iron filings, sand, copper sulfate and chalk dust?
- At least one experiment must be shown being performed in the segment — the experiment must be relevant to the segment and safe to perform (i.e. no explosions and no dangerous fumes produced).

The segment should be engaging and informative. It should have an introduction (either a scenario played out or a discussion between the presenters), an experiment to either test or demonstrate an idea, an explanation of the main concepts involved and a resolution that ties back into the original scenario or discussion. Remember: the main idea is to show that science is FUN!

7.11 Review

7.11.1 Study checklist

Elements and atoms

- describe some common chemical elements
- recall the chemical symbols of some common elements
- identify some of the dangers associated with some chemical elements
- model the structure of the atom and describe the characteristics of the three main particles
- recall that each chemical element is identified with a unique atomic number, which is equal to the number of protons in its nucleus
- distinguish between metals, non-metals and metalloids
- identify similar properties of groups of elements in the periodic table

Compounds and mixtures

- distinguish between elements, compounds and mixtures
- recall that the atoms in compounds are bonded very tightly together
- recall that elements can be separated from compounds only through a chemical reaction
- recognise that the properties of compounds are different from the elements that make them up
- use the formulas of simple compounds to identify the elements that make them up
- model the arrangement of atoms in the molecules of some compounds
- identify and describe some common compounds and their uses
- explain why elements and compounds can be represented by unique chemical formulas whereas mixtures cannot

Science as a human endeavour

- explain how the ideas about elements and the atom have changed over time
- describe the contributions of some of the scientists who have added to our knowledge of the atom and the elements
- recognise the impact of new scientific discoveries and technology on our understanding of the atom, elements and compounds

Activity 7.1	Activity 7.2	Activity 7.3
Investigating substances	Analysing substances	Investigating substances further
doc-6069	doc-6070	doc-6071

learn on ONLINE ONLY

7.11 Review 1: Looking back

To answer questions online and to receive **immediate feedback** and **sample responses** for every question, go to your learnON title at www.jacplus.com.au. *Note:* Question numbers may vary slightly.

Remember

1. About 2500 years ago, Democritus suggested what all substances were made up of.
   (a) In what way was Democritus' idea about substances the same as the model that scientists currently use to describe substances?
   (b) Suggest why most thinkers of the time disagreed with Democritus.
2. Copy and complete the following table, which describes the structure of atoms.

Part of atom	Location	Size and weight (relative)	Electric charge
		Large	Positive
Neutron			
	Outside the nucleus		

3. If a neutral atom has 12 protons, how many electrons does it have?
4. What takes up most of the space in an atom?
5. Identify the one feature that every single atom of the element sodium has in common.
6. What is the atomic number of each of the following elements?
   (a) Hydrogen   (b) Carbon   (c) Uranium
7. How many protons does each of the elements listed in question 6 have in its nucleus?
8. How many electrons does each of the elements listed in question 6 have in its nucleus?
9. Make a copy of the diagram of the atom at right and label an electron and the nucleus. Answer the following questions.
   (a) How many protons does this atom have?
   (b) How many neutrons does this atom have?
   (c) How many electrons does this atom have?
   (d) What is the atomic number of this atom?
   (e) Describe one use of the element that is made up of these atoms.
10. Complete the following table to summarise what you know about metals and non-metals.

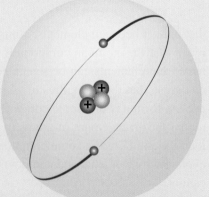

	Metals	Non-metals
Conduct electricity well		
Conduct heat well		
Surface features		

	Metals	Non-metals
State at room temperature		
Malleable		
Ductile		
Brittle		

11. Which of the elements iron, lead, hydrogen, oxygen, silicon, uranium and sodium are:
    (a) metals
    (b) metalloids
    (c) non-metals?
12. (a) Which element is used inside illuminated signs like the one below?

    (b) To which group in the periodic table does this element belong?
13. What event must take place in order to separate a compound into separate elements?
14. How are the molecules in polymers different from the molecules of other compounds?
15. Complete the table below to indicate whether the substances listed are elements, compounds or mixtures. Also indicate why you made that decision.

Substance	Element, compound or mixture	Why do you think so?
Gold		
Diamond		
Carbon dioxide		
Air		
Sea water		
Pure water		
Iron		
Ammonia		
Table salt (NaCl)		

16. Which of 'the bits that matter' is represented by each of the cartoons below?

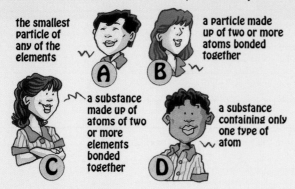

the smallest particle of any of the elements — A

a particle made up of two or more atoms bonded together — B

a substance made up of atoms of two or more elements bonded together — C

a substance containing only one type of atom — D

17. Explain why, unlike elements and compounds, mixtures cannot be represented by chemical formulas.

18. What do diamonds, the 'lead' in pencils and coal have in common?

19. Each of the diagrams below represents one of 'the bits that matter' that make up substances.

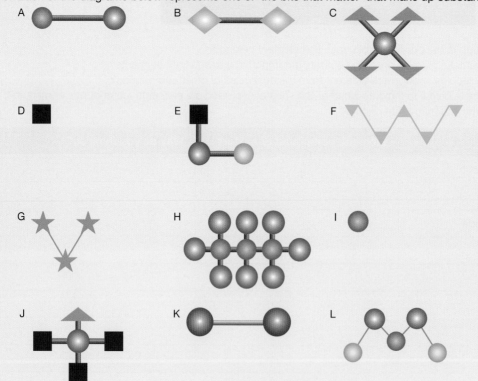

A   B   C

D   E   F

G   H   I

J   K   L

Which of the diagrams represents:
(a) an atom of an element
(b) a molecule of an element
(c) a molecule of a compound?

20. Most of the substances around you are compounds and mixtures.
    (a) What differences could be observed between a mixture of hydrogen and oxygen, and a compound of hydrogen and oxygen?
    (b) Explain the difference between a compound and a mixture in your own words.
21. Respiration is a chemical reaction in which carbon dioxide is produced.
    (a) Where in your body does respiration take place?
    (b) What is released during respiration apart from carbon dioxide?
    (c) Suggest how the carbon atoms in carbon dioxide enter your body.
22. Why doesn't water appear in the periodic table?

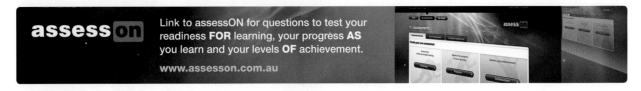

Link to assessON for questions to test your readiness **FOR** learning, your progress **AS** you learn and your levels **OF** achievement.

www.assesson.com.au

learn on RESOURCES — ONLINE ONLY

Complete this digital doc: Worksheet 7.5: Summing up (doc-18748)

TOPIC 8
Chemical change

8.1 Overview

Chemical reactions are happening everywhere. Chemical reactions in your body digest food, decay your teeth and much more. Chemical reactions occur in batteries to provide electricity, in the oven when you bake a cake, in your hair when it is bleached or coloured, and in your car when it burns fuel. Explosions are very fast chemical reactions.

8.1.1 Think about reactions

assess on

- Why does a half-eaten apple go brown?
- How is an explosion different from other chemical reactions?
- Why does a spoonful of sugar dissolve more quickly than a sugar cube?
- What makes a nail rust?
- Why is the Sydney Harbour Bridge continually being painted without a break?
- What is a backdraught and what causes it?
- What is the difference between soaps and detergents?
- What makes Lycra® so special?
- Why is recycling so important?

Numerous **videos** and **interactivities** are embedded just where you need them, at the point of learning, in your learnON title at www.jacplus.com.au. They will help you to learn the concepts covered in this topic.

8.1.2 Your quest

What is a chemical reaction?

What is a chemical reaction — and how do you know whether a chemical reaction has taken place?

Check out the images on this page and answer the questions based on what you already know about chemical reactions.

The boiling liquid in the pot at right is changing colour. It began as a mixture of reds, yellows and blues and, after stirring, is changing into a dangerous-looking green soup.

1. Write down your opinion about whether or not a chemical reaction is taking place.
2. Explain how you know whether a chemical reaction has taken place.
3. Is there a chemical reaction taking place underneath the pot? Explain your answer.
4. Clouds are forming above the pot. Is this evidence of another chemical reaction? Explain your answer.
5. Runners in long-distance races sweat heavily. The water lost due to sweating evaporates from the skin. Is this evaporation an example of a chemical reaction? Explain your answer.

It was a long, tough and hot cross-country race. Just as well this runner drank lots of water along the way.

6. Does a chemical reaction take place when you burn toast? What observations support your answer?
7. Does a chemical reaction take place when you toast bread without burning it? Explain your answer.
8. Is the frozen substance in the man's beard below the result of a chemical reaction? Explain your answer.

Oh, no! The toast is burnt again!

Antarctic weather is freezing — REALLY freezing!

8.2 Patterns, order and organisation: Physical and chemical properties

8.2.1 Physical or chemical?

Thousands and thousands of different substances are used in the objects that surround you. Each substance shown on the next page has physical and chemical properties that make it useful for a particular purpose.

The properties of most substances fall into two categories — physical or chemical. **Physical properties** are those that you can either observe using your five senses — seeing, hearing, touching, smelling and tasting — or measure directly. Examples include colour, size, shape, texture, temperature, malleability and ductility, but there are many, many more. **Chemical properties** are those that describe how a substance combines with other substances to form new chemicals or how a substance breaks up into two or more different substances. Examples of chemical properties include **flammability**, **reactivity** and **toxicity**.

- Flammability is an indicator of how easily a substance catches fire. When a substance burns, it creates new substances.
- Reactivity is a measure of how easily a substance combines with other substances to produce new substances.
- Toxicity refers to the danger to your health caused when poisonous substances combine with chemicals in your body to produce new substances and damaging effects.

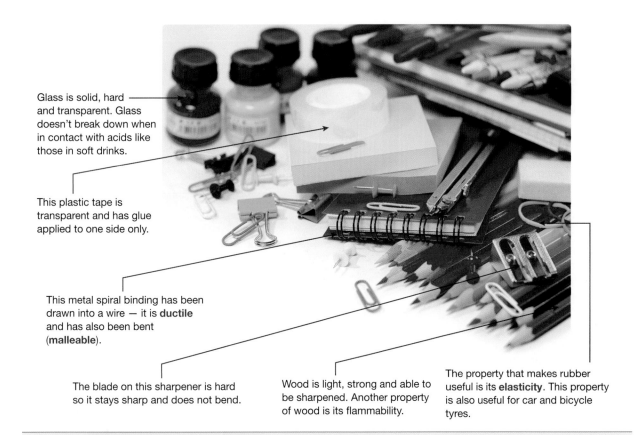

Glass is solid, hard and transparent. Glass doesn't break down when in contact with acids like those in soft drinks.

This plastic tape is transparent and has glue applied to one side only.

This metal spiral binding has been drawn into a wire — it is **ductile** and has also been bent (**malleable**).

The blade on this sharpener is hard so it stays sharp and does not bend.

Wood is light, strong and able to be sharpened. Another property of wood is its flammability.

The property that makes rubber useful is its **elasticity**. This property is also useful for car and bicycle tyres.

INVESTIGATION 8.1

Checking out properties

AIM: To describe the physical properties of a variety of substances

Materials:
a range of small items that might include a tennis ball, a table-tennis ball, a table-tennis paddle, a dishwashing sponge, assorted fabrics (for example, wool from a jumper, nylon socks and stockings, polyester, cotton), a magnifying glass or lens, a roll of sticky tape, a candle, paper clips, small springs, polystyrene cups, foam rubber, aluminium foil, a clear plastic bottle of dishwashing detergent, a bottle of perfume

Method and results

• Work in groups of three or four so that you can discuss the properties. Work on one item at a time.
1. For each item, list all its physical properties that you can think of. Some items will consist of more than one substance. In those instances, list the physical properties of each substance.
2. For each physical property of each substance in the item, explain how that property makes the substance useful for its purpose.

Discuss and explain

3. List tests that you could perform to discover some of the chemical properties of the substances in some of the items.

8.2 Exercises: Understanding and inquiring

To answer questions online and to receive **immediate feedback** and **sample responses** for every question, go to your learnON title at www.jacplus.com.au. *Note:* Question numbers may vary slightly.

Remember

1. Most metals can be described as ductile and malleable. How does each of these properties make metals useful?
2. Which of the two categories of properties can be directly observed with your five senses?
3. Some substances have the chemical property of toxicity. What does this mean?

▶

Think

4. What property of leather makes a soccer ball easy to grip?
5. List the properties of the packaging of potato chips.
6. List at least two physical properties that you can describe using your sense of hearing.
7. Three chemical properties are listed on the previous page — can you think of another one? Explain that chemical property using an example.
8. Look at the image on the previous page. Choose three items and explain how the properties of the materials make them suitable for their purpose.

Why are potato chip bags pumped full of air? Why is foil often used for the packaging? Could a different material be used?

Imagine

9. Imagine that you are designing a spacecraft that will take astronauts to the Moon and back. List the properties that the outer surface of the spacecraft would need to have. Include at least two chemical properties.

Investigate

10. Road bike frames for serious and competitive cyclists are made from aluminium, carbon fibre or a combination of both. Find out:
    (a) what properties both aluminium and carbon fibre have that make them suitable for the frames of road racing bikes
    (b) which properties make aluminium bikes more suitable than carbon fibre for some purposes. (Note that cost is not a property!)

 RESOURCES — ONLINE ONLY

📄 **Complete this digital doc:** Worksheet 8.1: Properties of materials (doc-18749)

8.3 Time for some changes

8.3.1 Chemical changes

When you hard-boil an egg, a **chemical change** takes place. At about 100 °C the eggwhite and yolk undergo chemical changes that alter their chemical make-up. Bonds between atoms or molecules are broken or new bonds between these particles are formed. Unlike cooling melted chocolate, which brings about another physical change, cooling the egg will not change it back to its raw state. In fact, most chemical changes are difficult to reverse.

When paper is burnt, it combines with oxygen to form ash and smoke. This is a chemical reaction, because new substances are formed. Burning gas in a Bunsen burner is also a chemical change. The methane gas burns with oxygen in the air to form two new substances: carbon dioxide and water vapour. During this chemical reaction heat is also produced.

8.3.2 How does a candle burn?

When you try to light a piece of solid wax it melts, but does not burn. If solid wax doesn't burn, how does a candle burn? Is it the string wick that is in the middle of the candle that burns? String will burn, but it doesn't burn like a candle does. How then does a candle burn?

When you light the wick of a candle, the wax at the top of the candle melts. The molten wax is drawn up the wick just as water soaks into a paper towel. As the liquid wax flows up the wick and gets closer to the heat of the flame it **evaporates**. The wax vapour mixes with oxygen in the air and burns.

When a glass is placed over a candle, initially it continues to burn, using up the oxygen in the glass and producing carbon dioxide.

A few seconds later, all the oxygen is used and the candle goes out.

If you enjoy eating chocolate you'll know that it's not so easy to eat on a hot summer's day. Energy transferred from the hot air surrounding the chocolate causes it to melt. The chocolate changes **state** from solid to liquid. The chocolate's change in state is reversible. The melted chocolate can be cooled and solid chocolate will form again.

INVESTIGATION 8.2

A burning candle

AIM: To observe and describe the changes that take place when a candle burns

Materials:

safety glasses   candle

jar lid   matches

heatproof mat

Method and results

- Place a jar lid on a heatproof mat.
- Light a candle and allow a drop of wax to drip onto the lid. Place the candle on the drop of wax and fix it to the lid.
1. Observe the candle and write down as many observations of the burning candle as you can.
   - Discuss your observations with others in your group.
   - Blow out your candle and you will see a white vapour rising from the top of the wick.

CAUTION

Do not smell the vapour directly. Fan the odour to your nose with your hand.

- To confirm that the white vapour is not smoke, carry out the following test: Relight the candle. Once it is burning properly, blow it out. Quickly light the top of the vapour trail. The flame should run down the vapour to the wick and relight the candle.

Discuss and explain

2. How far is the flame from the solid wax?
3. The solid wax forms a little pool of liquid wax around the wick. Why does this happen?
4. Describe the odour of the vapour that is present after the candle is blown out.
5. Draw a diagram of a candle and its flame. Label this diagram to explain how a candle burns.
6. Explain why lighting the wax vapour causes the candle to relight.

8.3.3 Physical changes

The changes to the chocolate described on the previous page are **physical changes**. Melting, evaporation, condensation and freezing are all physical changes. Changes of state are reversible physical changes.

Changes in the shape or size of a substance are also physical changes. These are not always reversible. For example, if you drop an egg, its shape is changed forever. But when you stretch an elastic band, it can quickly return to its original shape when you let it go.

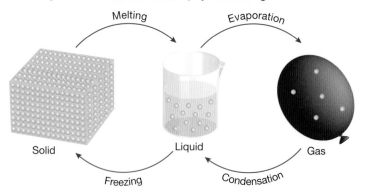

Changes of state are reversible physical changes.

A physical change does not break any bonds between the atoms of a substance, nor does it create any new bonds. No new substances are formed.

8.3.4 Describing change

In a burning candle, there are both physical and chemical changes. The melting of the solid wax to form liquid wax and the evaporation of liquid wax to form wax vapour are physical changes. The burning of the wax vapour is a chemical change. The wax vapour reacts with oxygen in the air to form new substances including carbon dioxide and ash.

Physical and chemical changes can be described using word equations.

Melting chocolate can be described by the equation:

$$\text{solid chocolate} \longrightarrow \text{liquid chocolate}$$

The burning of paper can be described by the equation:

$$\text{paper} + \text{oxygen} \longrightarrow \text{smoke} + \text{ash}$$

8.3 Exercises: Understanding and inquiring

To answer questions online and to receive **immediate feedback** and **sample responses** for every question, go to your learnON title at www.jacplus.com.au. *Note:* Question numbers may vary slightly.

Remember

1. What is the difference between a physical and a chemical change?
2. Describe two examples of a physical change.
3. Describe two examples of a chemical change.
4. Match the word with the definition.

Change from solid to liquid	Freezing
Change from gas to liquid	Melting
Change from liquid to solid	Condensation
Change from liquid to gas	Evaporation

5. Which type of physical change can always be reversed by heating or cooling?

Think

6. Copy and complete the table below.

Observation	Physical or chemical change
Water freezing to form snow	
A cake cooking	
Lighting the gas on the stove	
Petrol evaporating at the petrol pump	
Lighting a match	
Steam condensing on the bathroom mirror	
Melting gold to cast gold bars	
Dynamite exploding	
Bleaching a stain	
Dissolving eggshell in acetic acid	

7. Write two word equations to describe the changes of state that take place when a candle burns.
8. Write a word equation to describe the chemical change that takes place when a candle burns.
9. When you hard-boil an egg, the inside of the egg gets hard. Why is this a chemical change and not a physical change?

Create

10. Candles are a good example of both physical change and chemical change. Write a poem about a candle burning.

8.4 Chemical reactions

8.4.1 Chemical reaction

A **chemical reaction** is a chemical change in which a completely new substance is produced.

Almost all the products you use or wear each day are made by chemical reactions, from cosmetics to concrete, plastics to paper, glass to graphite, stainless steel to shampoo, fibres to food additives, margarine to medicines and many, many more.

8.4.2 Feeling hungry?

A hamburger is an incredible mixture of chemicals. Every part of it has been produced by chemical reactions. The most important chemical reaction in growing the lettuce is photosynthesis, in which the reactants are carbon dioxide and water. The products are glucose (a type of sugar) and oxygen. That chemical reaction cannot take place without light and a chemical called chlorophyll, which gives plants their green colour. In fact, none of the other components of the hamburger could be grown or produced without photosynthesis.

The substance used to make cheese is the product of a chemical reaction in which a protein in cow's milk called casein reacts with acetic acid when heated. Acetic acid is found in orange and lemon juice and is more commonly known as vinegar.

8.4.3 Reactants and products

The substances that you begin with in a chemical reaction are called the **reactants**; the substances that are produced are called the **products**. When you wash the dishes, a chemical reaction occurs between the detergent and the mess on the dishes. When you shampoo your hair, some of the chemicals in the shampoo react with the greasy substances on your scalp that contain dust, dirt and tiny organisms such as bacteria that can make your hair unhealthy.

> **WHAT DOES IT MEAN?**
> The word *product* comes from the Latin word *productum*, meaning 'thing produced'.

When water is added to dried copper sulfate, it turns blue — but has a chemical reaction taken place? No reaction has occurred since the solid turns white again when dried.

Where's the evidence?

You can usually tell whether a chemical reaction has taken place by identifying one or more of these clues:
- a **precipitate** (cloudiness caused by a solid substance) appears in a liquid or gas
- an odour is detected
- bubbles appear
- there is an increase or decrease in temperature

- light is emitted or a flame appears
- there is a change in colour.

However, the only way to be certain that a chemical reaction has taken place is to identify one or more new chemical products.

8.4.4 Chemical reaction experiments

Before you start each of the following four investigations, design a suitable table for recording your observations.

As you perform the experiments:
1. Make a note of the appearance of each of the reactants you start with.
2. Carry out the experiment and observe carefully to detect any changes that occur.
3. Describe the changes that take place and products of the reaction.

Safety glasses should always be worn during experiments involving chemical reactions.

INVESTIGATION 8.3

Heating copper carbonate

AIM: To observe and record the chemical reaction that occurs when copper carbonate is heated

Materials:
Bunsen burner, heatproof mat and matches
safety glasses
test tube, test-tube rack and test-tube holder
spatula
copper carbonate powder

Method and results
- Pour 2 spatulas of copper carbonate into the test tube.
- Using the test-tube holder, heat the test tube in the Bunsen burner flame. Remember to move the test tube in and out of the flame and point it away from people.
- Stop heating when the copper carbonate has changed colour.
1. Record your observations.

INVESTIGATION 8.4

Magnesium metal in hydrochloric acid

AIM: To observe and describe the chemical reaction between magnesium and hydrochloric acid

Materials:
heatproof mat
safety glasses
test tube and test-tube rack
1 cm piece of magnesium ribbon
dropping bottle of 0.5M hydrochloric acid

Method and results
- Put the magnesium in the test tube.
- Add 20 drops of hydrochloric acid to the test tube.

CAUTION

The test tube may become quite hot

---

1. Record your observations.

Sodium sulfate and barium chloride

AIM: To observe and describe the chemical reaction between sodium sulfate and barium chloride

Materials:
heatproof mat
safety glasses
test tube and test-tube rack
test-tube holder
dropping bottle of 0.1M sodium sulfate solution
dropping bottle of 0.1M barium chloride solution

Method and results

- Add 20 drops of the sodium sulfate solution carefully to the test tube.
- Add 20 drops of the barium chloride solution carefully to the test tube.

1. Record your observations.

Discuss and explain

2. What observation provides evidence that a chemical reaction has taken place? Explain your reasoning.

Steel wool in copper sulfate solution

AIM: To observe and record the chemical reaction between steel wool and copper sulfate

Materials:
heatproof mat
safety glasses
test tube and test-tube rack
glass stirring rod
1 cm ball of steel wool
dropping bottle of 0.5M copper sulfate solution

Method and results

- Put the steel wool in the test tube, using the glass stirring rod to push it gently to the bottom of the test tube.
- Add copper sulfate solution to the test tube to a depth of 2 cm.

1. Record your observations.

Discuss and explain

2. What observation provides evidence that a chemical reaction has taken place?

8.4.5 Writing word equations

Each of the chemical reactions in Investigations 8.3–8.6 can be described by a chemical word equation. In each case the reactants are on the left side of the equation and the products are on the right side.

1. When magnesium metal reacts with hydrochloric acid, hydrogen gas and magnesium chloride are formed:

   magnesium + hydrochloric acid $\longrightarrow$ hydrogen + magnesium chloride

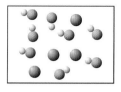

1. Magnesium is placed into hydrochloric acid.

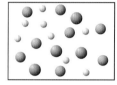

2. Bonds between hydrogen and chlorine atoms break.

3. New bonds form. Chlorine atoms bond to magnesium atoms to form molecules of magnesium chloride, while hydrogen atoms bond together to form molecules of hydrogen gas.

KEY

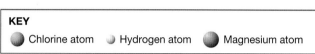

Chlorine atom   Hydrogen atom   Magnesium atom

2. Heating copper carbonate forms copper oxide and carbon dioxide:

$$\text{copper carbonate} \xrightarrow{\text{heat}} \text{copper oxide} + \text{carbon dioxide}$$

Although heat is required for this chemical reaction to take place, it is not a substance and therefore is not a reactant. It is written above the arrow for this reason.

3. Sodium sulfate and barium chloride in solution react to form solid barium sulfate and sodium chloride, which remains dissolved in the solution:

$$\text{sodium sulfate solution} + \text{barium chloride} \longrightarrow \text{solid barium sulfate} + \text{sodium chloride solution}$$

4. Steel wool (which is made of iron) dissolves in copper sulfate solution to form iron sulfate solution and copper metal:

$$\text{iron} + \text{copper sulfate solution} \longrightarrow \text{iron sulfate solution} + \text{copper}$$

8.4 Exercises: Understanding and inquiring

To answer questions online and to receive **immediate feedback** and **sample responses** for every question, go to your learnON title at www.jacplus.com.au. *Note:* Question numbers may vary slightly.

Remember

1. Write down four observations that could provide evidence that a chemical reaction has taken place.
2. When magnesium metal reacts with hydrochloric acid, hydrogen gas and magnesium chloride are formed.
   (a) What are the products?
   (b) What are the reactants?
3. What is the only real proof that a chemical reaction has taken place?

Think

4. Write word equations for the following chemical reactions.
   (a) Octane gas is burnt with oxygen in a car engine to produce carbon dioxide and water.
   (b) Sodium metal reacts with chlorine gas to form sodium chloride.
   (c) Hydrogen gas and oxygen gas combine to form water.
   (d) Zinc metal dissolves in hydrochloric acid to form hydrogen gas and zinc chloride.
5. Explain why the reaction that takes place when copper carbonate is heated is called a decomposition reaction.

6. Explain why the tomato, cheese, bread and meat in a hamburger cannot be grown or produced without photosynthesis.
7. Describe the evidence that one or more chemical reactions takes place when meat is grilled.

Create
8. Performing some chemical reactions can be dangerous. Design a safety poster for one of the experiments you have done.
9. Choose one of the materials below and find out how it is manufactured. Write a report about the chemical reactions used in its production.
   - Glass
   - Soap
   - Margarine
   - Paper
   - Nylon
   - Polyethylene

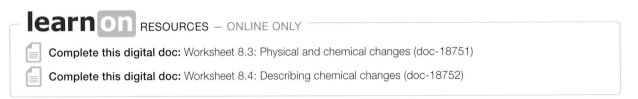

learn on RESOURCES — ONLINE ONLY

Complete this digital doc: Worksheet 8.3: Physical and chemical changes (doc-18751)

Complete this digital doc: Worksheet 8.4: Describing chemical changes (doc-18752)

8.5 Fast and slow reactions

8.5.1 Explosions

Explosions are chemical reactions that take place very quickly. Explosions also release a lot of heat, light and noise. In less than 10 milliseconds, a dynamite blast in a large mine can produce 5 billion litres of gas and release 20 billion joules of energy — enough energy to tear any rock apart.

In contrast, the chemical reactions that cause concrete to set are very slow. It can take several days for concrete to set hard. Rusting is another example of a slow chemical reaction.

The **reaction rate** is a measure of how quickly a chemical reaction occurs. How can the rate of a reaction be changed to make a slow reaction happen quickly or make a fast reaction slow down?

Explosions are fast chemical reactions.

WHAT DOES IT MEAN?
The word *explosion* comes from the Latin word *explosio*, meaning 'driven off by clapping or hooting'.

8.5.2 Speeding up a reaction with heat

Heating a substance adds energy to its particles. They move more rapidly and collide more frequently. When they collide, bonds between the particles are broken and new ones are more easily formed with the particles of other substances. Heating substances, therefore, usually causes the rate of a chemical reaction to increase.

8.5.3 Stay cool

Food 'goes off' because micro-organisms cause chemical reactions in the food that make it rot. These chemical reactions can be slowed by lowering the temperature of the food. Imagine what life would be like without a refrigerator or freezer.

8.5.4 Catalysts

A **catalyst** is a chemical that can speed up a chemical reaction but is still present once the reaction has finished. Catalysts are not reactants because they are not changed by the reaction.

Catalytic converters in car exhausts use a precious metal, such as platinum, as a catalyst. This enables nitrogen oxide to react with toxic gases, such as carbon monoxide, to form the less harmful carbon dioxide and nitrogen gases; this reaction would not occur in the absence of the catalyst.

This reaction in a catalytic converter can be shown as:

$$\text{carbon monoxide} + \text{nitrogen oxide} \xrightarrow{\text{platinum}} \text{carbon dioxide} + \text{nitrogen}$$

The catalysts in living things are called enzymes. Enzymes in the human body help to digest the food you eat more quickly.

Apples and other fruits go brown because chemicals in them, called phenolics, react with oxygen in the air. The brown chemical products are called quinones. Enzymes in the fruit speed up the reaction. The chemical word equation for this reaction is:

$$\text{phenolics} + \text{oxygen} \xrightarrow{\text{enzymes}} \text{quinones}$$

Apples go brown when phenolics react with oxygen in the air. Enzymes speed up the reaction.

8.5.5 Altering the reactants

No doubt you have been in situations where you wanted to increase the rate of a chemical reaction. Perhaps you wanted a camp fire to burn faster, a tablet to dissolve or a stain to be removed more quickly. What would you do in each case to make the reaction faster?

One solution is to add more reactant. In the case of the camp fire you can add more wood, or more oxygen by fanning the fire. To make the stain disappear more quickly, you could add more bleach.

Another solution is to increase the surface area of the reactants so that they can mix more easily. In the case of the camp fire, you can chop the wood or use smaller pieces of twigs and leaves. To help the tablet dissolve, you could crush it.

Granular sugar dissolves faster than a sugar cube because it has a larger surface area.

The effects of amount of reactants and surface area on the rate of reaction

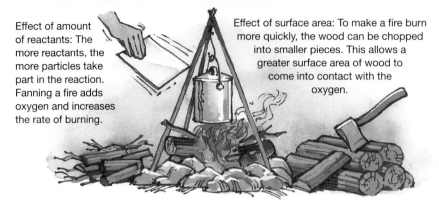

Effect of amount of reactants: The more reactants, the more particles take part in the reaction. Fanning a fire adds oxygen and increases the rate of burning.

Effect of surface area: To make a fire burn more quickly, the wood can be chopped into smaller pieces. This allows a greater surface area of wood to come into contact with the oxygen.

INVESTIGATION 8.7

The effect of temperature on a reaction

AIM: To investigate the effect of temperature on the rate of a chemical reaction

Materials:

safety glasses
test-tube holder
heatproof mat
Bunsen burner
matches

marble chips
dropping bottle of 1M hydrochloric acid
test tube
test-tube rack

Method and results

- Carefully slide one or two marble chips to the bottom of the test tube.
- Add 1M hydrochloric acid to half-fill the test tube.
- Observe the reaction.
- Now gently heat the test tube and observe the reaction.
1. Has a chemical reaction occurred? Describe the evidence that you observed.

Discuss and explain

2. What effect did heating the test tube have on the rate of this reaction?
3. In this chemical reaction, the calcium carbonate that makes up the marble chips reacts with the hydrochloric acid to produce calcium chloride, water and carbon dioxide gas.
   (a) List the reactants.
   (b) List the products.
   (c) Write a word equation for this chemical reaction.
4. Suggest a different method of increasing the rate of reaction.

HOW ABOUT THAT!

Have you ever had a composite resin filling in your tooth? The dentist uses blue light or ultraviolet (UV) radiation to set this type of filling. The visible or UV light speeds up the reactions that cause the materials in the filling to harden. Without the UV light, you would be waiting for hours for this type of filling to set.

UV light can speed up the setting of a composite resin filling.

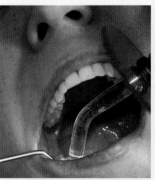

INVESTIGATION 8.8

Changing the rate of reaction

AIM: To investigate the factors that affect the rate of chemical reactions

Materials:

safety glasses
heatproof mat
test tubes and test-tube rack
white chalk
mortar and pestle

spatula
0.5M hydrochloric acid
1M hydrochloric acid
measuring cylinder

Method and results

Hydrochloric acid reacts with chalk to produce carbon dioxide gas, water and calcium chloride.

- Place a small amount of chalk in a test tube and add enough hydrochloric acid to cover it. Observe the chemical reaction.
- Discuss with your partner how you could use this reaction to demonstrate one of the hypotheses below.
  (a) Increasing the concentration or amount of reactants will speed up a chemical reaction.
  (b) Increasing the surface area of reactants will speed up a chemical reaction.
  (c) Decreasing the concentration or amount of reactants will slow a chemical reaction.
1. Design your experiment and write down the method.
2. Predict the results you would expect to obtain that would support the hypothesis you chose.
  - Perform the experiment.
3. Prepare a report of your findings.

8.5 Exercises: Understanding and inquiring

To answer questions online and to receive **immediate feedback** and **sample responses** for every question, go to your learnON title at www.jacplus.com.au. *Note:* Question numbers may vary slightly.

Remember

1. What is the rate of a reaction?
2. State five different methods of changing the rate of a reaction.
3. How does heating increase the rate of a reaction?
4. What is a catalyst?
5. How do you know that a catalyst is not a reactant?
6. What is an enzyme?

Think

7. Does a refrigerator stop food from rotting or does it just slow the rotting? Explain your answer.
8. What is the point of adding catalysts to washing powders?
9. Which will dissolve more quickly — a sugar cube or the same amount of sugar in a teaspoon? Explain your answer.

Create

10. Write and act out a short dramatic performance to show how a piece of chalk takes longer to react with an acid than the same amount of crushed chalk.

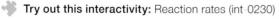

8.6 Rusting out

8.6.1 Corrosion

Rusting is an example of **corrosion**. Corrosion is a chemical reaction that occurs when substances in the air or water around a metal 'eat away' the metal and cause it to deteriorate.

There are many examples of corrosion: silver tarnish; the green film that forms on copper or brass objects; and, the most common one, the rusting of iron. Corrosion causes enormous damage to buildings, bridges, ships, railway tracks and cars.

8.6.2 Rust

Rust is the flaky substance that forms when iron corrodes. Iron reacts with water and oxygen in the air to form iron oxide and other iron compounds that make up the familiar red-brown substance known as rust. Rusting is a slow chemical reaction that can be represented by the following word equation:

The Sydney Harbour Bridge is continually painted to protect it from moisture and the air, which would cause its steel girders to rust.

iron + water + oxygen ⟶ rust

Even strong buildings and bridges that are made from steel, an alloy of iron, are weakened by rusting. The Sydney Harbour Bridge, for example, is continually painted to protect it from moisture and the air, which would cause its steel girders to rust. Ships and cars are also constructed largely of steel. Despite the strength of steel, it needs to be protected from the corrosive effects of the environment.

INVESTIGATION 8.9

Observing rusting

Steel wool is made from iron. You can observe rusting of the iron in steel wool by performing the following experiment.

AIM: To observe and describe the rusting of steel wool

Materials:
Petri dish
water
steel wool (without any soap)
small glass
permanent marker

Method and results
- Pour some water into the Petri dish.
- Place the steel wool in the middle of the Petri dish.
- Cover the steel wool by placing the glass over it upside-down.
- Mark the level of the water on the outside of the glass with a permanent marker.
- Leave for several days, adding water as required to keep the level at the mark on the glass.
1. Construct a table in which you can record your observations over several days.

Discuss and explain
2. What did you observe about the level of water inside the glass? Can you explain why this happened?
3. Write down a word equation for the chemical reaction that occurred inside the glass.

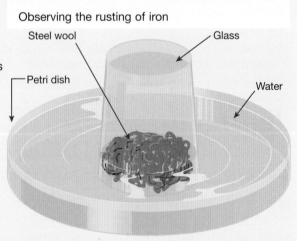

Observing the rusting of iron

Steel wool

Glass

Petri dish

Water

8.6.3 Speeding up rusting

Some substances in the environment make rusting happen much more quickly. One of the most effective of these is salt. Steel dinghies that are used in the ocean rust much faster than those that are used only in fresh water. This is because the salt in the sea water speeds up the reaction between oxygen in the air and the iron in the steel.

Some chemicals released from factories also increase the rate of rusting. A CSIRO study conducted in Melbourne found that rusting rates were high near airports and sewage treatment plants.

Rusting is much slower in dry environments like deserts, where the rainfall is nearly zero and there is very little water vapour in the air.

HOW ABOUT THAT!

In the Mojave Desert of Southern California, hundreds of unused aircraft are stored out in the open air. Due to the dry air, rusting occurs extremely slowly. As a result, some of the aircraft are still structurally sound after being exposed in the open air for about 20 years.

INVESTIGATION 8.10

Investigating the corrosion of different metals

AIM: To investigate the corrosion of a variety of metals

Materials:
small strips of a range of metals such as copper, aluminium, zinc and magnesium
sandpaper
other equipment approved by your teacher

Method and results

- Design and carry out an investigation to study the resistance of a selection of different metals to corrosion. Ensure that appropriate variables are controlled. Before commencing, clean the metal strips with sandpaper to ensure that any coatings already caused by corrosion are removed.
1. Write a report on your investigation that includes your aim, method, results (including a table), discussion and a clear conclusion listing the metals in order of resistance to corrosion, from most resistant to least resistant. Include the answers to the questions below in your discussion.

Discuss and explain

2. Identify the independent and dependent variables in your investigation.
3. Name the variables that you controlled.
4. Suggest how you might be able to improve or speed up the investigation.

8.6.4 Rust protection

The layer of rust that forms on an iron object flakes off the metal, allowing air and moisture to get through to the iron below. This causes more rusting to occur and eventually the iron becomes a heap of rust. It is important to protect iron and steel from corrosion, especially if they are part of a bridge or the hull of a ship.

There are several ways to protect iron and steel from rusting. One way is to prevent oxygen or moisture from contacting the metal. This is called **surface protection**. The metal can be protected by coating it with paint, plastic or oil. If the surface protection becomes scratched or worn off, the metal below can be attacked by moisture and oxygen, and rusting will occur. Examine the painted surface of an old car. Wherever the paint has chipped off you will find that corrosion has occurred and rust can be seen.

Another way to protect iron from rusting is to coat it with a layer of zinc. This is called **galvanising**. Zinc is a more reactive metal than iron, and in the presence of moisture and oxygen the zinc layer corrodes, leaving the iron unaffected. Many roofing materials and garden sheds are made from galvanised iron. You can also buy galvanised nails.

This wrecked car has rusted quickly because of its proximity to the sea.

INVESTIGATION 8.11

Rusting and salt water

AIM: To investigate the effect of salt water on the rate of rusting

Materials:
test tubes and test-tube rack
measuring cylinder
iron nails
water
salt (sodium chloride)

Method and results

- Design an experiment to test the effect of the saltiness of water on the time taken for an iron nail to rust.
- Propose a hypothesis.
- Discuss your experimental design with a partner. You will need to consider which conditions must be kept the same and which condition will be varied.
- You will need to set up a control test tube. Find out what the purpose of a control is.
1. Write down your method. It should be clear enough for someone else to follow without any help.
2. Construct a table in which to record your observations over the next few days.

Discuss and explain

3. What effect did salt have on the time taken for the iron nail to rust?
4. Was your hypothesis supported?
5. How do your results compare with those of others in your class?
6. Write a report of your findings.

8.6.4 Rusting can be useful

Not all rusting is bad. From a pharmacy you can buy hand warmers, which are commonly used by skiers and campers. These packages will produce heat when you shake them. The contents of the packet include powdered iron, water, salt and sawdust. When the packet is shaken vigorously, the iron rusts quickly, which produces heat.

HOW ABOUT THAT!

City councils face problems caused by the action of dogs on metal lampposts. The corrosive properties of the dogs' urine rusts the steel of the lampposts a few centimetres above the ground.

8.6 Exercises: Understanding and inquiring

To answer questions online and to receive **immediate feedback** and **sample responses** for every question, go to your learnON title at www.jacplus.com.au. *Note:* Question numbers may vary slightly.

Remember

1. What is corrosion?
2. What is rusting?
3. What is surface protection?
4. What is galvanised iron and what advantage does it have over iron?

Think

5. Why does rusting occur faster in coastal regions than in areas further away from the sea?
6. How can galvanising protect iron from rusting when the zinc coating corrodes more quickly than the iron?
7. Suggest why the powdered iron in hand warmers used by skiers and campers rusts much more quickly than an iron nail.

Investigate

8. Corrosion is found in many places. Survey your school for rust spots. Write a report about your findings.
9. If you have access to an old car, survey it carefully and record all its rust spots. Why are some parts of the car more likely to rust?
10. Aluminium corrodes quite quickly, yet it is used to make soft drink cans. Find out why aluminium cans are not corroded by the drinks they store.

 RESOURCES — ONLINE ONLY

Complete this digital doc: Worksheet 8.6: Rusting (doc-18754)

8.7 Burning is a chemical reaction

8.7.1 Oxidation reactions

Burning is a chemical reaction. It involves the combination of oxygen with a fuel and always produces heat and gases. Reactions that involve combination with oxygen are examples of **oxidation** reactions.

There are many other oxidation reactions. The rusting of iron to form iron oxide is an oxidation reaction. Rusting could correctly be described as a very slow type of burning reaction.

8.7.2 Burning fossil fuels

When a **fossil fuel** reacts with oxygen, heat is produced, along with carbon dioxide and water vapour. Fossil fuels are fuels formed from the remains of living things. Petrol, natural gas, coal, wood and even paper are fossil fuels.

The oxyacetylene torch

To obtain temperatures as high as 3000 °C — hot enough to melt iron and weld metals — acetylene fuel is mixed with pure oxygen in an oxyacetylene torch.

acetylene + oxygen ⟶ carbon dioxide + water

The car engine

Burning is also known as **combustion**. Car engines work by the combustion of petrol or gas in the cylinders. A mixture of air and fuel is drawn into each cylinder and ignited by a spark from the spark plug. The fuel reacts rapidly with oxygen in the air. The resulting explosion pushes the piston, which turns the drive shaft. The products of the reaction, carbon dioxide and water vapour, leave the car engine through the exhaust pipe.

An oxyacetylene torch is used in construction work.

HOW ABOUT THAT!

A backdraught occurs when a fire in a closed room dies down because it has been starved of oxygen, but flammable gases continue to stream out of the hot materials in the room. When a door to the room is opened, air is quickly drawn inside, restoring the supply of oxygen and allowing the fire to reignite. The resulting fire consumes all the flammable gases in a few seconds and produces sufficient heat to ignite any remaining materials in the room. This is very dangerous to firefighters.

Rocket fuels

Liquid and solid fuels are used in the NASA rocket program. When these fuels are burnt, they provide sufficient thrust to place a rocket in orbit hundreds of kilometres from Earth. Liquid hydrogen and liquid oxygen react to power the rocket's main engines.

hydrogen + oxygen ⟶ water

Most of the thrust required to place the rocket in its desired orbit comes from chemical reactions in the solid fuel, which is located in the solid rocket boosters. In space, liquid fuel such as hydrazine is oxidised to produce an enormous volume of gas. As the gas is released, the rocket is thrust forward. By controlling the direction of the thrust, it is possible to steer the rocket.

Oxidation reactions provide the thrust to launch a rocket.

INVESTIGATION 8.12

Burning magnesium

AIM: To observe and record the combustion of magnesium

Materials:

safety glasses
tongs
sandpaper
2 cm piece of magnesium ribbon
Bunsen burner, heatproof mat and matches

Method and results

- If the magnesium ribbon is dull, use the sandpaper to remove the dull layer.
- Hold the magnesium ribbon with tongs in the Bunsen burner flame.

CAUTION

Do not look directly at the flame — eye damage may occur.

- After burning the magnesium metal, observe the product that remains.

1. Describe the magnesium metal before burning.
2. During burning, the magnesium reacted with the oxygen in the air by combining with it to form magnesium oxide. Describe the magnesium oxide.

Discuss and explain

3. How do you know that a chemical reaction has taken place?
4. Write a word equation for the chemical reaction.

INVESTIGATION 8.13

Burning paper

AIM: To observe and record the combustion of paper

Materials:

safety glasses
Bunsen burner, heatproof mat and matches
tongs
gas jar
limewater
paper
deflagrating spoon

Method and results

- Place 10 mL of limewater in the bottom of the gas jar.
- Put a ball of scrunched-up paper into the deflagrating spoon.
- Light the paper and lower it into the gas jar.
- When burning has stopped, remove the deflagrating spoon and cover the jar.
- Shake the gas jar and observe the colour of the limewater.

1. What happened to the limewater?

Discuss and explain

2. What gas was given off by the burning paper?
3. Which other substance or substances were produced by the reaction?

8.7 Exercises: Understanding and inquiring

To answer questions online and to receive **immediate feedback** and **sample responses** for every question, go to your learnON title at www.jacplus.com.au. *Note:* Question numbers may vary slightly.

Remember

1. Define the term 'burning'.
2. What evidence is there that burning is a chemical reaction?
3. What is a fossil fuel? List three examples of fossil fuels.
4. List and describe three examples of useful oxidation reactions.
5. Write a word equation for each of the three examples listed in question 4.
6. Is rusting an example of burning? Explain.

Think

7. Complete this word equation:

   fuel + _____ ⟶ _____ + water vapour
8. Name at least one fuel that is not a fossil fuel.
9. What are the three different one-word names given to the chemical reactions in which fuels react with oxygen?
10. Space agency scientists and engineers are constantly searching for better rocket fuels. List the properties that they are looking for. Also, list the properties that are undesirable.

Create

11. Choose one fuel from the list below and prepare a poster on its use. Include in your poster details of where the fuel comes from, what it is used for, and a word equation for its oxidation reaction.
    - methane
    - ethanol
    - butane
    - propane
    - kerosene
    - lignite
    - diesel
    - acetylene

Investigate

12. Fire has always been present in Australia and is a major cause of change in the environment. Find out how Australian Aboriginals have traditionally used fire to benefit themselves and the environment.

learn on RESOURCES — ONLINE ONLY

📄 **Complete this digital doc:** Worksheet 8.7: Combustion (doc-18755)

8.8 A new breed of materials

8.8.1 Fantastic plastic

The scientists and engineers who develop new plastics for spacesuits that allow astronauts to walk in space need a knowledge of chemistry to create materials that are strong, light and heat resistant. Developing new materials for a particular purpose requires an assessment of the required properties and an understanding of chemical reactions.

Metals, paper and ceramics have been used for thousands of years. But plastics have been around for less than 100 years. Plastics are synthetic (manufactured) materials that can be easily moulded into shape. Some plastics are flexible and soften when they are heated. They can be easily moulded into products such as milk and fruit juice containers, rubbish bins, spectacle lenses, electrical insulation and laundry baskets. Others are quite hard and rigid. These plastics are used to make items such as toilet seats, electrical switches, bench tops and outdoor furniture. Most plastics are the products of chemical reactions with crude oil, from which petrol and bitumen are also produced, as the main reactant.

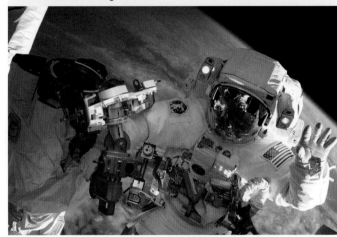

The spacesuits worn by astronauts when they are walking in space contain many layers of materials developed by scientists and engineers.

WHAT DOES IT MEAN?
The word *plastic* comes from the Greek word *plastikos*, meaning 'able to be moulded'.

HOW ABOUT THAT!
Australia was the first country in the world to use only plastic notes for currency. The notes are more difficult to forge and last much longer than the old paper notes.

8.8.2 The clothes you wear

Until the development of nylon in 1938, just in time to make parachutes for World War II, the world relied almost completely on fabrics made from **natural fibres** such as wool, cotton, linen and silk.

Animal-based products include wool from sheep and silk from silkworms. Cotton is derived from cotton bushes and linen comes from flax plants. Today, it would be impossible to provide clothing and bedding for the world's population with purely natural fibres because of the amount of land and water that would be needed for crops and sheep.

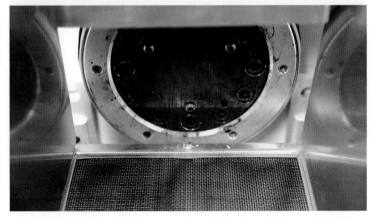

Synthetic fibres form when soft plastic is forced through the holes of a spinneret.

Synthetic fibres such as those used in compression sports gear have many desirable qualities that natural fibres lack, including easy care, colour-fastness and light weight.

Of the many synthetic fibres, the most widely used are **nylon** and **polyester**. Synthetic fibres are made by pushing softened plastic materials through tiny holes in a nozzle called a **spinneret**, which looks a little like a shower head.

Each fibre, whether natural or synthetic, has advantages and disadvantages. Some of these are outlined in the table below.

Fibre	Advantages	Disadvantages
Wool	Warm in cold weather Crease resistant Burns slowly Retains its shape well	Shrinks when washed Turns yellow in sunlight
Cotton	Absorbs moisture Soft Cool in hot weather	Creases easily Burns quickly
Nylon	Dries quickly Light Strong Elastic	Builds up static electricity Melts rather than burns Turns yellow in sunlight
Polyester	Dries quickly Crease resistant Resistant to many chemicals	Builds up static electricity Melts rather than burns

learn on RESOURCES — ONLINE ONLY

▥ **Watch this eLesson:** The future of clothing (eles-0859)

HOW ABOUT THAT!

The spinneret gets its name from the organ used by spiders to spin their webs. Liquid silk flows through the spider's spinneret. It hardens into a fibre as it passes through. Most spiders have six spinnerets.

The spider pictured here has used the fibre to build an intricate web.

INVESTIGATION 8.14

Putting fibres to the test

AIM: To observe and describe the combustion of a range of fibres

Materials:

a range of threads of different fibres, e.g. cotton, polyester, wool, nylon, rayon
uniformly sized fabric samples made from different fibres or blends of fibres
equipment decided upon by the group

Method and results

- Work in groups of three or four to complete this investigation.
1. Start by listing the properties of either the fabric samples or the fibres that can be tested by experiments. Some examples to help get you started include flammability, elasticity and the ability to absorb water.
   - Devise an experiment that will allow you to compare one property of threads of either different fibres or different fabric samples.
2. Make a list of the equipment you will need.
3. Have your experiment plan and equipment list checked by your teacher.

CAUTION

Obtain your teacher's approval before carrying out any tests. Synthetic fibres or blends should be burned only in a fume cupboard.

---

4. Carry out your experiment and keep a record of your measurements and observations.
5. Write a report about your experiment.

Discuss and explain

Include answers to the following questions in the 'Discussion' section of your report.
6. List the variables that you were able to control in your experiment.
7. Identify any variables that you didn't feel the need to control.
8. Identify the most useful properties of each of the fibres or fabrics that you tested.
9. Suggest at least one improvement that you could have made to your experiment.

8.8.3 The best of both worlds

Many of today's fabrics are made from blends of natural and synthetic fibres to make the best of the properties of each fabric in the blend. A blend of polyester and cotton is commonly used for shirts and dresses. The cotton helps keep the wearer cool, while the polyester reduces creasing.

Rayon is shiny, easy to dry and cool in summer. On the 'down' side it has low **durability** and is not **elastic**. Elasticity describes the ability of a material to return to its original size after being stretched. Rayon is neither a natural nor a synthetic fibre. To make it, cellulose fibres from spruce and eucalyptus trees are mixed with chemicals that soften them. The mixture is then passed through a spinneret.

8.8.4 A new breed of fibres

When you watch the feats of Olympic athletes, cyclists, skiers and skaters, it's almost certain that they are wearing Lycra. Lycra is not a fabric. Lycra is the registered trademark of a synthetic fibre called spandex. Spandex was invented in 1958. Spandex is lightweight, durable, retains its shape and fits snugly. It even pulls moisture away from the wearer's skin. Spandex is very elastic. It can be stretched to up to seven times its normal length and spring back to its initial length when released. Spandex is always blended with other fibres. As little as 2 per cent of this material in a blend makes a difference to the properties of the fabric. Lycra suits usually consist of between 3 per cent and 10 per cent spandex.

8.8.5 Cleaning up

Whether you're washing dishes, clothes or yourself, it's obvious that water alone is not good enough. That's because dirt, grease and oil don't dissolve in water. You need to use soaps or detergents which, when mixed with water, loosen or remove soil and greasy and oily substances from surfaces. The dirt and oil can then be rinsed away by the action of running water or a wash cloth.

Cleansing agents like hair shampoo release soil, grease and oil from the scalp. Almost all hair shampoos are detergents.

Comparing soaps and detergents

Soaps and detergents share many properties. Both dissolve grease and oil (a chemical property), both can be found as solids and liquids (a physical property) and both are very, very slippery (another physical property). Most soaps and detergents are biodegradable (a chemical property), which means that they will break down naturally after their journey down the plughole.

The biggest difference between soaps and detergents is the way they are made. Soaps are made from natural fats and oils. No-one is certain about when the first soap was made, but a recipe for a substance that would have resembled soap has been found in the Middle East, dating back to the Babylonian civilisation about 4800 years ago. The ingredients included ashes from fires, oils from trees and sesame seed oil. Detergents, like most plastics, are made from chemicals obtained from crude oil. They are heavily scented with fragrances to disguise the odour, and preservatives are added so that they don't spoil.

8.8.6 The perils of packaging

Just about everything you buy at the supermarket comes in a package. Even if it doesn't, you usually put it in a bag to take it home. The type of packaging needed depends on the properties of the product inside. For example, you can't package tomato sauce in a paper bag. The most commonly used materials in packaging are paper (or cardboard), plastic, metal and glass. For a consumer, it's not just the properties of the packaging that are important. At least two questions should be asked when you make a choice about buying a product:

• Is the packaging recyclable?
• Is the packaging biodegradable?

If the packaging is glass, aluminium or steel, it is probably recyclable, which can save energy and water. If it is a plastic bottle, it is also likely to be recyclable. If the packaging is not recyclable, think about whether it is **biodegradable**; that is, can it be broken down by natural chemical reactions in the bodies of worms or other small **organisms** that live in the soil? Plastics, metals and glass are not biodegradable. If they are thrown out with other household rubbish such as food waste, they end up in rubbish tips and will not break down. This creates the need for more rubbish tips. Of course, there is a limit to how much land can be used for rubbish tips in or near major towns and cities.

Paper is mostly biodegradable. Paper packaging that has been contaminated by food or oils cannot be recycled. But at least when it gets to the rubbish tip it can be broken down in the soil. If you have a choice, choose items with packaging that is either recyclable or biodegradable.

8.8 Exercises: Understanding and inquiring

To answer questions online and to receive **immediate feedback** and **sample responses** for every question, go to your learnON title at www.jacplus.com.au. *Note:* Question numbers may vary slightly.

Remember

1. Identify the single property that all plastics have.
2. From which substance found beneath the ground are most plastics made?
3. Where do all natural fibres come from?
4. Why is woollen clothing popular in winter whereas cotton clothing is popular in summer?
5. Why are cotton and polyester blends so commonly used for shirts and dresses?
6. From which group of materials are synthetic fibres made?
7. What is a spinneret used for?
8. Which fibre do Lycra suits always contain?
9. How do soap and detergents remove grease, oil and dirt from surfaces?
10. What is the major difference between soap and detergents?
11. Identify two properties that soap and detergents have in common.
12. Plastics, metals and glass are not biodegradable.
    (a) What does this mean?
    (b) If non-biodegradable rubbish cannot be recycled, what happens to it?

Think

13. Which properties make plastic more suitable for use in outdoor furniture than:
    (a) wood
    (b) metal?
14. Which properties of plastic currency notes make them more suitable than the old paper ones?
15. Which properties of nylon made it suitable for making parachutes during World War II?
16. Explain why rayon is neither a natural fibre nor a synthetic fibre.
17. Explain why Lycra is not a fabric.
18. How would blending with spandex change the properties of a pure cotton fabric?
19. Which properties make Lycra suitable for:
    (a) the clothes of speed skaters
    (b) toddlers' clothing
    (c) underwear?
20. Why is it important to rinse your hair thoroughly after shampooing?
21. State the properties of the packaging that make it suitable for each of the following supermarket items.
    (a) Tomato sauce in a glass bottle
    (b) Tomato sauce in a red plastic bottle
    (c) Lemonade in an aluminium can
    (d) Lemonade in a clear plastic bottle

Imagine

22. Make a list of at least ten items in your house made from plastic. Imagine that they could no longer be made from plastic. For each item, write down:
    (a) the most important properties that the items must have
    (b) which other material could be used to make them.

Investigate

23. Find out why the group of chemicals known as phosphates is no longer used in most detergents.
24. Research and report on how each of the following materials was discovered or developed, by whom and what they are used for.
    (a) Polyester
    (b) Kevlar®
    (c) PSZ (partially stabilised zirconia)

8.9 Use it again

Science as a human endeavour

8.9.1 Recycling

The material that you throw out as household rubbish is buried in landfill tips.

The food scraps that make up almost half of your household rubbish are biodegradable. They will be broken down by microbes and other decomposers in the soil such as worms. Chemical changes take place when these organisms digest the scraps, returning nutrients to the soil. You can use compost bins or compost heaps to allow this to happen in your own backyard. Even paper and cardboard break down fairly quickly in the soil.

Look closely at this photo. This is not garbage. All these things can be recycled, including cardboard, paper, egg cartons, steel cans, plastic and glass.

However, materials such as plastic, glass and metals take hundreds or even thousands of years to break down. The properties of these materials allow them to be **recycled**.

8.9.2 Looking after your PET

There are two very good reasons for recycling plastics:

- Plastics are non-biodegradable. That is, they are not broken down naturally by micro-organisms. Plastics add thousands of tonnes of new rubbish to the environment every year.
- Plastics are made from oil — a resource that is expensive and dwindling. The continued production of new plastic is not **sustainable**.

Household waste contains many different types of plastic, which need to be separated. The plastics industry has introduced a code system to help consumers identify recyclable plastics. The symbols shown on the next page make the sorting of plastics before recycling easier and cheaper. Some plastics are more easily recycled than others because of differences in the structure of the chains of molecules of which they are made.

PET or PETE (polyethylene terephthalate) is used to make plastic soft-drink bottles. Most commonly known as PET, this plastic is recycled to make carpet fibre and flower tubes. Empty PET bottles completely and remove the lids before placing them in a home recycling bin. The lids are recyclable, but their small size makes the sorting process awkward.

HDPE (high-density polyethylene) is used to make plastic milk and fruit juice bottles. This plastic is recycled to make bottles, crates, pipes, wheelie bins and playground equipment. Empty HDPE bottles completely and remove the lids before placing them in a home recycling bin.

Polyvinyl chloride, more commonly known as PVC, is used to make pipes, fencing and bottles containing substances other than food. It is not easily recycled and can be dangerous to your health and the environment. PVC bottles can be placed in your home recycling bin as long as the lids are removed. They are separated from the more easily recyclable plastics and sent to a separate plant for processing.

LPDE (low-density polyethylene) is used to make wash bottles and other containers. These are recyclable and can be placed in your recycling bin. LPDE is also used to make supermarket plastic bags, which should not be placed in your home recycling bin. These bags interfere with the automatic sorting machines in recycling plants. It is best to avoid using them by using reusable bags for shopping.

PP

PP (polypropylene) is used in synthetic fibres to make clothing, industrial fibres, car batteries, bumper bars and other car parts. Although these products are not appropriate for your recycling bin, polypropylene can be recycled for use in carpet, furniture, white goods and even polymer bank notes. Most car workshops, scrap metal dealers and service stations will accept used car batteries for recycling.

PS

Polystyrene is a very light plastic used in solid form to make plastic cutlery, toys and cases for CDs and DVDs. In its softer foam form, it is used for disposable drinking cups and packing materials. Polystyrene products should not be placed in home recycling bins. The lightness of polystyrene foam makes it difficult to sort and recycle. Although all polystyrene can be recycled, it is a very expensive process.

OTHER

Other plastics, including nylon, fibreglass and polycarbonate, are not generally recycled and should not be placed in your home recycling bin.

8.9.3 See-through recycling

About 45 per cent of the glass packaging used in Australia is recycled. Used glass bottles, known as **cullet**, are collected and melted down in a furnace to produce new products. The overall energy saving is only 8 per cent of that used in making new glass. This is because of the high cost of collecting and melting down the bottles. In some countries, milk is sold in bottles that can be sterilised and reused up to 50 times before they need melting down, which saves a large amount of energy.

8.9.4 Saving trees

Over a million tonnes of paper, about a third of our annual consumption, is recycled in Australia. Paper is made out of fibres of the chemical cellulose and is relatively easy to recycle. Waste paper is first mixed with water to separate the fibres. Additives such as ink and adhesives are then removed, producing low-quality fibres that can be used to make cardboard and other products. Steam rollers are used to improve the quality of the finished paper. Recycling paper reduces the amount of new paper needed, saving millions of trees.

8.9.5 And metal too!

Metals such as steel and aluminium are easily recycled as long as they can be cheaply separated from other rubbish. Steel cans, aerosol containers, jar lids and bottletops can be recycled. The recycling of aluminium cans saves huge amounts of energy. Twenty aluminium cans can be recycled with the same amount of energy needed to produce just one new can.

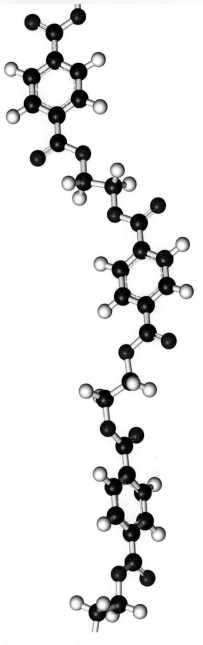
Plastics are made from chains of molecules and are also called polymers (poly means 'many'). This chain of molecules has two repeated units.

8.9.6 Sorting it out

The separation of the items in your recycling bin relies on differences in their physical properties including size, weight, magnetic properties and even colour. For example, items of different weights can be separated using blasts of air or a centrifuge that works like the spin dryer of a washing machine. Steel can be separated from other metals by a large magnet.

Special recycling programs

There are separate recycling programs for some products that cannot be placed in home recycling bins. These recycling programs are generally used to collect products containing substances that would endanger the environment or the community if they were dumped in landfill tips. For example, printer cartridges can be placed in recycling boxes at many Australia Post outlets and retail stores that sell computers and printers. Mobile phones can be left at most mobile phone outlets for recycling. Use the **Recycling** weblink in your Resources tab to find out where computers and other electronic equipment, white goods such as fridges and washing machines, corks, light globes and many other items are collected for recycling. This site also provides information about how to dispose of chemical wastes from home, school or industry. Oil, paints and unused medicines should not be placed in rubbish bins or flushed down the sink.

8.9.7 You can make a difference

The three bin collection system used by many city and shire councils throughout Australia makes it very easy for you to make a difference to the environment by recycling.

During the 12 months to June 2009, Hornsby Shire in NSW recycled 18 000 tonnes of paper. That's the equivalent of:
- saving over 299 000 gigajoules of energy. That's enough to power 13 870 homes for a year!
- preventing 9032 tonnes of greenhouse gas from entering the atmosphere. That's like taking 2169 vehicles off our roads — permanently!
- saving over 298 million litres of water. That's enough to fill 119 Olympic-sized swimming pools!

Imagine the combined effect of all cities and shires Australia wide!

This compost bin is made from recycled polypropylene (PP). The compost decreases in volume as it breaks down. Almost 50 per cent of domestic waste in Australia is suitable for composting.

8.9 Exercises: Understanding and inquiring

To answer questions online and to receive **immediate feedback** and **sample responses** for every question, go to your learnON title at www.jacplus.com.au. *Note:* Question numbers may vary slightly.

Remember

1. How are biodegradable substances different from those that are not biodegradable?
2. Where do the chemical changes that break down biodegradable waste take place?
3. What are the main benefits of recycling plastics?
4. Why are plastics such as PET now identified by a code?
5. What is cullet?

Think

6. The first part of the process of recycling paper involves mixing it with water. What is the major purpose of mixing the paper with water?
7. List three problems associated with the disposal of waste in landfill sites.
8. Why is only a small amount of energy saved when glass is recycled?
9. List two benefits of recycling aluminium.
10. What factors influence the decision as to whether it is worth the trouble of recycling a resource?
11. Ink is removed from paper when it is recycled. Is the combination of ink and paper a compound or a mixture? Explain your answer.
12. If a plastic bag manufacturer claimed that the bags it produced were biodegradable, what evidence would you need to be satisfied that the claim was correct?

Create

13. Draw a poster that could be used to encourage people to do one of the following.
    (a) Recycle all plastic products.
    (b) Reduce their use of LDPE plastic bags.
    (c) Separate household rubbish into different bins for more efficient recycling.
    (d) Recycle paper used at school.
    (e) Recycle newspaper.
    (f) Recycle aluminium cans.

 **RESOURCES** — ONLINE ONLY

 Explore more with this weblink: Recycling

8.10 Target maps and single bubble maps

8.10.1 Target maps and single bubble maps

1. Draw three concentric circles on a sheet of paper.
2. Write the topic in the centre circle.
3. In the next circle, write words and phrases that are relevant to the topic.
4. In the outer circle, write words and phrases that are not relevant to the topic.

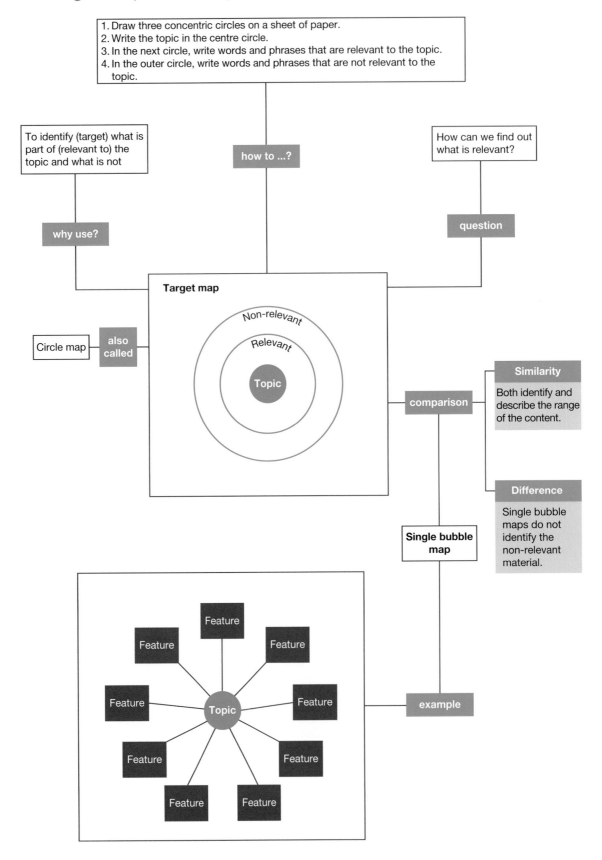

To identify (target) what is part of (relevant to) the topic and what is not

how to ...?

How can we find out what is relevant?

why use?

question

Target map

Non-relevant

Relevant

Topic

Circle map

also called

comparison

Similarity

Both identify and describe the range of the content.

Difference

Single bubble maps do not identify the non-relevant material.

Single bubble map

Feature

Feature

Feature

Feature

Topic

Feature

Feature

Feature

Feature

Feature

example

8.10 Exercises: Understanding and inquiring

To answer questions online and to receive **immediate feedback** and **sample responses** for every question, go to your learnON title at www.jacplus.com.au. *Note:* Question numbers may vary slightly.

Think and create

1. Use the words listed below to construct a target map about physical change.

2. The single bubble map at right identifies some of the ideas associated with a burning candle.
   (a) Draw your own single bubble map about the topic 'a burning candle', adding as many additional bubbles as you can.
   (b) Construct a single bubble map that identifies clues that provide evidence that a chemical reaction has taken place.
3. (a) In teams, brainstorm as many single words as you can that are associated with chemical reactions.
   (b) Use the brainstorm list to create a team single bubble map on chemical reactions.
   (c) Compare your list with those of other teams and then work together to construct a class single bubble map on chemical reactions. Suggest other types of thinking tools that could be useful.
4. Suggest a topic for each of the target maps below.

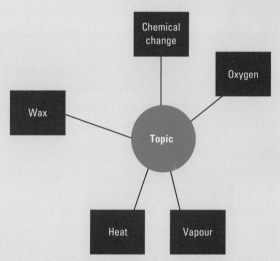

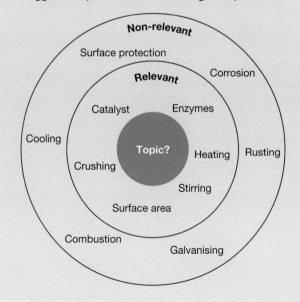

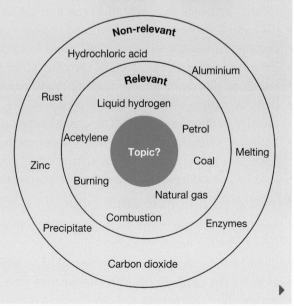

5. The single bubble map below identifies some of the ideas associated with rusting.
   (a) Complete the bubble map by adding as many ideas as you can.
   (b) Use the ideas in your single bubble map to create a target map on rusting.

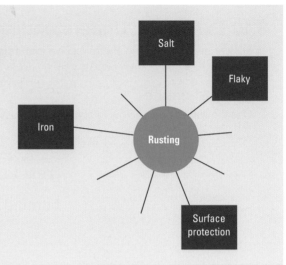

8.11 Review

8.11.1 Study checklist

Physical and chemical properties

- distinguish between the physical and chemical properties of substances
- outline some examples of physical and chemical properties
- recognise that the chemical properties of a substance affect its use
- outline the benefits and disadvantages of plastics
- compare the properties and method of manufacturing soaps and detergents
- explain the difference between natural and synthetic fibres, and discuss their advantages and disadvantages

Physical and chemical change

- define chemical change as a change in which the bonds between atoms or molecules are broken or new bonds between these particles are formed
- distinguish between physical changes and chemical changes

Chemical reactions

- define a chemical reaction as a chemical change in which a new substance is produced
- identify evidence that a chemical reaction has taken place
- distinguish between the reactants and products of a chemical reaction
- describe simple chemical reactions using word equations
- describe a variety of methods of speeding up or slowing down chemical reactions
- recognise that corrosion and burning (combustion) are chemical reactions
- describe rusting as an example of corrosion
- identify burning in oxygen as an oxidation reaction
- outline examples of the use of fuels in combustion reactions
- recall that most plastics are made from crude oil
- recall that biodegradability depends on chemical reactions within living organisms

Science as a human endeavour

- investigate the use of fire by traditional Aboriginal people
- investigate the development by scientists of a variety of new materials
- distinguish between biodegradable and non-biodegradable materials
- evaluate the suitability of different types of packaging for recycling
- outline the requirements and development of systems for the collection and recycling of household waste

8.11 Review 1: Looking back

To answer questions online and to receive **immediate feedback** and **sample responses** for every question, go to your learnON title at www.jacplus.com.au. *Note:* Question numbers may vary slightly.

1. List two useful properties of:
   (a) glass
   (b) metals
   (c) plastics
   (d) paper.

2. Explain, using examples, the difference between physical and chemical properties.

3. In your own words, express the meaning of each of these terms that describe the properties of substances.
   (a) Elastic
   (b) Ductile
   (c) Reactive
   (d) Malleable
   (e) Lustrous
   (f) Toxic
   (g) Transparent
   (h) Flammable
   (i) Melting point

4. Identify the following as either chemical or physical changes.
   (a) The wax on a burning candle melts.
   (b) The wax vapour at the top of a candle wick burns with oxygen to produce carbon dioxide, water vapour and heat.
   (c) Calcium carbonate is dissolved by hydrochloric acid to form calcium chloride, water and carbon dioxide gas.
   (d) Hydrogen gas explodes with oxygen gas to form water.

5. Write word equations for each of the changes in question 4.

6. How do you know that:
   (a) toasting bread is not a physical change
   (b) rusting of a nail is not a physical change?

7. Some chemical reactions can be useful. Write down three examples of useful chemical reactions.

8. Catalysts are sometimes added to the reactants taking part in a chemical reaction.
   (a) What is a catalyst?
   (b) When a word equation is written to describe a chemical reaction, catalysts are not included either as reactants or products. Why?

9. When a lead nitrate solution is added to a potassium iodide solution, a chemical reaction takes place. A bright yellow solid appears. It is the compound lead iodide. Another compound, potassium nitrate, remains in the solution and is not visible.
   (a) Name the reactants in this chemical reaction.
   (b) What are the products of the reaction?

(c) The yellow lead iodide will eventually settle to the bottom of the flask. What 11-letter word beginning with 'p' is given to a substance that behaves like the lead iodide?

(d) Write a chemical word equation for the reaction.

10. Rusting is an example of a slow chemical reaction.
    (a) What are the three reactants of rusting?
    (b) What is the product of the rusting reaction?

11. For each of the reactions below, suggest ways that the reaction could be made to happen more quickly.
    (a) Burning a pile of dry leaves
    (b) Cooking potatoes
    (c) Dissolving marble chips in acid
    (d) Removing a stain using bleach
    (e) Making an iron nail go rusty
    (f) Milk going sour

12. Some chemical reactions can be destructive. Write down three examples of harmful chemical reactions.

13. Children's steel swing sets in beachside towns and suburbs rust much faster than those further from the coast.
    (a) Explain why this happens.
    (b) Suggest two methods of slowing down or preventing the rusting of steel swing sets.

14. The oxyacetylene torch shown at right is used to melt metals to allow them to be joined together.
    (a) What type of chemical reaction takes place in the oxyacetylene torch?
    (b) What evidence is there in the photo that a chemical reaction has taken place?

15. Just as chemicals can be grouped or classified, so can chemical reactions. What name is given to the following chemical reactions?
    (a) The corrosion of iron
    (b) The reaction of substances with oxygen
    (c) Burning

16. The illustration below shows a camper boiling water in a billy over a camp fire.
    (a) What physical change in the wood has taken place during the preparation of the camp fire?
    (b) What physical change is shown taking place?
    (c) What chemical change is shown taking place?
    (d) List two ways in which the chemical reaction taking place has been sped up.

17. Which properties of the plastic used to make light switches and power points make it right for the job?

18. What is the difference (other than their properties) between natural and synthetic fibres?

19. Identify each of the fibres listed below as natural, synthetic, or neither natural nor synthetic.
    (a) Nylon
    (b) Cotton

(c) Rayon

(d) Lycra

(e) Wool

(f) Polyester

20. How are synthetic fibres such as nylon made?

21. Why are many garments made of a blend of two or more different fibres?

22. Soap and detergents can be used for the same purpose — including washing your own body.

   (a) What is the main difference between soap and detergents?

   (b) What properties do soap and detergents have in common?

23. Which properties are essential for the packaging of the following products?

   (a) Pool chemicals

   (b) Eggs

   (c) Soft drink

   (d) Peanuts

24. Some plastic containers are marked with this symbol at right.

   (a) What substance would you expect to find in bottles made from this type of plastic?

   (b) What two things should you do before placing bottles made from this type of plastic in a recycling bin?

   (c) State two uses for this type of plastic after it has been recycled.

PETE

25. Describe the 'three-bin system' used by many cities and shires, and explain how it helps the environment.

TOPIC 9
Sedimentary, igneous and metamorphic rocks

9.1 Overview

The Earth's surface is constantly changing. Volcanoes and earthquakes can cause quick changes, but most of the changes to the Earth's surface happen slowly. Rocks on and below the surface of the Earth are slowly and constantly being changed by natural events. Rocks also provide a valuable record of past events.

9.1.1 Think about rocks

assess[on]

- Which rock is light enough to float on water?
- Which rocks are formed from the remains of living things?
- What do butterflies, frogs, Mr Hyde, werewolves and metamorphic rocks have in common?
- How do we know what living things that have not existed for millions of years looked like, how they walked and what they ate?
- How can whole skeletons of animals be fully preserved for millions of years?
- What can you learn from a dinosaur footprint?
- Why did the dinosaurs vanish from the Earth 65 million years ago?

Numerous **videos** and **interactivities** are embedded just where you need them, at the point of learning, in your learnON title at www.jacplus.com.au. They will help you to learn the concepts covered in this topic.

9.1.2 Your quest

Bathroom rocks

When you last used the bathroom, you probably weren't thinking about rocks. After all what does a bathroom have to do with rocks?

But where did the materials to make the shower recess come from? What about the taps and pipes that delivered the water? Where do the materials to make tiles come from? And what about the soap, talcum powder and toothpaste — where do they come from?

The answers to all of these questions lead back to rocks. For example, metals are extracted from rocks. Talcum powder is made from a mineral called talc, which comes from a rock. Some of the ingredients of toothpaste come from rocks.

Even food and clothes can be traced back to rocks. Plants grow in soil, which is made up mostly of weathered rock. Glass is made using sand, which is weathered rock.

Think

Work in small groups to answer the following questions.

1. What materials are mirrors made from?
2. Where does the metal used to make bathroom taps come from?
3. What are bathroom tiles made from?
4. List some building materials that are:
   (a) made directly from rocks
   (b) not made directly from rocks but can be traced back to rocks.
5. The terms 'igneous', 'sedimentary' and 'metamorphic' are used to describe the three main groups of rocks. Just from looking at the words, suggest how each of these groups of rocks is formed.
6. Make a list of the names of rocks that you know. Attempt to classify them as igneous, sedimentary or metamorphic.

Taps, tiles, mirrors and even soap. Where do the materials needed to produce these come from?

9.2 Rocks and minerals

9.2.1 How rocks are formed

Rocks are made up of substances called **minerals**. Any naturally occurring solid substance with a definite chemical composition is called a mineral.

All rocks are formed in the Earth's **lithosphere**, which includes the Earth's crust and the top part of its mantle, where partially molten rock called **magma** flows very slowly under the crust. Some rocks are formed when magma gets close to the surface and slowly cools. Some of that red-hot magma breaks through the Earth's crust to form fiery volcanoes, releasing lava to cool quickly on the surface or even underwater. Other rocks form as a result of the weathering of older rocks and erosion, creating layers of **sediments**, which are eventually buried under more sediments and changed by heat and pressure. Some rocks are even formed from the remains of living things.

All rocks are formed in the Earth's lithosphere.

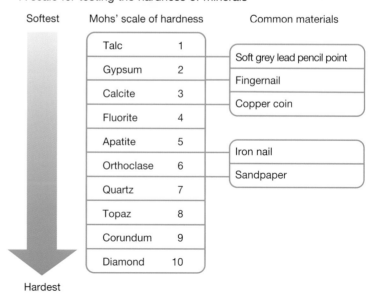

9.2.2 What's in a rock?

Elements found naturally in their uncombined form are also minerals. These elements, called **native elements**, include diamonds (pure carbon) and gold.

Most minerals in rocks are compounds with one or more metal elements together with the elements oxygen and silicon. The colours, shapes and textures of the minerals in rocks tell us what they are made of and how they were formed, as well as providing clues about the past.

Cool shapes

The atoms that join together to form minerals make up regular geometric shapes in particles called **crystals**. The way crystals grow when a mineral is formed depends on the speed of the cooling process of the molten material from which they form and how much space is available. The quartz crystals shown on the next page have cooled slowly and have had a lot of time and space to grow. Quartz, one of the most common minerals, consists of hexagonal-shaped crystals of silicon dioxide (SiO_2).

Identifying minerals

Although colour might seem to be the quickest way to identify a mineral, it is

A scale for testing the hardness of minerals

Softest

Mohs' scale of hardness		Common materials
Talc	1	
Gypsum	2	Soft grey lead pencil point
Calcite	3	Fingernail
Fluorite	4	Copper coin
Apatite	5	
Orthoclase	6	Iron nail
Quartz	7	Sandpaper
Topaz	8	
Corundum	9	
Diamond	10	

Hardest

not reliable. Many different minerals have similar colours. Some minerals, even though they have the same chemical composition, can have different colours. Quartz, for example, can be colourless like glass, or may be pink, violet, brown, black, yellow, white or green. Other properties can be used to identify minerals.

Quartz is one of the most common minerals in the Earth's crust.

The **lustre** of a mineral describes the way that it reflects light. Minerals could be described, for example, as dull, pearly, waxy, silky, metallic, glassy or brilliant.

The **streak** is the powdery mark left by a mineral when it is scraped across a hard surface like an unglazed white ceramic tile.

The **hardness** of a mineral can be determined by trying to scratch one mineral with another. The harder mineral leaves a scratch on the softer mineral. Friedrich Mohs' scale of hardness is a numbered list of ten minerals ranked in order of hardness. Higher numbers correspond to harder minerals. The hardness of a mineral is determined by comparing it with the minerals in Mohs' scale. For example, a mineral that can be scratched by quartz but not by orthoclase has a hardness between 6 and 7.

The diagram at the bottom of the previous page shows that some more common materials can be used to determine the hardness of a mineral if the minerals in Mohs' scale are not available.

INVESTIGATION 9.1

Which mineral is it?

AIM: To observe the properties of a range of minerals

Materials:
mineral kit
common materials to substitute for unavailable Mohs' scale minerals
hand lens
white ceramic tile

Method and results

1. Construct a table like the one shown below and use it to record your observations as you work through the following steps for each mineral.
2. Describe the colour and lustre of the mineral.

Mineral	Colour	Lustre	Crystal shape and size	Streak	Hardness

3. Use the magnifying glass to look closely at the mineral, and describe the shape and size of its crystals.
4. Scrape the mineral across the unglazed side of a white ceramic tile. Record the colour of the streak.
5. Use Mohs' scale minerals or the common materials to estimate the hardness of the mineral by trying to scratch it. An approximate range, such as 5–6, is sufficiently accurate.

Discuss and explain

6. Other than those already described, what additional properties of minerals could be used to identify them?
7. If two unlabelled mineral samples have the same colour and lustre, can you be sure that they are the same mineral? Explain how you would find out.

9.2 Exercises: Understanding and inquiring

To answer questions online and to receive **immediate feedback** and **sample responses** for every question, go to your learnON title at www.jacplus.com.au. *Note:* Question numbers may vary slightly.

Remember

1. Which parts of the Earth make up the lithosphere?
2. List three ways in which rocks can be formed in the Earth's lithosphere.
3. What is a mineral?
4. What is a native element? List two examples.
5. Which minerals are present in granite?
6. List at least five properties that you could observe to help you identify an unknown mineral.
7. What is the approximate hardness on Mohs' scale (to the nearest whole number) of a mineral that can be scratched by sandpaper but not by an iron nail?

Think

8. Explain the difference between a rock and a mineral.
9. You have two samples, each of a different mineral, but no other equipment to test them for hardness. How could you tell which mineral is harder?
10. A mineral can be scratched by a copper coin but not by a fingernail. You know that the mineral is quartz, fluorite or calcite. Which is it?
11. Is table salt a mineral? Think carefully about your answer and suggest reasons for and against classifying it as a mineral.

Create

12. Find out how crystals can be artificially grown and then grow a crystal garden.

learnon RESOURCES — ONLINE ONLY

Complete this digital doc: Worksheet 9.1: Identifying and classifying minerals (doc-18760)

9.3 'Hot' rocks

9.3.1 Extrusive rocks

Lava is released from erupting volcanoes at temperatures of 1000 °C or more. At that temperature, it could take weeks for the lava to cool down to become solid rock. But lava surging violently into the air from explosive volcanoes cools much faster. The lava erupting from underwater volcanoes on the ocean floor also cools quickly.

Rocks that form from the cooling of magma below the Earth's crust or lava are called **igneous rocks**. Igneous rocks that form from red-hot lava above the Earth's surface are called **extrusive rocks**. Igneous rocks that form from the lava spilling from underwater volcanoes are also classified as extrusive rocks.

The appearance of all extrusive igneous rocks depends on two major factors:

• how quickly the lava or magma cooled
• what substances it contains.

WHAT DOES IT MEAN?

The word *igneous* comes from the Latin word *ignis*, meaning 'fire'. The words ignite and ignition also come from the same Latin word.

Fast or slow?

The size of the crystals that make up extrusive igneous rocks depends on how quickly the lava cooled. When it cools quickly, there is not enough time for large crystals to form. The rocks formed from lava that cools more slowly have larger crystals.

Frothy rocks

Some violent volcanic eruptions shoot out lava filled with gases. The lava cools quickly, while it is still in the air, and traps the gases inside. Rocks that form this way are full of holes. Two examples of this type of rock are **pumice** and **scoria**.

Pumice

Pumice is a pale-coloured rock. It is very light because it is full of holes. Pumice floats on water and sometimes washes up on beaches. Powdered pumice is used in some **abrasive** cleaning products.

Scoria

Scoria is heavier than pumice, and darker because it contains more iron. It is usually found closer to the volcano's crater than pumice. Scoria is a reddish-brown or grey rock that can be crushed and used in garden paths or as a drainage material around pipes.

Basalt

Basalt is an extrusive rock that can take on many appearances. One big difference between samples of basalt is the size of the crystals that make up the rock. For example, basalt formed from lava cooling in cold ocean water has much smaller crystals than basalt formed by lava cooling on the ground. In fact, basalt that formed under water has crystals so small they are difficult to see.

The basalt in the photograph below has large crystals because it formed from lava on the ground. The crystals had time to grow before the rock became solid. It also has large bubble-like holes. That's because the lava was filled with gases when it began to cool. The gases have since escaped.

Pumice

Scoria

Basalt with bubbles

Obsidian

Obsidian is a smooth, black rock that looks like glass. It is formed when lava cools almost instantly. This rock is different from basalt because it cooled so quickly that no crystals formed. Sometimes very fine air bubbles are trapped in the rock, which give it a coloured sheen.

Obsidian

learn on   RESOURCES — ONLINE ONLY

Watch this eLesson: Volcanoes (eles-0130)

Does fast cooling make a difference?

AIM: To investigate the effect of cooling rate on the size of crystals

Materials:
freshly made saturated solution of potassium nitrate
potassium nitrate
spatula
250 mL beaker
3 test tubes and test-tube rack
test-tube holder
Bunsen burner, heatproof mat and matches
crushed ice
safety glasses
hand lens

CAUTION
Safety glasses must be worn during this experiment.

Method and results
- Half-fill a beaker with crushed ice.
- Quarter-fill a clean test tube with saturated potassium nitrate solution. Add a spatula of potassium nitrate.
- Gently heat the solution over a Bunsen burner flame until the added potassium nitrate has dissolved or until the solution starts to boil.
- Pour half the warm solution into each of two clean test tubes.
- Place one test tube in the beaker of crushed ice and the other test tube in the rack to cool.
- When crystals have formed in each test tube, examine them with a hand lens.
1. Draw a labelled diagram of some crystals in each test tube, concentrating on their shape and size.

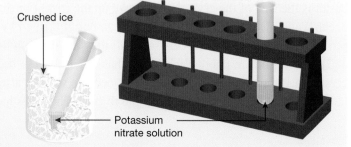

Cool one solution quickly and the other one slowly.

Crushed ice

Potassium nitrate solution

Discuss and explain
2. Which test tube contained the larger crystals — the one that cooled quickly or the one that cooled slowly?
3. Which type of extrusive rock would you expect to have the larger crystals — those that cool slowly on the surface or those that cool quickly underwater?
4. Why do safety glasses need to be worn during this experiment?

9.3.2 Cooling underground

Igneous rocks can form below the surface of the Earth. Those that form from magma that cooled below the surface are called **intrusive rocks**. They cool very slowly and become visible only when the rocks and soil above them erode. Large bodies of intrusive rock are called **batholiths**. They can stretch over distances of up to 1000 kilometres.

Intrusive rocks (sometimes called plutonic rocks) have larger crystals than extrusive rocks because the crystals had more time to grow.

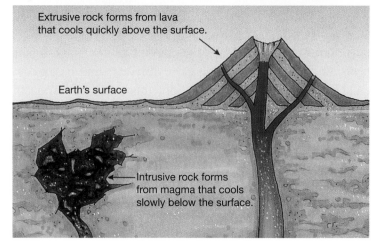

Igneous rocks can form below or above the Earth's surface.

Extrusive rock forms from lava that cools quickly above the surface.

Earth's surface

Intrusive rock forms from magma that cools slowly below the surface.

If a batholith is exposed to the environment, it will start to wear away along the cracks. Over time, the batholith may break down completely. The breakdown of rocks is called weathering.

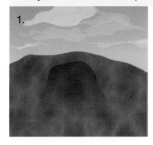 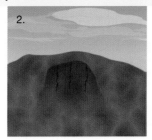 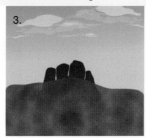 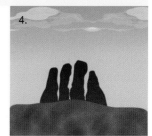

Granite

Granite is a common intrusive rock. The crystals in granite form over long periods of time and grow large enough to be easy to see with the naked eye. Granite is very hard and can be used for building. Headstones and other monuments are often made from granite that has been polished to give it a glossy finish.

The crystals found in granite are a mixture of white, pink, grey, black and clear minerals. These are quartz (clear to grey), feldspar (white and pink) and mica (black). Feldspar is made of aluminium silicate, and black mica is aluminium silicate combined with potassium, magnesium and iron.

Granite

9.3 Exercises: Understanding and inquiring

To answer questions online and to receive **immediate feedback** and **sample responses** for every question, go to your learnON title at www.jacplus.com.au. *Note:* Question numbers may vary slightly.

Remember
1. What causes the frothy appearance of pumice and scoria?
2. Why are the crystals in basalt that formed under water smaller than those in basalt that formed on the ground?
3. Batholiths form well below the ground. Explain how they become visible on the Earth's surface.
4. Distinguish between the ways extrusive and intrusive igneous rocks are formed.
5. Describe two major differences between the appearance of granite and basalt.
6. List the three minerals found in granite.

Think
7. Explain how you would decide whether an igneous rock formed from a volcanic eruption.
8. Rhyolite is an extrusive rock that contains the same minerals as granite. In what ways would you expect it to be different from granite?

Investigate
9. Locate a building, statue or memorial in your area that is made from granite. Describe the granite in the structure, and suggest why it was the chosen material.

learn on RESOURCES — ONLINE ONLY

Complete this digital doc: Worksheet 9.2: Igneous rocks (doc-18756)

9.4 Sedimentary rocks

9.4.1 Sedimentary rocks

Rocks that are formed from the particles of sediments are called **sedimentary rocks**.

Sediments are deposited when weathered rock is moved from one place to another by the wind, running water, the sea or glaciers. That process is called **erosion**. Deposits of dead plants and animals are also called sediments.

Sand deposited by the wind forms sand dunes, especially in coastal areas where sand is picked up and blown inland until it is stopped by obstacles such as rock or vegetation.

A fast-moving river is likely to carry with it sand, gravel and smaller particles. As it slows down on its path to the sea, the river loses energy and particles are deposited, forming sediments. The larger particles, such as gravel and sand, settle first. By the time the river reaches the sea, it is usually travelling so slowly that the very fine mud particles begin to settle.

During floods when rivers break their banks, sediments are deposited on flat, open land beside the river. These plains are called **floodplains**.

Many sedimentary rocks form in this way.

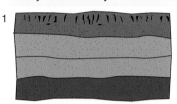

1 Sediments are laid down by ice, wind or water, in horizontal layers called beds.

2 Within each bed, the sediment grains are squashed together so that they are in close contact.

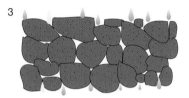

3 Water seeps in between the grains, bringing with it many dissolved chemicals.

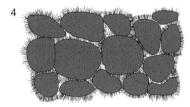

4 When the water evaporates, these chemicals are left behind as crystals around the edges of the grains. These crystals cement the grains of sediment together to form rock.

The water in fast-moving rivers, along with the weathered rock it takes with it, can carve out deep valleys in the Earth's surface. One of the most spectacular examples of this is the Grand Canyon in Arizona, USA.

In the coldest regions of the Earth, especially at high altitudes, bodies of ice called **glaciers** slowly make their way down slopes. They generally move between several centimetres and several metres each day. Being solid, glaciers can push boulders, rocks, gravel and smaller particles down the slope, and deposit them on curves beside the glacier or at the end of the glacier. These deposits are called **moraines**.

HOW ABOUT THAT!

Chalk is a sedimentary rock. It is similar to limestone, but not as hard. Chalk is formed from very fine grains of calcium carbonate that separate from sea water and settle to become a white, muddy sediment on the sea floor. The sediment hardens over time to form chalk. This process takes millions of years. The remains of shellfish and other sea animals are also found in the sediment that forms chalk, but most of these remains are microscopic.

The white cliffs of Dover that overlook the English Channel are composed of chalk.

9.4.2 Clastic sedimentary rocks

Most sedimentary rocks are formed from weathered rock. Grains of sediment are cemented together over a long period of time to form solid rock. The process is shown in the diagram below.

1. Fast-flowing water can move sand, soil and even big rocks over long distances. All creeks and rivers flow to the sea or to inland lakes, but, by the time they reach the seas or lakes, the water flows much more slowly.
2. As the water slows down, the bigger rocks are deposited.
3. By the end of the river's journey, all but very fine sediments have been deposited.
4. Coastlines can change quite quickly as a result of weathering, erosion and the deposition of sediments.
5. Ocean waves wear away the rocks that make up cliff faces. The waves pound rocks, smashing them into smaller and smaller pieces.
6. Sand is picked up by currents in the waves along one beach and deposited on other beaches. Strong winds have enough energy to pick up sand and carry it inland.

Sandstone is formed from grains of sand that have been cemented together over a period of time. **Mudstone** and **shale** are formed from finer grains of sediment deposited by calm water in the form of mud. **Siltstone** has grains slightly larger than those of mudstone. **Conglomerate** contains grains of different sizes that have been cemented together.

Conglomerate is formed from sediments that might be deposited by a fast-flowing or flooded river.

WHAT DOES IT MEAN?

The word *conglomerate* comes from the Latin word *conglomerare*, meaning to 'roll together'.

9.4.3 Rocks from living things

Limestone is a sedimentary rock that is formed from deposits of the remains of sea organisms such as shellfish and corals. The hard parts of these dead animals contain calcium carbonate. These deposits are cemented together over a period of time in very much the same way as other sedimentary rocks form from weathered rock.

Coal is formed from the remains of dead plants that are buried by other sediments. In dense forests, layers of dead trees and other plants build up on the forest floor. If these layers are covered with water before rotting is completed, they can become covered with other sediments. The weight of the sediments above compacts the partially decayed plant material. Over millions of years the compacting increases the temperature of the sediment and squeezes out the water, forming coal.

The Grand Canyon is a spectacular example of erosion by a fast-flowing river.

INVESTIGATION 9.3

Sediments and water

AIM: To investigate the order in which different sediments are deposited

Materials:
mixture of garden soil, gravel, sand and clay
large jar with lid
watch or clock

Method and results
- Before commencing this experiment, form your own hypothesis about the order in which the different types of particles will settle. Give reasons for your hypothesis.
1. Draw a diagram to illustrate your hypothesis.
   - Place enough of a mixture of garden soil, gravel, sand and clay in a large jar to quarter-fill it.
   - Add enough water to three-quarters fill the jar and place the lid on firmly. Shake the jar vigorously.

- Put the jar down and watch carefully as particles begin to settle. Note the time taken for each layer of sediment to settle completely.
- Leave the jar for a day or two. Then compare your observations of the jar with your diagram.
2. Which type of sediment settled first?
3. Where are the other particles of sediment while the first layers are settling?
4. Draw a labelled diagram showing clearly any layers that form. Identify the layers if you can.
5. Which sediments settled after a day or two?

Discuss and explain
6. Why did the last sediments take so long to settle?
7. Was your hypothesis supported by your observations?
8. What is the relationship between the size of sediment particles and the time taken to settle?

9.4.4 Rocks from chemicals

Some sedimentary rocks form when water evaporates from a substance, leaving a layer that is compressed after being buried by other sediments. **Rock salt** is an example of a rock formed in this way. It forms from residues of salt that remain after the evaporation of water from salt lakes or dried-up seabeds and can form beds that are hundreds of metres thick. Rock salt is used on roads and driveways in very cold areas to combat ice. Gypsum is another mineral that is formed in this way.

9.4.5 Rocks in layers

Layers of sedimentary rock are often clearly visible in road cuttings and the faces of cliffs. The limestone in the photograph at right was formed on the ocean floor. Layers of sediments and sedimentary rocks can be pushed upwards by the same forces below the Earth's surface that form mountains. Those forces can also bend and tilt the rock layers.

When fossils are found in sedimentary rock, the layer they are found in can be used to figure out how old the fossil may be.

9.4.6 Using sedimentary rocks

Sandstone and limestone are often used as external walls of buildings. These sedimentary rocks are well suited to carving into bricks of any shape. Shale can be broken up and crushed to make bricks.

Limestone is broken up to produce a chemical called lime. Lime is used to make mortar, cement and plaster, and is also used in the treatment of sewage and on gardens to neutralise acid in the soil.

This limestone, rich in corals and shells, is many metres above sea level. How did it get there?

Layers of sedimentary rock can be pushed upwards, bent and tilted by forces beneath the Earth's surface.

Coal is another useful sedimentary rock. It is used as a fuel and burned in electric power stations to boil water. The steam is then used to drive the turbines that produce electricity. In some countries, coal is burned in home heaters, although this can cause air quality problems.

INVESTIGATION 9.4

Identifying sedimentary rocks

AIM: To use a key to identify a variety of sedimentary rocks

Materials:

several examples of unlabelled sedimentary rocks, including limestone
dropping bottle of dilute hydrochloric acid

Method

- Use the key below to identify the samples of sedimentary rocks you have been given.
- To do the acid test, just add one drop of dilute hydrochloric acid onto the sample.

Discuss and explain

1. How many of the unlabelled rocks did you identify?
2. Which of the rock samples were the most difficult to identify?

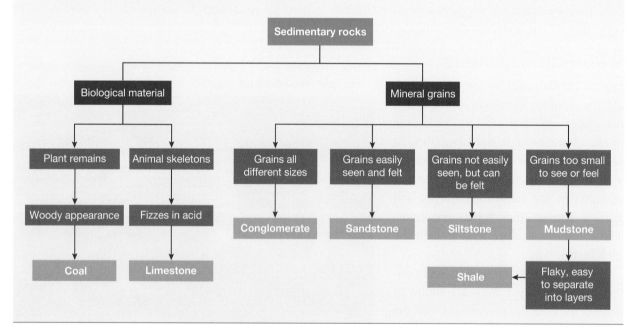

9.4 Exercises: Understanding and inquiring

To answer questions online and to receive **immediate feedback** and **sample responses** for every question, go to your learnON title at www.jacplus.com.au. *Note:* Question numbers may vary slightly.

Remember

1. What are most sediments composed of before they eventually form sedimentary rocks?
2. As a flooded river slows down, which particles are likely to settle first — gravel, sand or fine clay?
3. Explain how a floodplain is created.
4. From what are all sedimentary rocks formed?
5. Explain, with the aid of a diagram, how grains of sand can become part of a sedimentary rock.
6. Sedimentary rocks formed from the erosion of weathered rock cannot be identified by their colour. What feature allows you to identify them?
7. How is coal formed?
8. Explain why sedimentary rocks are found in layers.

Think

9. What type of sediment would you expect to find on the bed of the Yarra River in Melbourne?
10. A road cutting reveals a layer of sandstone beneath a layer of mudstone. Between them is a much thinner layer of conglomerate.
    (a) Which layer would have formed from sediments beneath the sea?
    (b) Which layer would have formed while the area was flooded by a swollen, fast-flowing river?
    (c) Which layer would have formed while the land was the floor of a still lake?
    (d) Which layer was formed most recently?
11. Explain why limestone and coal are sometimes referred to as biological rocks.
12. In which type of sedimentary rock would you be most likely to find embedded seashells?

Investigate

13. What do peat, brown coal and black coal have in common? How are they different from each other?

learn **on** RESOURCES — ONLINE ONLY

 Complete this digital doc: Worksheet 9.3: Sedimentary rocks (doc-18757)

9.5 Stability and change: Metamorphic rocks

9.5.1 Metamorphic rocks

The thing that never changes about the Earth is that it never stops changing. Igneous rocks and sedimentary rocks are constantly undergoing change. If they are on the surface, they are weathered, cracked or even bent.

Igneous and sedimentary rocks deep below the Earth's surface are buried under the huge weight of the rocks, sediments and soil above them. They are also subjected to high temperatures. The temperature increases by about 25°C for every kilometre below the surface. This heat and pressure can change the composition and appearance of the minerals in rocks.

The process of change in the rocks is called **metamorphism** and the rocks that are formed by these changes are called **metamorphic rocks**.

WHAT DOES IT MEAN?

The word *metamorphic* comes from the Greek words *meta*, meaning 'change', and, *morph* meaning 'form'.

Shale (pictured below) is a common type of sedimentary rock. It has fine grains and crumbles easily along its layers. When shale is exposed to moderate heat and pressure, it forms **slate**.

Marble (pictured below) forms from limestone under heat and pressure. It contains the same minerals as limestone.

Shale

Marble

 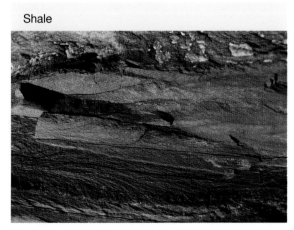

The changes that take place during the formation of metamorphic rocks depend on:
- the type of original rock
- the amount of heat to which the original rock is exposed
- the amount of pressure caused by the weight of the rocks above
- how quickly the changes take place.

Metamorphic rocks that are mainly the result of great pressure can often be identified by bands or flat, leaf-like layers. These bands are evident in the sample of **gneiss** (pronounced 'nice') pictured at right.

The diagram at right shows how rocks can be changed by the high temperatures that result from contact with hot magma.

Other common examples of the formation of metamorphic rocks are:

Shale (sedimentary)	mainly pressure ⇒	Slate
Sandstone (sedimentary)	mainly heat ⇒	Quartzite
Limestone (sedimentary)	mainly heat ⇒	Marble

Gneiss is formed mainly as a result of great pressure on granite.

The formation of metamorphic rock by contact with hot magma

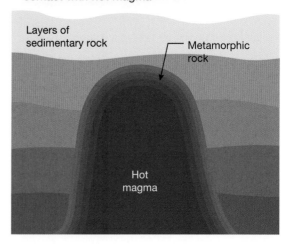

9.5.2 Clues from metamorphic rocks

The nature of metamorphic rocks above and below the ground can provide clues about the history of an area. Think about why the presence of quartzite or marble high in a mountain range would suggest that the area was once below the sea.

The presence of slate might suggest that the area was once the floor of a still lake or river mouth. The sediments were probably buried under many other sediments, and cemented together to form shale. The shale was transformed, or metamorphosed, into slate as a result of new rock formed above it.

9.5.3 Uses of metamorphic rock

Marble's strength, appearance and resistance to weathering make it suitable for use in statues and the walls and floors of buildings (inside and outside). It is usually highly polished. The hardness, flat structure and strength of slate make it ideal for use in buildings, especially in roofing and floor tiles. The sedimentary rocks from which marble and slate are formed could not be used for these purposes.

HOW ABOUT THAT!

A tadpole grows into a frog, female frogs lay eggs, and eventually more tadpoles emerge from the eggs. That's a life cycle. Some of the changes in rocks can be described as cycles too. Weathered rock is moved by erosion and the particles form sediments, which can be cemented together to form sedimentary rocks, which in turn may eventually change into metamorphic rocks. Once those rocks are exposed at the surface the weathering starts all over again. A complete cycle normally takes millions of years, but sometimes never takes place at all. Why?

There are many cycles in nature. Some happen faster than others.

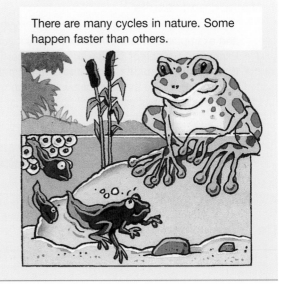

9.5.4 The rock cycle

The rock cycle below describes how rocks can change from one type to another. Weathering, erosion, heat, pressure and remelting are processes that help change rocks. The rock cycle is different from other cycles because there is no particular order in which the changes happen. Some rocks have been unchanged on Earth for millions of years and may not change for millions more. Some rocks change very quickly, especially near the edges of the plates that make up the Earth's crust.

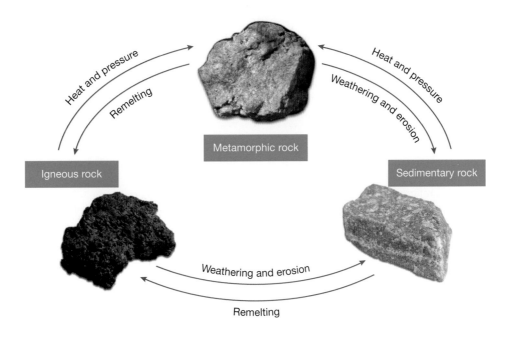

Rocks — the new generation

AIM: To examine and compare a selection of metamorphic rocks and their corresponding 'parent' rocks

Materials:

labelled samples of granite, gneiss, limestone, marble, sandstone, quartzite, shale and slate
hand lens

Method and results

- Try to sort the rocks into pairs of 'parent' rock and corresponding metamorphic rock. Use the descriptions and examples on the previous two pages if you have trouble pairing the rocks.
- Examine each pair of rocks with a hand lens. Take particular note of grain or crystal size and banding.
- If necessary, re-sort the rocks into different pairs.
1. Copy and complete the table below by noting the similarities and differences between the 'parent' and metamorphic rock of each pair.

Comparing 'parent' and metamorphic rocks

'Parent' rock	Metamorphic rock	Similarities	Differences	Main cause of metamorphism
Shale				
	Gneiss			
Sandstone				
	Marble			

Discuss and explain

2. Why is the term 'parent' rock used to describe the rock before metamorphosis?
3. Use the last column of your table to suggest whether the main cause of metamorphism was heat or pressure.

9.5 Exercises: Understanding and inquiring

To answer questions online and to receive **immediate feedback** and **sample responses** for every question, go to your learnON title at www.jacplus.com.au. *Note:* Question numbers may vary slightly.

Remember

1. Rocks are classified into three groups. Metamorphic rocks are one of these. Identify two groups of rock from which metamorphic rocks are formed.
2. What can cause rocks to change form and become metamorphic rocks?
3. Describe the differences between gneiss and granite.
4. What causes granite to be transformed into gneiss?
5. Slate is commonly used in floor tiles. Why?

Think

6. If an igneous or sedimentary rock gets so hot that it melts completely, it does not become a metamorphic rock. Explain why.
7. Why is limestone referred to as the 'parent' rock of marble?
8. Metamorphic rocks are generally formed deep below the surface of the Earth. However, they are often found above the ground — even high in mountain ranges. How can this be so?
9. Devise a 'buildings trail' in your city or town to locate buildings made of different kinds of rock. Draw a map to show the location of the buildings and the type of rock used in constructing them.

Investigate

10. Find out more about the uses of marble and slate. Where are they obtained? What are they used for? Why are they expensive?

9.6 Mining for metals

Science as a human endeavour

9.6.1 Minerals

Metals play an important part in our lives every day. You probably slept on a mattress with metal inner springs. The alarm clock that woke you up has metal parts. You use metal cutlery to eat food. The bus, car or bike that you may have used to get to school is made from metal.

The metal elements used to make these things are all found in minerals within rocks in the Earth's crust. The pie chart at right shows that almost three-quarters of the Earth's crust (by weight) is made up of the non-metals oxygen and silicon. Most of the metal elements are combined with other elements in compounds, usually with oxygen, silicon or other non-metals. It is these compounds that are the minerals.

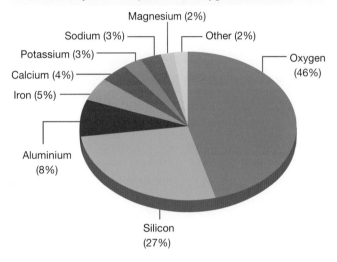

The elements in the Earth's crust. The metal elements are relatively rare compared with oxygen and silicon.

Magnesium (2%)
Sodium (3%)
Other (2%)
Potassium (3%)
Calcium (4%)
Iron (5%)
Oxygen (46%)
Aluminium (8%)
Silicon (27%)

Minerals containing metals of value are called **mineral ores**. It takes a lot of time, effort and money to get the rocks that contain the mineral ores out of the ground, separate the mineral ores from the rock and extract the metal element from the mineral ore. The **mining** of a mineral ore can take place only if enough of it is found at a single location.

The mining industry makes a major contribution to Australia's economy. Apart from the profits that go to shareholders in mining companies and to the government in taxes, the mining industry employs many thousands of Australians. Scientists and engineers are involved at every stage of the mining process.

9.6.2 Mineral exploration

Finding minerals below the Earth's surface where you can't see them is an expensive business. Before any money or effort is spent on using helicopters and analysing samples, geologists have to know where to look. Their knowledge of sediments, rocks and minerals of all types, and the clues they provide, helps them to predict where precious mineral ores are likely to be found.

Geologists also make use of satellites equipped with cameras, radar and other sensors to search for geological features that are likely to contain high concentrations of minerals. Minerals in the crust dissolve in rain and running water and get washed into creeks and rivers. A chemical analysis of the sediments and surface water of lakes and streams, therefore, provides evidence of the presence of minerals in the area.

The magnetic properties of large bodies of rocks containing some minerals can be detected from aircraft or by geological surveyors on the ground. Samples of soil and rocks are taken using portable equipment. On average, only one in 1000 sites that are sampled is eventually mined.

If there is sufficient evidence of useful mineral deposits that might be worth mining, a licence must be obtained before any clearing is done or heavy drilling equipment is brought in. Helicopters are sometimes used to bring in heavy equipment to protect sensitive ecosystems. The drilling allows mining companies to have a detailed look at what lies beneath the surface. Mining companies are required by law to clean up exploration drill sites and ensure they are left as they were found.

9.6.3 You can't start until …

In the past, mining was often carried out without considering its long-term effect on the environment and the people who live and work in the area. Today, however, an **environmental impact statement (EIS)** must be prepared before a mining operation can commence. An EIS outlines how the mining company intends to manage all environmental aspects of the proposed mine. It also outlines how the land will be **rehabilitated** or reconstructed, so that it can be used again after the mining is completed.

The environmental impact statement, along with any other relevant information, is studied by the government before permission to proceed is granted.

INVESTIGATION 9.6

Searching without disturbing
AIM: To model the search for minerals below the ground

Materials:

a tray of sand	10 paperclips
blindfold (optional)	compass
paper and clipboard	ruler

Method

- Find a partner. Each of you should then draw identical maps of the sand tray. Use a ruler to construct a grid on each map. Label the grids across the top and down the side (e.g. A–J across the top, 1–15 down the side). Each grid should consist of at least 100 equal-sized rectangles or squares.
- Without showing your partner, hide the paperclips in the tray of sand and mark the location of the ten clips on your map.
- Your partner's task is to locate the ten paperclips and mark them on the map without disturbing the sand. You might wish to set a time limit.
- Swap roles and repeat the steps above.

Discuss and explain

1. What property of the paperclips allowed them to be located?
2. How could your predictions of the location be checked with a pencil?
3. After checking, can the sand be restored to its initial condition?

The EIS reports on:
- existing flora, fauna and soils
- existing towns and roads in the area
- proposed new towns, roads and other developments
- how the new development might affect the local community and environment
- alternative plans to complete the development that might have less impact on the environment
- measures that will be put in place to monitor and control air, water and noise pollution during the project and while rehabilitation is undertaken
- rehabilitation proposals for the area.

9.6.4 Taking out the mineral ore

To obtain mineral ore from the ground, it is often necessary to remove large amounts of rocks and soil. The way this is done depends on how close the mineral ore is to the surface. If it is close to the surface, first the vegetation and topsoil are removed. Then waste rock from beneath the topsoil, called **overburden**, is removed. The removed topsoil and overburden are used to fill areas that have already been mined, or are left in a pile to restore the newly mined area when mining is completed. This method of mining is called **open-cut mining**.

If the mineral ores are deep below the surface, miners use **underground mining**. This mining method is more dangerous and expensive than open-cut mining. Shafts and tunnels are dug up to four kilometres into the ground to reach the rocks containing the mineral ore. The development of open-cut and underground mining is overseen by mining engineers.

9.6.5 Getting the metal

Obtaining the metal element takes place in two stages:
1. Mineral extraction separates the mineral ore from the rock taken from the ground. This involves crushing, grinding and washing the rock to separate the minerals from the unwanted rock.
2. Metal extraction separates the desired metal element from the mineral. This always involves chemical reactions. The nature of these reactions depends on a number of factors, including the chemical composition of the mineral ore. Chemical engineers are involved in the design of this process.

9.6.6 Rehabilitation

Before mining of a new site begins, seeds of the natural vegetation of the area are collected so that seedlings can be cultivated at a later stage. The seedlings are grown in special nurseries until they are mature enough to return to the site of the mine.

During open-cut mining, the overburden (the material removed from the site to expose the mineral ore or coal) is used to fill holes left from earlier stages of the mining operation. Fresh topsoil is used to cover the overburden to ensure that new vegetation will grow. The soil surface is shaped to fit in with the surroundings, fertilised and sown with seeds or planted with seedlings. Care is taken to shape the new surface to prevent the newly sown soil from being eroded or washed away by wind or rain.

Resurfacing and replanting a former open-cut iron mine on Koolan Island, Western Australia.

9.6 Exercises: Understanding and inquiring

To answer questions online and to receive **immediate feedback** and **sample responses** for every question, go to your learnON title at www.jacplus.com.au. *Note:* Question numbers may vary slightly.

Remember

1. Where are all mineral ores found?
2. In the Earth's crust, where are the metal elements found?
3. Describe the method of open-cut mining for removing mineral ores from the ground.
4. Outline the two stages involved in obtaining a metal element from rock.
5. What is an EIS?
6. Outline the information that is included in an EIS.
7. How do mining companies rehabilitate the land used for mining?

Think

8. The most common element in the Earth's crust is oxygen. This element is a gas except at extremely low temperatures. In what form is it found in the Earth's crust?
9. Explain why it is important to recycle metals as much as possible.
10. In a table like the one below, make a list of the benefits and disadvantages of mining.

Benefits	Disadvantages

11. Summarise the reasons for and against allowing mining to take place in Australia's national parks.

Brainstorm

12. In a small group, discuss and list:
    (a) the factors a mining company should consider when it decides whether to start a mining project
    (b) the different tasks that scientists and engineers might perform from the beginning of mining exploration until mining rehabilitation is complete.
    Compare the lists of your group with those of others in your class.

9.7 Rock technology

Science as a human endeavour

9.7.1 Back to the Stone Age

Rock technology began about two million years ago when early humans started using rocks to make simple chopping tools. This was the beginning of the period known as the **Stone Age**. For the great civilisations of Asia, Europe and North Africa, the Stone Age ended around 3000 BC with the discovery of bronze, an **alloy** of copper and tin.

The most commonly used resource in the Stone Age was a fine-grained sedimentary rock called **flint**. When flint breaks, it leaves a razor-sharp edge, so it was ideal for making sharp tools like knives, axes and spearheads.

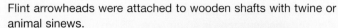
Flint arrowheads were attached to wooden shafts with twine or animal sinews.

Small tools were made by striking tool stones like flint or the glass-like igneous rock obsidian with harder stones, such as quartzite, a metamorphic rock. To remove large flakes from the tool stone, a sharp blow was delivered by the harder rock. If the tool stone was struck correctly, a flake sheared from it. This process is called **percussion flaking**. The toolmaker continued to remove flakes from the stone until the desired shape was obtained. The flakes were then used to make tools such as knife blades, scrapers and engravers.

Larger items such as axeheads and spearheads were made with a combination of techniques, such as percussion flaking, grinding stones against each other and chiselling against the edge of a stone with tools made of bone or wood.

9.7.2 Indigenous ingenuity

Aboriginals and Torres Strait Islanders were still using Stone Age tools when Europeans began to settle in Australia in 1788. They were highly skilled at working with stone. In fact, Indigenous Australians were the first people to use ground edges on cutting tools and to grind seed.

Their stone axes and other sharp tools were used to cut wood, shape canoes, chop plants for food, skin animals and make other tools out of stone or wood. The sharpened stones were often attached to wooden handles with twine from trees or with animal sinews.

Grinding stones are slabs of stone used with a smaller, harder top stone to grind seeds such as corn and wheat, berries, roots, insects and many other things to prepare food for cooking. Leaves and bark were sometimes ground to make medicines. Aboriginals also used grinding stones to grind various types of soil and rock to make the powders used to paint shields and other wooden implements with traditional patterns.

Hand axes made and used by the Ngadjonji people of the tropical rainforests of northern Queensland

Grinding stones were rough and usually made from sandstone, basalt with large crystals or quartzite. The smaller top stone was usually a hard, smooth river pebble.

The tools and the type of stone used to make them varied from group to group, depending on the location. Aboriginal people were skilled at making good use of the available resources. Apart from grinding stones, axes and other cutting tools, they made items such as bowls, cups and food graters out of stone.

An Aboriginal grinding stone with a top stone, or muller. The grinding stone is 40 cm long, 35 cm wide and 10 cm high. It is made from sandstone. The top stone is a hard, smooth river pebble.

A food grater made from stone by the Ngadjonji people of northern Queensland

9.7 Exercises: Understanding and inquiring

To answer questions online and to receive **immediate feedback** and **sample responses** for every question, go to your learnON title at www.jacplus.com.au. *Note:* Question numbers may vary slightly.

Remember

1. List one example of each of the following types of rock that were used in the Stone Age to make tools.
   (a) Igneous
   (b) Sedimentary
   (c) Metamorphic
2. Which alloy replaced stone as the substance chosen to make tools when the Stone Age ended?
3. What role did animal sinews play in toolmaking by Indigenous Australians?
4. List three different uses of grinding stones.

Think

5. What properties of flint made it so useful during the Stone Age?
6. List some properties that you would look for when selecting a suitable top stone for a grinding stone.
7. Suggest how the process of percussion flaking got its name.

Investigate

8. Research and report on a range of tools, weapons and other devices made from rocks or other natural materials that Aboriginal and Torres Strait Islander peoples used in their daily lives.

9.8 Every rock tells a story

9.8.1 Clues in rocks

If only rocks could talk! They would have so much to say. They would tell us about the Earth's history — about prehistoric creatures whose fossils lie within them, about explosive volcanoes, earthquakes, about flooded rivers that washed them away and about what it is like inside the Earth.

Although rocks can't talk, geologists are able to use them to answer questions such as:

- What did the first humans on Earth eat?
- How has the Earth's climate changed over millions of years?
- What caused the extinction of the dinosaurs?

The clues lie in the appearance of the rocks, how they feel and the different layers of rock beneath the Earth's surface. There are also clues in fossils. A **fossil** is evidence of living things preserved in rocks.

9.8.2 Layers of clues

Over very long periods of time, rivers change their course, mountains form where seas once existed and the climate changes. As these changes take place, different layers of sediment can be deposited at the same location. Some layers will be thicker than others. Sedimentary rocks, which are formed by the different layers of sediments, provide many clues about the order in which events took place. Sudden events, such as erupting volcanoes or earthquakes, are also recorded in the layers.

Slow movements caused by the forces beneath the surface can tilt, curve and push up the layers. Weathering and erosion can expose some layers that were below the ground millions of years ago.

7. These layers were deposited last. They have started to weather and erode.
6. A long period of weathering and erosion left the layer of limestone with a flat surface. When a volcano then erupted nearby, lava from the volcano cooled to form basalt on the flat surface.
5. A sudden event, such as an earthquake, has occurred to break the layers of rocks like this. This event took place after the lower layers were folded. A break like this is called a fault.
4. A slow event has caused the lower levels to buckle. This is called **folding**. Folding can occur when rock layers are under pressure from both sides.
3. The third event to occur was the deposition of limestone. It tells us that there were probably marine organisms present in the area during this time.
2. This is the second layer deposited. Shale is a fine-grained rock that is deposited in a quiet environment such as a swamp, lake or the slow-flowing part of a river.
1. Conglomerate was deposited first in this rock sample. This layer was deposited by a glacier or an active environment — such as a very fast-flowing river.

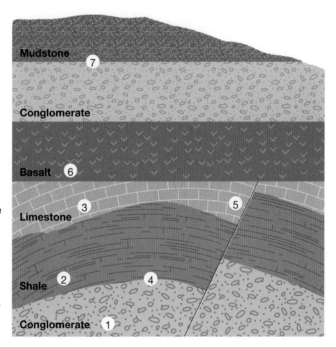

9.8.3 It's all relative

Fossils provide a way of finding out how living things have changed over time. Evidence of the very oldest living things is buried within the deepest and oldest layers of rock. Scientists who study fossils are called **palaeontologists**.

By comparing fossils found in rocks in different areas, including different continents, it is possible to compare the **relative age** of rocks throughout the world. The relative age of a rock simply indicates whether it was formed before or after another rock. In any particular location it is almost certain that a layer of sedimentary rock is older than the rocks above it and younger than those below it. The relative ages of some igneous rocks and metamorphic rocks can be determined in the same way. It can also be assumed that the fossils in lower layers are older than those in the layers above.

Fossils provide clues about life in the past. This is a fossil of an ancient reptile.

9.8.4 How fossils form

The remains of most animals and plants decay or are eaten by other organisms, leaving no trace behind. However, if the remains are buried in sediments before they disappear, they can be preserved, or fossilised. Fossils can take several forms.

Hard parts

The hard parts of plants and animals are more likely to be preserved than the softer parts. Wood, shells, bones and teeth can be replaced or chemically changed by minerals dissolved in the water that seeps into them. Fossils formed in this way are the same shape as the original remains but are made of different chemicals; petrified wood is an example. Animal bones and shells can be preserved in sediments or rock for many years without changing. The types of bones, shells and other remains found in the layers of sedimentary rock provide clues about the environment, behaviour and diets of ancient animals.

These insects were trapped in the resin of a tree millions of years ago.

INVESTIGATION 9.7

Making a fossil
AIM: To model the formation of a fossil

Materials:
small seashell
fine sand
small box (shoebox or milk carton)
plaster of Paris

CAUTION

Do not put plaster of Paris down the sink.

Method

- Half-fill the box with fine, damp sand.
- Make a clear imprint of a small seashell in the sand.
- Mix some plaster of Paris and pour it carefully into the imprint.
- Once the plaster has set, remove the plaster cast carefully from the sand.

You have two records of the seashell — the mould or imprint in the sand and the plaster cast.

Discuss and explain

1. Which parts of animals are most likely to be preserved as casts?
2. Is the fossil of a fern leaf more likely to be found as a cast or a mould? Why?
3. Dinosaur fossils are found in casts and moulds. What evidence of dinosaurs is likely to be found as a mould?

Whole bodies

Sometimes, fossils of whole organisms, including the soft parts, are preserved. Such fossils are rare and valuable. Insects that became trapped in the resin of ancient trees (the fossilised resin is called amber) have sometimes been wholly preserved. Similarly, if the remains of animals or plants are frozen and buried in ice, they can be fully preserved. Whole bodies of ancient woolly mammoths (including skin, hair and internal organs) have been found trapped in the ice of Siberia and Alaska. These remains provide clues to the way in which living things have changed since ancient times. Whole bodies and preserved skulls of animals can even reveal evidence of their last meal before death.

An ancient woolly mammoth. Whole bodies of these ancient animals have been discovered in the ice of Siberia and Alaska.

DINOSAURS PRESERVED IN ROCK

1. After the death of a dinosaur, its body would usually be eaten by meat-eating animals (**carnivores** or **scavengers**). Its bones would be crushed or weathered, leaving no remains. If, however, the remains of a dinosaur were buried in sediment, the bones could be preserved.
2. If a dinosaur died near a muddy swamp, shallow lake or riverbed, its remains sank in the mud or were washed into a river in a flood. The bones were quickly buried in sediment.
3. Over millions of years, more layers of sediment were deposited on top of the buried remains. Chemicals dissolved in the water that seeped into the remains changed their colour and chemical composition. The shape, however, was preserved. The sediments were gradually transformed into sedimentary rock.

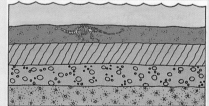

4. The layers of rock containing the fossilised remains were pushed upwards, bent and tilted by forces beneath the Earth's surface. Weathering and erosion by the wind, sea, rivers or glaciers might expose one or more of the bones or teeth. If the exposed fossils were discovered before being buried again, palaeontologists might discover the remains.

Making an impression

The remains of animals or plants sometimes leave an impression, or imprint, in hardened sediments or newly formed rock. It is also possible for remains trapped in rock to be broken down by minerals in water, leaving a **mould** in the shape of the organism.

Just a trace

Some fossils, called **trace fossils**, provide only signs of the presence of animals or plants. For example, footprints preserved in rock can provide clues about ancient animals, including dinosaurs, and how they lived. By studying the shape, size and depth of footprints, hypotheses can be made about the size and weight of **extinct** animals as well as how they walked or ran. Plant, leaf and root imprints, and feather impressions are other examples of trace fossils.

The imprint of the leaf of an ancient fern left in stone is a trace fossil.

9.8.5 Delving into dinosaurs

It is about 65 million years since the last non-flying dinosaurs existed on the Earth.
- What did they look like?
- What colour were they?
- How fast could they move?
- How did they behave?
- What did they eat?

Palaeontologists use fossils to try to answer all of these questions and more!

Not just a pile of bones

Dinosaur fossils are not all bones. They may include the following.

- Fossilised teeth: The shape of the teeth and the way they are arranged provide vital clues about the diets of dinosaurs.

  Flat-surfaced grinding teeth would have belonged to a dinosaur with a plant diet. When fossilised teeth like these are examined under a microscope, scratches caused by the grinding of the teeth are sometimes visible. Sharp-pointed teeth suited to tearing flesh would have belonged to a meat-eating dinosaur.

- Footprints: Dinosaur footprints are often preserved in rock. Footprints from a single dinosaur provide clues about its size and weight. They also indicate whether the dinosaur walked on two legs or four, and how its weight was spread. The distance between footprints enables palaeontologists to estimate how fast the dinosaur moved. Footprints also provide clues about the behaviour of dinosaurs and whether they lived in herds or alone.

- Impressions of skin may be left in mud that has hardened.

9.8 Exercises: Understanding and inquiring

To answer questions online and to receive **immediate feedback** and **sample responses** for every question, go to your learnON title at www.jacplus.com.au. *Note:* Question numbers may vary slightly.

Remember

1. A road cutting reveals the layers of rock shown in the diagram on the next page. Which of the rocks in the cutting is:
   (a) the oldest rock
   (b) the youngest rock
   (c) evidence of volcanic activity?
2. Explain why some layers of sedimentary rock are tilted, even though the sediments that formed them were laid in horizontal beds.
3. What does a palaeontologist study?
4. What clues about life in the past do fossils provide?
5. Under what circumstances can whole ancient living things be preserved as fossils?
6. Describe trace fossils and how are they useful.
7. What is the difference between a cast and a mould?
8. List the information about dinosaurs that can be obtained from fossils.
9. Fossils of dinosaurs form when their remains are buried under many layers of rock. Explain why fossils are often discovered in rocks and soil on the surface.

Think

10. In which rocks shown in the diagram below would you be most likely to find the fossil of:
    (a) a seashell
    (b) the leaf of a fern usually found in swamps?

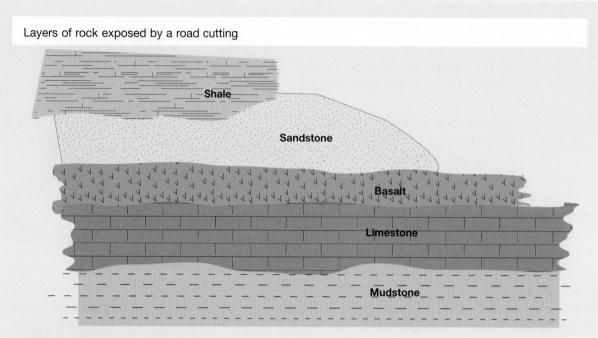

Layers of rock exposed by a road cutting

Shale

Sandstone

Basalt

Limestone

Mudstone

11. Why are some layers in the diagram below thicker than others?
12. Explain why the hard parts of plants and animals are more likely to be preserved than the softer parts.
13. Explain how it is possible to use preserved dinosaur footprints to form hypotheses about:
    (a) whether dinosaurs lived alone or in herds
    (b) the way that dinosaurs walked
    (c) the weight of different kinds of dinosaurs
    (d) the walking or running speed of dinosaurs.

Create and explore

14. Use plasticine to construct a sample of sedimentary rocks. Apply a gentle force to the sides of the layers. Describe how the layers fold under gentle pressure.

Investigate

15. Find out how the actual age of a rock in years is determined. This actual age is known as the absolute age.
16. Even an animal's droppings can become fossilised. Use the internet or books to research and report on the following.
    (a) Which animal was responsible for a huge fossilised dropping found in Canada in 1998?
    (b) How long was the dropping?
    (c) What can palaeontologists find out from it?

learn on RESOURCES — ONLINE ONLY

⚡ **Try out this interactivity:** Relative age of rocks (int-0233)

📄 **Complete this digital doc:** Worksheet 9.6: Tracking changes in the rock (doc-18761)

9.9 Questioning and predicting

9.9.1 What happened to the dinosaurs?

Two of the inquiry skills that geologists and other scientists use are questioning and predicting. The question of how the dinosaurs died out has intrigued scientists for many years. In answering this question, scientists use scientific knowledge to make 'predictions' about what happened many millions of years ago.

9.9.2 Solving the dinosaur riddle

Between about 250 million and 65 million years ago, dinosaurs were the most successful animals on Earth. In fact, those years are known as 'the age of the dinosaurs'. Dinosaurs thrived and dominated the land while mammals lived in their shadow. Fossil evidence indicates that the last of the dinosaurs died about 63 million years ago. There are several theories about the extinction of the dinosaurs. Scientists and others argue about whether the end of the dinosaurs was sudden or gradual. Scientists do generally agree that the riddle of the dinosaur extinction remains unsolved. Palaeontologists and other scientists continue to look for clues that might provide the final solution.

The asteroid theory

The most widely accepted solution to the dinosaur riddle is that an asteroid collided with the Earth around 65 million years ago. The asteroid's impact threw billions of tonnes of dust into the air, blocking out sun-

light and plunging the Earth into darkness for two or three years. Plants stopped growing but their seeds remained intact. The temperature dropped. The large plant-eating dinosaurs would have died quickly of starvation. The meat-eating dinosaurs would probably have died next, having lost their main food supply but surviving for a while by eating smaller animals. Many smaller animals would have survived by eating seeds, nuts and rotting plants.

As the debris began to settle and sunlight filtered through the thinning dust clouds, many of the plants began to grow again. The surviving animals continued to live as they did before the impact. The surviving mammals were no longer competing with dinosaurs for food. It was the beginning of the age of mammals.

The volcano theory

The eruption of Mount Pinatubo in the Philippines in June 1991 showed that ash and gases from volcanoes could reduce average temperatures all over the world. The average global temperatures during 1992 and 1993 were almost 0.2°C less than expected. While this is not a large drop in temperature, the size of the eruption of Mount Pinatubo was very much smaller than those of many ancient volcanoes.

The ash from a large volcano could have the same effect on sunlight and the Earth's temperature as an asteroid impact. If there was an unusually large amount of volcanic activity about 65 million years ago, the extinction of the dinosaurs could be explained. The largest known volcanic eruption occurred about 250 million years ago in what is now Siberia. It is believed that many types of marine animals became extinct at about the same time.

The cooling climate theory

The gradual cooling of the Earth's climate due to changes in the sun's activity could have caused the extinction of the dinosaurs. Dinosaurs, with no fur or feathers, had less protection from cold weather than mammals and birds. The larger dinosaurs would have found it very difficult to shelter from the cold conditions. Many smaller animals could burrow below the ground or shelter in the hollow trunks of trees or in caves. Many mammals and birds would have been able to migrate to warmer regions closer to the equator.

The emerging plants theory

During the Cretaceous period (140 million to 65 million years ago), new types of plants began to appear. Flowering plants evolved, competing with the more primitive plants such as ferns for nutrients, water and sunlight. The plant-eating dinosaurs did not eat flowering plants. According to this theory, as their traditional food supply became more scarce, the plant-eating dinosaurs could not survive, and the meat-eating dinosaurs that preyed on them starved as well.

Cold-blooded or warm-blooded?

Until recently, it was believed that dinosaurs were **ectothermic**. Ectothermic animals have body temperatures that depend on the temperature of their surroundings. As the surrounding temperature decreases, their body temperature decreases and they become less active.

Mammals are **endothermic**. Endothermic animals are able to maintain a constant body temperature that is usually above that of their surroundings. They are able to remain warm and active in lower surrounding temperatures.

If dinosaurs were in fact ectothermic, a cooler climate would have made it more difficult for them to compete with other animals for food. However, many scientists now believe that dinosaurs may have been endothermic.

The question of whether dinosaurs were cold-blooded or warm-blooded needs to be answered before the riddle of the dinosaurs can be solved.

9.9 Exercises: Understanding and inquiring

To answer questions online and to receive **immediate feedback** and **sample responses** for every question, go to your learnON title at www.jacplus.com.au. *Note:* Question numbers may vary slightly.

Remember

1. What is the most widely accepted theory about the extinction of the dinosaurs?
2. Why would smaller animals be more likely to survive the effects of an asteroid impact or large volcanic eruption than larger animals?
3. What is the difference between an ectothermic animal and an endothermic animal?

Think

4. In what ways were the dinosaurs different from mammals?
5. How could volcanic eruptions affect life throughout the whole world?
6. How could meat-eating dinosaurs be endangered by the evolution of new types of plants?
7. Which group of animals benefited the most as a result of the extinction of the dinosaurs?
8. List as many weaknesses as you can in each of the four theories about the dinosaur extinction presented.
9. Which theory of the extinction of the dinosaurs do you think is most likely to be correct? Explain your answer.

Imagine

10. Imagine what it would have been like 65 million years ago if an asteroid plunged into the Earth. Write a story about the first 24 hours after the impact.
11. Which animals and plants do you think would be most likely to survive if an asteroid struck central Australia now? Explain your answer.

9.10 Fishbone diagrams and tree maps

9.10.1 Fishbone diagrams and tree maps

1. Think of an event that you do not know the causes of.
2. Brainstorm as many possible causes as you can for this event.
3. In pairs or teams of four, use an affinity diagram to organise your list of causes into groups.
4. Write the event that you are analysing as the 'fish's head' of a fishbone diagram. Your groups of causes then become the main 'bones' of the diagram, one bone for each group.
5. Write the title for each of your groups of causes on its relevant 'fishbone'.
6. Write the causes on the smaller 'fishbones' that are joined to the sides of the main bones. (You can attach causes to more than one bone or group of causes.)

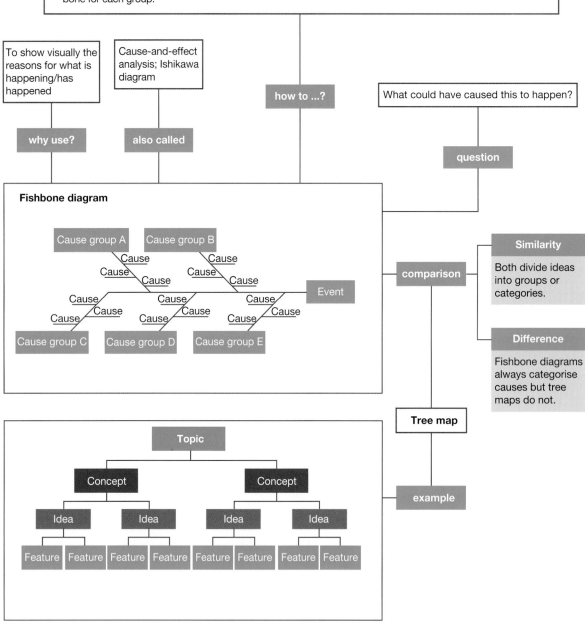

9.10 Exercises: Understanding and inquiring

To answer questions online and to receive **immediate feedback** and **sample responses** for every question, go to your learnON title at www.jacplus.com.au. *Note:* Question numbers may vary slightly.

Think and create

1. Create a fishbone diagram that shows the possible causes of the extinction of the dinosaurs. Use the one below as a template, adding the smaller 'bones' and causes yourself. The four theories about the cause discussed in this subtopic are already on the template. Include a separate theory of your own on the diagram. It could be a combination of the four other theories, or something completely different.

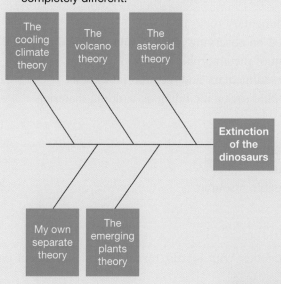

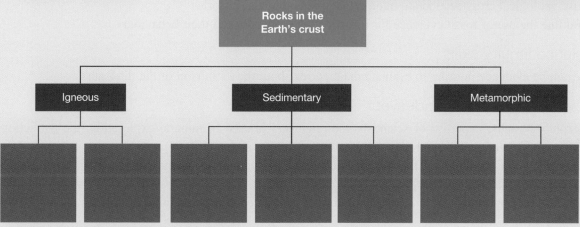

2. Working with a partner, copy the tree map above onto A3 paper and complete it to represent the three main types of rock found in the crust of the Earth. Add further branches below the existing ones if you can.
3. Create a fishbone diagram to represent the formation of rocks. Use the cause categories to the right, and add any others that you think should be there.

CAUSE CATEGORIES

weathering   erosion   volcanoes
pressure   heat

9.11 Review

9.11.1 Study checklist

Classifying rocks

- describe the formation of igneous rocks
- distinguish between extrusive and intrusive igneous rocks
- explain how cooling rate affects crystal size
- describe the formation of sedimentary rocks
- explain the role of water in the formation of many sedimentary rocks
- identify a range of sedimentary rocks using a key based on observing physical and chemical properties
- describe the roles of heat and pressure in the formation of metamorphic rocks
- identify the cyclic nature of the formation of igneous, sedimentary and metamorphic rocks
- outline the uses of igneous, sedimentary and metamorphic rocks, including as building materials

Minerals

- recall that all rocks are made of substances called minerals
- describe the physical properties of a variety of minerals
- recognise that the minerals in some rocks provide valuable resources
- describe the processes involved in mining mineral ores

Reading the earth's history in rocks

- explain how layers of sedimentary and other rocks, together with fossils, reveal information about past environments and life on Earth
- use geological cross-sections to interpret simple geological histories
- outline the use of fossil evidence for investigating dinosaurs and their behaviour

Science inquiry skills

- examine evidence in order to evaluate the theories about the extinction of the dinosaurs

Science as a human endeavour

- explain how the expertise of scientists and engineers is used in mineral exploration, the extraction of mineral ores and metals, and the rehabilitation of mining sites
- describe some examples of the use of traditional rock technology in the daily lives of Aboriginal and Torres Strait Islander peoples
- discuss the environmental and community issues associated with the mining of mineral resources

Individual pathways

ACTIVITY 9.1	ACTIVITY 9.2	ACTIVITY 9.3
Investigating rocks and minerals doc-6057	Analysing rocks and minerals doc-6058	Investigating rocks and minerals further doc-6059

learn on ONLINE ONLY

9.11 Review 1: Looking back

To answer questions online and to receive **immediate feedback** and **sample responses** for every question, go to your learnON title at www.jacplus.com.au. *Note:* Question numbers may vary slightly.

1. In which parts of the Earth are rocks formed?
2. How are all igneous rocks formed?
3. Explain the difference between the ways in which extrusive igneous rocks and intrusive rocks are formed.
4. List three examples of extrusive igneous rocks.
5. What clues does the size of the crystals in an igneous rock provide about how the rock was formed?
6. What are sediments?
7. Describe the three different ways in which sedimentary rocks can be formed.
8. Suggest a way of checking that a rock sample that appears to have seashells embedded in it is limestone.
9. Explain why some layers of sedimentary rocks are tilted or bent.
10. While studying sedimentary rocks in a railway cutting, a geologist discovers a bed of rock with ripple marks in its surface. How could the ripple marks have been made in the rock?
11. Describe two ways in which igneous and sedimentary rocks can be transformed into metamorphic rocks.
12. What is a parent rock?
13. Copy and complete the table below to summarise what you know about igneous, sedimentary and metamorphic rocks.

Type of rock	How it is formed	Special features	Example	Uses
Igneous				
Sedimentary				
Metamorphic				

14. Copy and complete the diagram at right to show how some common metamorphic rocks are formed.

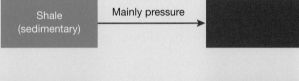

15. The changes that lead to the formation of the three main groups of rocks can be drawn as a cycle, as shown below.

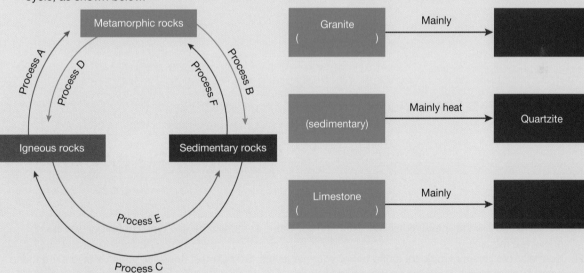

Which of processes A–F involve:
(a) weathering and erosion
(b) heat and pressure
(c) remelting?

16. Explain why the crystals in granite are larger than those in basalt.
17. What characteristic of minerals do the following terms describe?
(a) Lustre
(b) Streak
(c) Hardness

18. What property of minerals does Mohs' scale provide an approximate measure of?

19. If you were given a sample of each of two different minerals, how could you tell which one had the greater hardness?

20. Which sedimentary rock was the most important material for toolmaking in the Stone Age? What properties made it so useful?

21. In the Stone Age, tools were made by a process called percussion flaking.
    (a) Describe the process of percussion flaking.
    (b) Which property of the rock used to make the tool must be different from those of the stone from which the tool is formed?

22. What is the most common element in the Earth's crust and where is it found?

23. One factor that determines the way in which mineral ores are mined is their depth. Compare the mining processes used for mineral ores located near the Earth's surface with those used for mineral ores located deeper in the Earth's crust.

24. According to many geologists, parts of Antarctica are rich in mineral resources, similar to those found in Australia. Use a two-column table to list reasons why these mineral resources should be mined and why they should not be mined.

25. The mining industry provides employment for many Australians. Make a list of occupations that are involved in the mining industry. (*Hint*: Think about what happens before, during and after mining is undertaken.)

26. Imagine that the set of fossilised dinosaur footprints shown in the illustration below were found in a layer of sedimentary rock.
    (a) Use the footprints to write a description of what might have happened millions of years ago.
    (b) Compare your interpretation of the footprints with others.
    (c) Does each person interpret the evidence in the same way?
    (d) If there are differences of opinion about what happened, is there any way of knowing who is right?
    (e) List as many differences as you can between the two types of dinosaurs making these footprints.

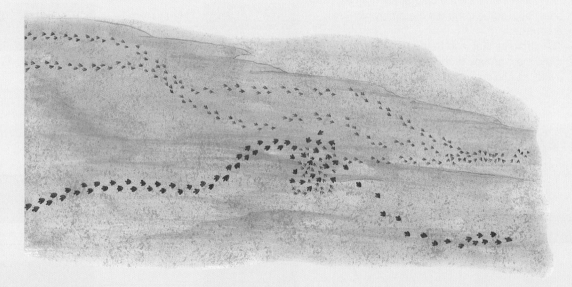

27. Not all fossils are the actual remains of living things. Name and describe two types of fossils that are not preserved remains.

28. Normally, old layers of rock are found below younger layers. Sometimes, however, younger layers are found beneath older layers. Explain how this could happen.

29. The photograph on the next page is of dinosaur footprints that have been preserved in rock at Gantheaume Point near Broome
    (a) What type of fossil is it?
    (b) Why is it classified as a fossil even though it could be described as a dent in a rock?
    (c) Have all dinosaur footprints been preserved? Why have these been preserved for hundreds of millions of years?
    (d) What can be learned about the features of the dinosaur that left these footprints?
    (e) What forms of evidence, apart from preserved footprints, can be used to gather knowledge about dinosaurs?

This dinosaur footprint has been preserved in rock for hundreds of millions of years at Gantheaume Point near Broome.

TOPIC 10
Transferring and transforming energy

10.1 Overview

A fireworks display is one of the most spectacular energy transformations; you can not only see it but also hear, feel and smell it. When fireworks are ignited, the energy stored in the substances inside them is quickly transformed into movement (kinetic energy), light energy, sound energy and thermal energy (more commonly called heat). Energy that is stored is known as potential energy.

10.1.1 Think about energy

assess on

- Which type of energy do you find in chocolate?
- When you drop a tennis ball to the ground, why doesn't it return to its initial height?
- How much electrical energy is wasted as heat by an incandescent light globe?
- How does ceiling insulation keep your house warmer in winter?
- Where does a firefly get the energy to light up?
- How do glow-in-the-dark stickers work?
- How does a didgeridoo player create its unique sound?

Numerous **videos** and **interactivities** are embedded just where you need them, at the point of learning, in your learnON title at www.jacplus.com.au. They will help you to learn the content and concepts covered in this topic.

10.1.2 Your quest

Potential energy and kinetic energy

All substances and objects possess **potential energy**. But you can't tell unless something happens to transform the potential energy into a different type of energy. In the case of fireworks it's obvious when they explode. When a diver dives from a platform or diving board, the **kinetic energy** they gain on the way down is transformed from the energy stored in them because of their height above the ground. And the energy stored in the stretched string of a bow is transformed into the kinetic energy of the arrow when it is released.

THINK

1. Copy and complete the table below. One example has been completed for you.

Object	What to do to release the stored energy	Potential energy is usefully transformed into ...
Torch battery	Switch it on	electrical energy and light energy
Chocolate		
Petrol		
Dynamite		
Olympic diver on platform		
Match		
Stretched elastic band		

2. Answer the following questions about the wind-up toy shown below.
   (a) Where is the energy stored when it is wound up?
   (b) What do you have to do to allow the stored energy to be transformed into different forms?
   (c) Name two forms of energy into which the potential energy is transformed.
   (d) From where does the energy come that allows the user to wind up the toy?

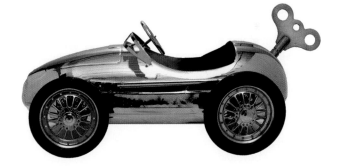

10.2 Matter and energy: Making things happen

10.2.1 What is energy?

Energy is a word that you sometimes use to describe how active you feel. Sometimes you don't seem to have any energy. At other times you feel like you have enough energy to do just about anything. Energy is defined as 'the ability to do work'. That is, it is the ability to make something observable happen.

We know that:
– all things possess energy even if they are not moving
– energy cannot be created or destroyed. This statement is known as the **Law of Conservation of Energy**. It means that the amount of energy in the universe is always the same.

– energy can be transferred to another object (for example, from a cricket bat to a ball) or transformed into a different form (for example, from electrical into sound)
– energy can be stored.

10.2.2 Types of energy

Types of energy	
Potential energy (stored energy that, when released, is converted to other forms such as kinetic, sound, heat or light energy)	**Other types of energy** (often converted from potential energy, these are more easily observed by our senses)

Gravitational (potential energy of an object elevated above the ground)		**Kinetic** (energy possessed by objects that are moving)	
Elastic (energy stored by an elastic object that is stretched, such as a spring or rubber band)		**Heat** (energy that causes objects to gain temperature)	
Chemical (energy stored in chemicals that, when reacted together such as in burning reactions, release heat, sound or light)			
Nuclear (energy stored in the nucleus of atoms that can release energy slowly, such as in a nuclear reactor, or quickly, such as in a nuclear explosion)		**Light** (energy that may be released, for example, when an object is hot or by a nuclear reaction in a star)	
Electrical (energy supplied to homes by powerlines and available to your appliances via power outlets in the home)		**Sound** (energy carried by the air in a room and detected by the ear)	

Light energy, sound energy, thermal energy and kinetic energy are all very easily observed. All objects that move have kinetic energy. **Electrical energy** can be seen if there is a spark or a lightning strike, but you can't see it when it's moving in wires. You become aware of it when it is changed into other forms, for example into light in a fluorescent tube or into sound in an iPhone.

Types of energy changes involved in bouncing on a trampoline

| Gravitational potential energy | → | Kinetic energy | → | Elastic potential energy | → | Kinetic energy | → | Gravitational potential energy |

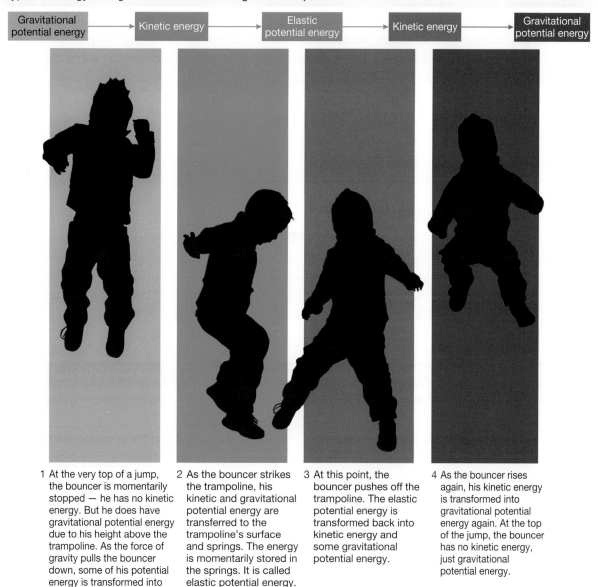

1 At the very top of a jump, the bouncer is momentarily stopped — he has no kinetic energy. But he does have gravitational potential energy due to his height above the trampoline. As the force of gravity pulls the bouncer down, some of his potential energy is transformed into kinetic energy.

2 As the bouncer strikes the trampoline, his kinetic and gravitational potential energy are transferred to the trampoline's surface and springs. The energy is momentarily stored in the springs. It is called elastic potential energy.

3 At this point, the bouncer pushes off the trampoline. The elastic potential energy is transformed back into kinetic energy and some gravitational potential energy.

4 As the bouncer rises again, his kinetic energy is transformed into gravitational potential energy again. At the top of the jump, the bouncer has no kinetic energy, just gravitational potential energy.

Stored energy is known as potential energy because it has the 'potential' to make something happen. There are several different forms of potential energy, some of which are described in the following examples.

- A ball held above your head has **gravitational potential energy**. This form of energy becomes noticeable when you drop the ball and its stored energy is transformed into kinetic energy.
- A battery contains **chemical energy** but this is not noticeable until the battery is connected in an electric circuit. When that happens the chemical energy is transformed into electrical energy, which in turn is transformed into other types of energy — to make things glow, get hot, produce sounds or move. It is the chemical energy stored in food and drinks that gives you the energy to live and be active. The chemical energy in fuels is transformed to operate cars and other vehicles, keep you warm and generate electricity.
- The **elastic potential energy** stored in a stretched elastic band is released when you let go of one end. The stored energy is transformed into kinetic energy.

- **Nuclear energy** is the energy stored at the centre of atoms, the tiny particles that make up all substances. The energy we receive on Earth from the sun has been transformed from nuclear energy. Under the right conditions, nuclear energy can be transformed into electrical energy in a nuclear power station. Unfortunately it can also be transformed into thermal energy in nuclear weapons.
- Electrical energy can also be stored. For example, if you rub a plastic ruler with a cloth it can become charged. You can't see the stored electrical energy but you can tell it's there when the ruler bends a slow stream of water from a tap.

learn on RESOURCES — ONLINE ONLY

Watch this eLesson: Energy in disguise (eles-0063)

10.2.3 An unavoidable loss

Every electrical appliance you use, whether powered by batteries or plugged into a power point, converts electrical energy into other forms of energy. Most of that energy is usually converted into useful energy — but some is converted into forms of energy that are wasted or not so useful. But all of the electrical energy is converted — that's the Law of Conservation of Energy in action. The table below shows some examples of energy conversion by electrical appliances. None of the wasted energy is actually lost.

Energy conversion by appliances

Appliance	Electrical energy usefully converted to ...	Electrical energy wasted ...
Microwave oven	thermal energy of food	heating air in the oven, plates and cups etc.
Television	light and sound	heating the television and the surrounding air
Hair dryer	thermal energy and kinetic energy of air	as sound
Electric cooktop	thermal energy of food	as light and heating the surrounding air

This loss of useful energy is also apparent when you step on the brake pedal in a car — not all the energy you transfer to the pedal is used to stop the car. Much of it is lost in the brakes and to the surrounding air as heat. The same applies to using the brakes of a bicycle. Also, when you drop a tennis or cricket ball it never bounces back to its original height because some energy is lost as heat. On a larger scale it is seen in power stations, where the fuel, falling water, solar energy or any other energy source is used to produce electricity; some of the energy of the source is transformed to heat, warming the power equipment, the surrounding air and the water used as coolant. The 'loss' of useful energy is unavoidable.

Some types of lighting waste more energy than others. Old-fashioned incandescent light bulbs convert more energy to wasted heat than to light. They emit light only when the filament inside gets white hot. Fluorescent lights and LEDs (light-emitting diodes) waste very little energy. Almost all of the electrical energy is converted to light, so you use much less energy to produce the same amount of light than you would using an incandescent bulb.

HOW ABOUT THAT!

In old-fashioned incandescent light bulbs, electricity passes through a thin filament in the bulb, causing it to glow white hot. The light is a useful form of energy, but about 90 per cent of the electrical energy is wasted as heat. Compact fluorescent lights (CFLs) offer a more energy-efficient form of lighting, but light-emitting diodes (LEDs) are even more efficient.

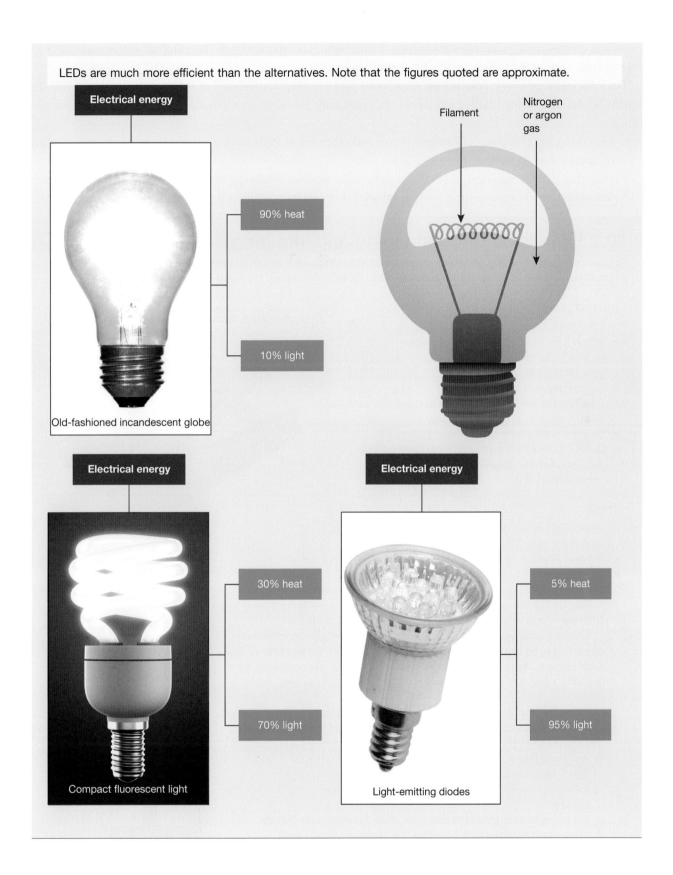

LEDs are much more efficient than the alternatives. Note that the figures quoted are approximate.

Electrical energy

90% heat

10% light

Old-fashioned incandescent globe

Filament

Nitrogen or argon gas

Electrical energy

30% heat

70% light

Compact fluorescent light

Electrical energy

5% heat

95% light

Light-emitting diodes

10.2.4 Efficiency

The **efficiency** of a car, light bulb, gas heater, power station, solar cell or any other energy converter is a measure of its ability to provide useful energy. Efficiency is usually expressed as a percentage. The efficiency of the incandescent light globe pictured on this page is 10 per cent because 10 per cent of the total electrical energy input is usefully transformed into light. The efficiency of the compact fluorescent light is 70 per cent, and the LED light is 95 per cent efficient.

The efficiency of every device that uses fossil fuels is very important for the environment and life on Earth. Scientists and automotive engineers are constantly working on methods of reducing fuel consumption by:

- increasing the efficiency of burning petrol and other fossil fuels such as diesel by reducing the amount of energy wasted as heat
- changing the external design of cars to reduce the amount of energy needed to overcome air resistance
- searching for alternative fuels such as ethanol that can be produced from sugar cane and grain crops.

10.2 Exercises: Understanding and inquiring

To answer questions online and to receive **immediate feedback** and **sample responses** for every question, go to your learnON title at www.jacplus.com.au. *Note:* Question numbers may vary slightly.

Remember

1. State the Law of Conservation of Energy.
2. Which form of energy is observed when:
   (a) an athlete runs
   (b) a spring is squashed?
3. List five types of stored energy.
4. Describe the difference between gravitational potential energy and elastic potential energy.
5. Define the efficiency of an energy converter.
6. Outline three of the methods being used by scientists and automotive engineers to reduce the fuel consumption of cars.

Think

7. Identify four types of energy that are present during a lightning strike.
8. From which type of energy does the sound of a cymbal come?
9. From which two types of energy does the sound of a bass drum come?
10. How can you tell that a high diver has gravitational potential energy?
11. A catapult like the one top right was used by the Romans more than 2000 years ago to attack castles, cities and invading armies.

     The long arm was held in its usual vertical position with rope twisted around its base in what is known as a torsion bundle. The arm was pulled back towards the ground using a second rope so that the bucket could be loaded with a missile. This causes the torsion bundle to twist more tightly.

     When the arm was released, the torsion bundle quickly untwisted and it returned to its vertical position, releasing the missile from the bucket at high speed towards the target. The missiles fired included rocks, burning tar and even human corpses.

     Use flowcharts to show:
     (a) the energy transfers that take place during the loading and firing of the missile
     (b) the energy transformations that take place from the time that the missile is loaded until the time that the missile finds its target.
12. Construct a table similar to the one on the next page and use it to list the useful energy and the wasted energy converted by the following devices.
     (a) A torch
     (b) A wind-up toy
     (c) A pop-up toaster
     (d) A gas cooktop
     (e) A car engine

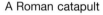
A Roman catapult

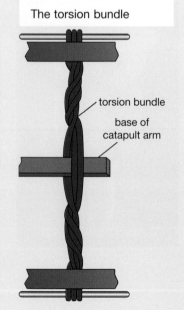
The torsion bundle

torsion bundle

base of catapult arm

Device	Source of energy	Energy usefully converted to ...	Forms of energy wasted

13. When a tennis ball is bounced on the ground, it never returns to its original height. Does this break the Law of Conservation of Energy? Explain your answer.
14. Outline at least three reasons why efficiency is important for devices that use fossil fuels.
15. Suggest some methods that drivers could use to increase the fuel efficiency of their vehicles.

Create

16. Illustrate the energy transfers and transformations of the person on the trampoline shown on page 391 in the form of a flowchart.
17. Create a poster-sized flowchart to show the energy transformations that take place to produce lightning and thunder. (Think first about how the clouds become electrically charged during an electrical storm.)

Investigate

18. Water wheels have been used in the past (and are still being used) to convert the energy of moving water to other useful forms of energy. Research and report on one example of the use of a water wheel. In your report, use flowcharts to illustrate the transformations and transfers of energy that take place.
19. Are solar-powered cars a realistic alternative to cars that run on fossil fuels or biofuels like ethanol? Find out what scientists, engineers and members of the public have contributed to the design of solar-powered vehicles.
20. Find out the purpose of the Australian International Solar Challenge and how you could become involved in it.

---

learn on RESOURCES — ONLINE ONLY

Watch this eLesson: Australian International Model Solar Challenge (eles-0068)

Try out this interactivity: Coaster (int-0226)

Complete this digital doc: Worksheet 10.1: Skateboard flick cards (doc-18764)

Complete this digital doc: Worksheet 10.2: Types of energy (doc-18765)

10.3 Hot moves

10.3.1 Heat transfer

If you accidentally touch a hotplate you'll find out quickly — and painfully — that heat travels from warm objects to cooler objects.

It is the rapid transfer of energy into your hand that causes the pain. Sports people sometimes use ice baths to assist with injury. The body heat is transferred quickly to the cold ice. If you touch something that has the same temperature as your hand, you won't feel any sensation of heat transfer into or away from your hand.

Heat is energy in transit from an object or substance to another object or substance with a lower temperature. Heat can move from one place to another in three different ways — by conduction, convection or radiation.

10.3.2 Conduction

If you've ever picked up a metal spoon that has been left in a hot saucepan of soup you will know that heat moves along the spoon and up to the handle. This is an example of **conduction** of heat. Metals are very good conductors of heat. Like all substances, metals are made up of tiny particles. The particles in all solid substances are vibrating. Of course, you can't see the vibrations because the particles are far too small to see — even with a microscope. When you heat a solid object, its temperature increases. The particles vibrate faster and bump into each other. The vibrations are passed from particle to particle along the object until the whole object is hot.

Not all solids conduct heat at the same rate. Metals, for example, are much better conductors than most other solids. Some solid substances are very poor conductors of heat. Glass, wood, rubber and plastic are all poor conductors of heat, and are called **insulators**. Many metal saucepans have a plastic or wooden handle. Suggest a reason for this.

10.3.3 Convection

The particles that make up solids are close to each other and held together tightly. They can vibrate faster only when heated. However, in liquids and gases the particles are further apart and can move around. So when they are heated, rather than the vibration passing between particles, the particles themselves can move. Heat can travel through liquids and gases by a process called **convection**.

The figure on the right shows how convection takes place. Heat causes the particles of air to gain energy, move faster and spread out. This warmer air is less dense than the air around it, so it rises. As it rises it begins to cool. The particles lose some of the energy gained, slow down and move closer

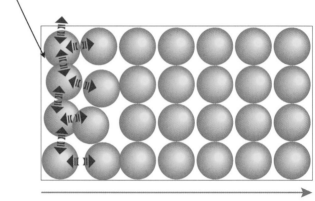

Conduction of heat occurs as a result of vibrating particles.

These particles vibrate faster.

The vibrations are passed along the object.

Flame or heat

Modelling heat transfer in air by convection

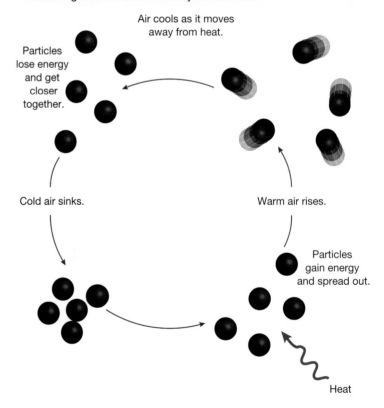

Air cools as it moves away from heat.

Particles lose energy and get closer together.

Cold air sinks.

Warm air rises.

Particles gain energy and spread out.

Heat

together. This cooler air is denser than the air around it, so it falls. The whole process then starts again, creating a pattern of circulation called a convection current.

Gas wall heaters create convection currents with the aid of a fan that pushes warm air out near floor level so that it heats the entire room as the air rises.

10.3.4 Radiation

Heat from the sun cannot reach Earth by either conduction or convection because there are not enough particles in space to transfer heat by moving around or passing on vibrations. Heat from the sun reaches Earth by **radiation**. Heat transferred in this way is called **radiant heat**. Heat transfer by radiation is much faster than heat transfer by conduction or convection.

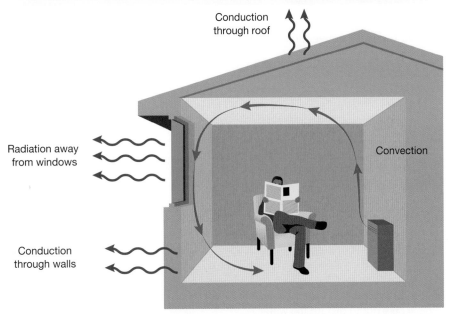

The transfer of heat in a house by conduction, convection and radiation

A camp cookout — heat is transferred by radiation, conduction and convection.

Sea breezes caused by convection currents

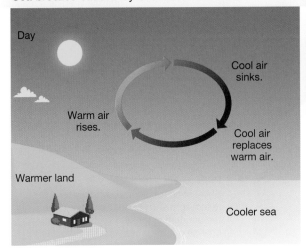

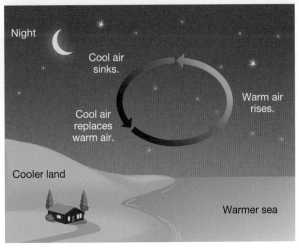

Moving particles

AIM: To model convection currents in a liquid

Materials:
250 mL beaker   *tweezers*
single small crystal of   *drinking straw*
    potassium permanganate
Bunsen burner and
    heatproof mat
tripod and gauze mat

Method and results

- Fill the beaker with water and place it on a gauze mat and tripod.
- Use the tweezers to drop a crystal of potassium permanganate down the drinking straw into the water at the bottom of the beaker.
- Slowly remove the straw, taking care not to disturb the water.
- Light the Bunsen burner and turn it to a blue flame. Be sure not to disturb the beaker.
1. Draw a diagram to show the movement of colour through the beaker. This will show the currents within the beaker.

Discuss and explain

2. Explain why the colour moved in the way it did.
3. Is this experiment successful at modelling convection? Explain why or why not.

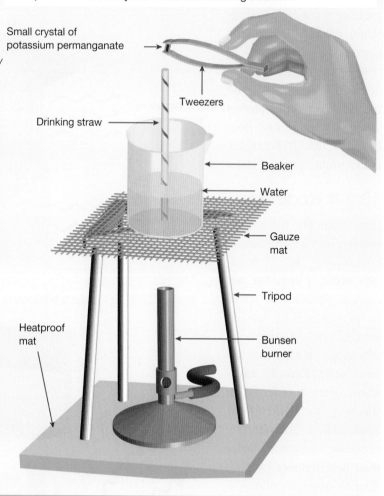

Drop the coloured crystal down the drinking straw.

Small crystal of potassium permanganate

Tweezers

Drinking straw

Beaker

Water

Gauze mat

Tripod

Heatproof mat

Bunsen burner

10.3.5 Transmission, absorption and reflection

When radiant heat strikes a surface, it can be **reflected**, **transmitted** or **absorbed**. Most surfaces do all three; some surfaces are better reflectors, others are better absorbers and some transmit more heat. The diagrams on the right show how different surfaces are affected by radiant heat.

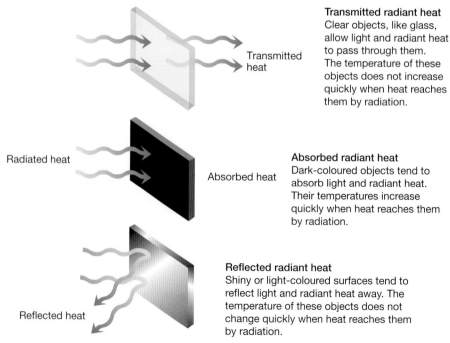

Transmitted heat

Radiated heat

Absorbed heat

Reflected heat

Transmitted radiant heat
Clear objects, like glass, allow light and radiant heat to pass through them. The temperature of these objects does not increase quickly when heat reaches them by radiation.

Absorbed radiant heat
Dark-coloured objects tend to absorb light and radiant heat. Their temperatures increase quickly when heat reaches them by radiation.

Reflected radiant heat
Shiny or light-coloured surfaces tend to reflect light and radiant heat away. The temperature of these objects does not change quickly when heat reaches them by radiation.

During cold weather, snakes lie against rocks that have absorbed some heat from the sun. The fast-moving particles in the rocks transfer some of their energy to the snake, warming it up.

10.3 Exercises: Understanding and inquiring

To answer questions online and to receive **immediate feedback** and **sample responses** for every question, go to your learnON title at www.jacplus.com.au. *Note:* Question numbers may vary slightly.

Remember

1. Copy and complete the table below.

Type of heat transfer	Describe briefly how heat moves	Substances in which heat moves in this way
Conduction		
Convection		
Radiation		

2. What is an insulator? Name three different materials that can act as insulators.
3. Heat can travel through empty space (for example, between the sun and Earth). How does the heat move?
4. What three things can happen to radiated heat when it arrives at any surface?

Think

5. Conduction occurs in solid materials like metals but is not an effective way of transferring heat in liquids and gases. Explain why this is so.
6. Draw a diagram similar to the one on page 397 (top right) to show how air-conditioners push cool air out near the ceiling to create convection currents that cool rooms in hot weather.
7. When you hold a mug of coffee or hot soup, your hands feel warm. How is the heat transferred to your hands? Use a storyboard, cartoon or flowchart to illustrate your response.
8. Would it be hotter to sit in a black or a white car during summer? Why?

Investigate

9. Compare the advantages and disadvantages of evaporative and refrigerated air-conditioners.
10. How quickly do things cool? The rate at which substances cool is determined by many factors. A cup of hot chocolate will cool more rapidly than the same cup filled with thick vegetable soup. The material in the cup is one variable that affects how quickly cooling takes place. The size of the container, the temperature around the outside of the container, and the type of container are other variables that affect the rate of cooling.

   Choose one variable to investigate. All other variables must remain the same so that the test is fair. If, for example, you decide to investigate the effect of the shape of the cup, you must make sure that nothing but the shape changes. The two or three shapes of cup you choose to investigate would need to contain the same amount of liquid, start at the same temperature, be made from the same materials, and be in the same surroundings.
   - Write down the aim of your investigation and state your hypothesis.
   - List the set of steps that you will follow.
   - Decide what equipment is needed and make a list of it.
   - Decide how your results will be recorded and draw up any necessary tables.
   - Check with your teacher before beginning.
   - Use your results to write a conclusion. State whether your hypothesis was supported.

10.4 A costly escape

10.4.1 Saving energy

Knowledge of how heat moves from a warm place to a cooler place can help you to save on the energy that is used to heat and cool your home.

Using less energy for heating and cooling also conserves valuable resources such as coal and natural gas that are used to generate electricity.

10.4.2 Staying warm

In winter, heat leaves the inside of a warm, cosy home by conduction, convection and radiation. New homes are designed to reduce heat losses by all three methods. However, there are also measures that occupants can take to reduce heat losses (and the bills that go with them).

Heat can escape from many different places.

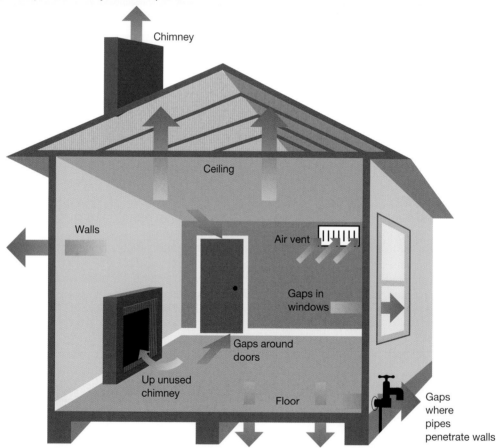

Using the sun

The direction that a house faces, positioning of windows and skylights, and the types of trees planted around the house all affect the amount of sunlight and radiated heat that enter a home. **Deciduous** trees planted near north-facing windows allow radiated heat from the sun through in winter but block it out in summer.

Insulation

Heat loss by conduction occurs through the ceiling, walls, windows and floor. Since air is a very poor conductor of heat, materials containing air reduce heat loss. However, if the air is free to circulate, it can move away, taking heat with it. The best insulators, therefore, are those that contain air that is restricted from moving. Woollen clothes, birds' feathers and animal fur are all good insulators because they restrict heat loss by both conduction and convection.

Some ways in which insulation is used in the home include:

- ceiling insulation such as fibreglass batts and loose rockwool that can be blown in. These materials contain pockets of air that provide insulation, and reduce the loss of warm air from the roof by convection.
- cavity wall insulation, a foam that can be sprayed in between the inside and outside walls
- heavy curtains, which trap a still layer of air between them and windows
- double glazing — the use of two sheets of glass in windows with a narrow gap of air between them
- cavity bricks, which have holes in them. The still air in the holes reduces heat loss by conduction and convection.

10.4.3 Do you feel a draught?

Preventing draughts is the easiest way to reduce heat loss in winter. There are many products available from hardware stores designed to seal small cracks and gaps to stop draughts. Draughts from chimneys and exhaust fans are difficult to control, but some exhaust fans have automatic shutters that close when the fan is not in use. Chimneys may have a metal plate to seal off air when there is no fire alight.

10.4.4 Radiation

A warm house radiates heat in all directions. Heat loss by radiation can be reduced with shiny foil that reflects radiated heat. Foil can be added to insulation in the ceiling and is also used in external walls.

INVESTIGATION 10.2

Investigating insulators
AIM: To investigate the insulating ability of a range of materials

Materials:

6 empty soft drink cans	polystyrene foam and sticky tape, or foam
6 thermometers	drink can holder
newspaper	foam rubber
woollen cloth	hot water
cottonwool	measuring cylinder
	sticky tape (to tape on the materials)

Method and results

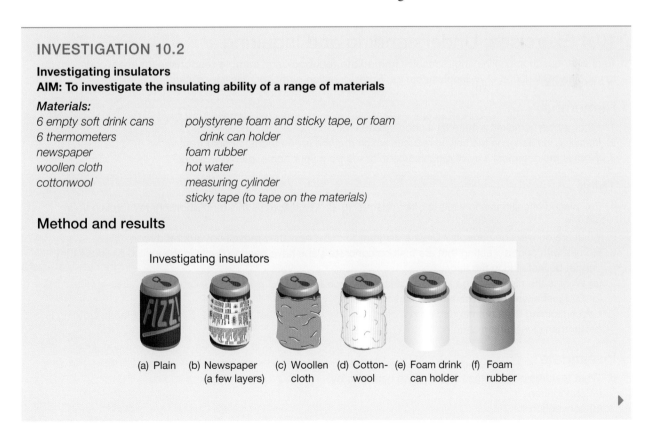

Investigating insulators

(a) Plain   (b) Newspaper (a few layers)   (c) Woollen cloth   (d) Cotton-wool   (e) Foam drink can holder   (f) Foam rubber

---

- Surround each can except one with a different material.
- Copy the table below into your workbook and use it to record your measurements.
- Measure out and pour 100 mL of hot water into each of the cans.
- Measure the temperature of the water in each can. Repeat the measurement of temperature every 5 minutes for 20 minutes.

Temperature of water in cans (°C)

Can covering	Time (minutes)				
	0	5	10	15	20
None					
Newspaper					
Woollen cloth					
Cottonwool					
Foam can holder					
Foam rubber					

1. Draw a bar graph that will allow you to compare the drop in temperature of the water in the cans after 20 minutes.

Discuss and explain

1. Which covering appears to be the most effective insulator?
2. Which one or more of the three methods of heat transfer does the most effective insulator reduce?
3. Use your data to suggest a good container for a mug of hot chocolate.
4. Why was one can left without a covering?
5. Are your conclusions reliable? Discuss the difficulties encountered in making sure that the comparison of insulators was fair.

10.4 Exercises: Understanding and inquiring

To answer questions online and to receive **immediate feedback** and **sample responses** for every question, go to your learnON title at www.jacplus.com.au. *Note:* Question numbers may vary slightly.

Remember

1. What property makes a material a good insulator?
2. Installing insulation in the ceiling reduces which method (or methods) of heat transfer?
3. What is the cheapest way of reducing heat losses from your home in cold weather?

Think

4. Foil placed in ceilings and walls is often referred to as 'insulation'. Is this term appropriate? Explain your answer.
5. What are convection currents? Draw a diagram to show how they move heat around a room.
6. Homes with central heating that are built on concrete slabs have heating ducts in the ceiling because they cannot be installed in the floor.
   (a) What is the disadvantage in having ducts in the ceiling?
   (b) Suggest a way of overcoming this disadvantage.
7. Loose clothing is recommended on hot days as it allows body heat to escape. Explain why loose clothing is better than close-fitting clothing for this purpose.

Investigate

8. What features of a thermos flask reduce heat loss by:
   (a) conduction
   (b) convection
   (c) radiation?

10.5 Light energy

10.5.1 The importance of the sun

Like all stars, the sun changes some of the energy stored inside it into light energy. A burning candle converts some of the chemical energy stored in wax into light energy. Some living things are also able to change chemical energy stored in their bodies into light energy.

Without light from the sun, the world would be in darkness. Plants would not grow and no other life on Earth would exist. However, light makes up only a very small fraction of the energy that comes to us from the sun.

Each of the light sources shown here is luminous.

Light travels through space at about 300 000 kilometres per second, taking almost 10 minutes to get here.

10.5.2 Sources of light

The sun is not the only source of light. Any objects or substances that give off their own light are said to be **luminous**. Examples of some other sources of light are shown above.

Most of the light sources shown are **incandescent**. They emit light because they are hot. The sun and all other stars, light bulbs and flames are incandescent. Other sources, such as fluorescent tubes, the paint on the hands and numerals of clocks and watches, fireflies, glow-worms and some deep-sea fish, emit light without getting hot — they are not incandescent. Living things that emit light without heat are referred to as **bioluminescent**.

Most things that you see are not luminous. We see **non-luminous** objects because light from luminous objects is reflected from them. They do not emit their own light. Light from luminous objects, such as the sun, light globes or fluorescent tubes, strikes them and is reflected into your eyes. The moon is not a luminous object. Its surface reflects light from the sun.

10.5.3 The deep black sea

Light from the surface does not reach deep below the ocean. From a depth of about 1000 metres downwards, the ocean is in complete darkness. Imagine the problems the fish that live there have in finding food. Some deep-sea fish swim closer to the surface to get their food, but others spend all of their time in the dark. The angler fish wiggles a luminous lure to attract its prey. The viperfish uses bioluminescent lights in its open mouth to entice prey directly into its stomach. The black dragonfish produces red light from a spot just beneath its eye. This allows the dragonfish to see its prey without being seen itself, as most of its prey can't see red light.

10.5.4 Seeing the light

Light is not normally visible between its source and any surface that it strikes. You can see a beam of light only if there are small particles in its path. The light is then **scattered** in many directions by the particles, some of it reaching your eye.

The angler fish, living in darkness about 4000 metres below the ocean surface, uses a luminous lure to attract its prey.

Light beams are visible only when there are particles in the air to scatter the light into your eyes. Light from a spotlight can be seen if there is smoke or fog in the air.

Observing a radiometer

A radiometer consists of four vanes, each of which is black on one side and silver on the other. The vanes are balanced on a vertical support so that they can turn with very little friction. The mechanism is encased inside a glass bulb from which air has been pumped out, making it almost a vacuum.

AIM: To observe the effect of sunlight on a radiometer

Materials:
radiometer

Method and results

1. Put the radiometer in direct sunlight. Record your observations.
2. Put it in the shade. Record your observations again.

Discuss and explain

3. What effect does sunlight have on a radiometer?
4. How does this experiment demonstrate that sunlight is a form of energy?
5. Research a scientific theory to explain the effect of sunlight on the radiometer.

Seeing the light

AIM: To observe and explain the scattering of light

Materials:
moderately dark room
torch or projector
matches or a well-used chalk duster

Method

- Shine the torch or projector on a nearby wall.
- Now hit the chalk duster with your hand, or light and blow out a match, so that chalk dust or smoke falls between the light source and the wall.

Discuss and explain

1. Can you see the light beam between the light source and the wall without the chalk dust or smoke?
2. What changes when the chalk dust or smoke is present?
3. Explain what happens to the light from the source to make it visible.

WHAT DOES IT MEAN?

The word *absorb* comes from the Latin word *sorbere*, meaning 'to suck in'.

10.5.5 Meeting new substances

When light energy travels from one substance to another, three things can happen to it.

1. It can be transmitted; that is, the light energy can travel through the substance. For example, light is transmitted through clear glass.

2. It can be absorbed; that is, the light energy can be transferred to particles inside the substance. For example, the tinted glass in many cars contains a substance that absorbs some of the light energy passing through it.

3. It can be reflected from the surface of the substance or reflected (scattered) by small particles inside the substance. For example, light is reflected from opaque objects like a piece of wood and scattered by particles of water in fog. This is how you are able to see them.

You can't see the people in this car because most of the light energy coming from inside the car is absorbed by the tinted glass.

10.5.6 The visible spectrum

Light reaching us from the sun is known as white light.

Household lighting and torches are almost always designed to produce white light. By observing a rainbow, you can see that white light consists of many different colours. This set of colours is called the **visible spectrum.**

The colours of the visible spectrum are usually described as red, orange, yellow, green, blue and violet. However, there is no sharp boundary between the colours. They merge into each other.

White light can also be separated into the colours of the visible spectrum by passing a narrow beam through a triangular glass prism. This separation of colours by droplets of water to form a rainbow or by a triangular prism is called **dispersion**.

Photo of a rainbow

10.5.7 Meet the rest of the family

Light is just one part of a 'family' of forms of energy known as **electromagnetic radiation** emitted by the sun and other stars. Together, this family makes up the **electromagnetic spectrum** and you are probably already familiar with most, if not all of the members, including microwaves, ultraviolet radiation and X-rays. All types of electromagnetic radiation can be produced artificially here on Earth.

All electromagnetic radiation travels through air at 300 000 000 metres per second. Unlike sound waves, electromagnetic radiation can travel through a vacuum.

Radio waves

Radio waves include the low-energy waves that are used to communicate over long distances through radio, television. mobile phones and wi-fi. Radar uses radio waves to detect objects from a very long distance. The microwaves used in microwave ovens for cooking are also radio waves.

Infra-red radiation

Infra-red radiation, invisible to the human eye, is emitted by all objects and is sensed as heat. It has less energy than visible light. The amount of infra-red radiation emitted by an object increases as its temperature increases. When you push a button on a remote control to operate a television, PlayStation® or other electronic device, a beam of infra-red radiation is sent towards the device. A detector in the device converts the infra-red energy into electrical energy, which operates the controls.

A gentle push of a button sends infra-red radiation to an electronic device at 300 million metres per second.

Ultraviolet radiation

Like infra-red radiation, **ultraviolet radiation** is invisible to the human eye. It is needed by humans to help the body produce vitamin D; however, too much exposure to ultraviolet radiation causes sunburn. Ultraviolet radiation has more energy than visible light.

X-rays

X-rays have enough energy to pass through human flesh. They can be used to kill cancer cells, find weaknesses in metals and analyse the structures of complex chemicals. X-rays are produced when fast-moving electrons give up their energy quickly. In X-ray machines, this happens when the electrons strike a target made of tungsten.

Some parts of the human body absorb more of the energy of X-rays than others. For example, bones absorb more X-ray energy than the softer tissue around them. This makes X-rays useful for obtaining images of bones and teeth. To obtain an image, X-rays are passed through the part of the body being examined. The X-rays that pass through are detected by photographic film on the other side of the body. Because bones, teeth and hard tissue such as tumours absorb more energy than soft tissue, they leave shadows on the photographic film, providing a clear image.

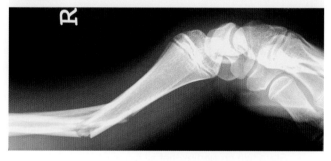

X-ray showing fractures of a radius and an ulna in a forearm

CT scanners (or CAT scanners) are X-ray machines that are rotated around the patient being examined.

Visible light represents only a very small part of the electromagnetic spectrum.

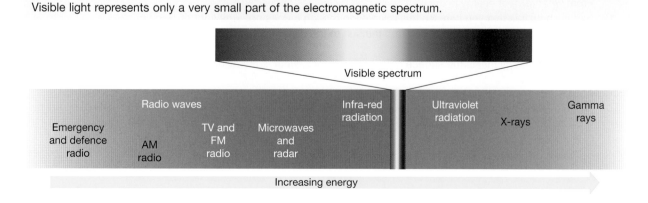

Gamma rays

Gamma rays are emitted by radioactive substances and larger stars. Gamma rays have even more energy than X-rays and can cause serious damage to living cells. They can also be used to kill cancer cells and find weaknesses in metals. Gamma rays are produced when energy is lost from the nucleus of an atom. This can happen during the radioactive decay of nuclei or as a result of nuclear reactions.

Gamma cameras are used in PET scans to obtain images of some organs. To obtain a PET scan, a radioactive substance that produces gamma rays is injected into the body (or, in some cases, inhaled). As it passes through the organ being examined, it produces gamma rays, which are detected by the camera.

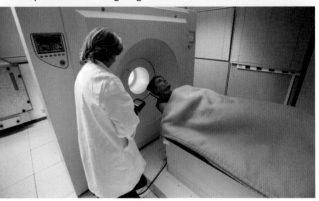

Cameras that detect gamma rays are used in PET scans. This patient is undergoing a PET scan of her brain.

10.5 Exercises: Understanding and inquiring

To answer questions online and to receive **immediate feedback** and **sample responses** for every question, go to your learnON title at www.jacplus.com.au. *Note:* Question numbers may vary slightly.

Remember

1. What is light and how fast does it travel through space?
2. (a) What does 'incandescent' mean?
   (b) List two examples of light sources that are incandescent.
   (c) List two examples of light sources that are not incandescent.
3. Why do you see the beam of light from a torch if it is foggy?
4. Describe what can happen to light energy travelling through the air when it meets a new substance.
5. List the six commonly known colours of the visible spectrum.
6. How do you know that white light consists of different colours?
7. Apart from visible light, what other forms of energy come to Earth from the sun?
8. State two properties that all forms of electromagnetic radiation have in common.
9. List the forms of electromagnetic radiation that have more energy than visible light.

Think

10. Which of the following objects are luminous?
    (a) The sun
    (b) The moon
    (c) The stars
    (d) A burning candle
    (e) This page
11. From what form of energy is the light produced by fireflies converted?
12. Explain how it is that you can see this page even though it does not emit light of its own.
13. How long does it take light to travel from the sun to the distant dwarf planet Pluto when it is 6000 million kilometres from the sun?
14. When light energy meets the surface of your sunglasses, what is the evidence that some of it has been:
    (a) transmitted
    (b) reflected
    (c) absorbed?

Investigate
15. Find out how the energy that reaches the Earth is produced by the sun.
16. Find out how a rainbow is formed.

learn on RESOURCES — ONLINE ONLY

Complete this digital doc: Worksheet 10.4: Light energy (doc-18767)

10.6 Light forms images

10.6.1 Rays

Light is especially important to us because of its ability to form images. The reflection of light allows you to see images in mirrors. When you take a photo, the image is created when light is bent by a lens. In fact, everything that you see is an image, created at the back of your eye when light energy entering your eye bends. To understand how images are formed, we need to have a closer look at how light energy behaves.

Light travels in straight lines as it travels through empty space or through a uniform substance like air or water. The lines that are used to show the path of light are called **rays**. A beam is a stream of light rays.

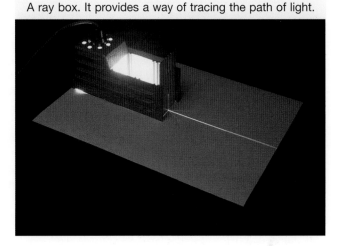

A ray box. It provides a way of tracing the path of light.

10.6.2 Tracing the path of light

The ray box shown in the photograph above right provides a way of tracing the path of light. It contains a light source and a lens that can be moved to produce a wide beam of light that spreads out, converges or has parallel edges. The light box is placed on a sheet of white paper, making the beam visible as some of the light is reflected from the paper into your eyes.

Black plastic slides can be placed in front of the source to produce a single thin beam or several thin beams. The beams are narrow enough to trace with a fine pencil onto the white paper. The fine pencil line can be used to represent a single ray.

10.6.3 Reflections

When you look in a mirror you see an image of yourself. If the mirror is a plane or flat mirror, the image will be very much like the real you. If the mirror is curved, the image might be quite strange, like the one in the photograph at right.

The images in mirrors are formed when light is reflected from a very smooth, shiny metal surface behind a sheet of glass. Images can also be formed when light is reflected from other smooth surfaces, such as a lake.

Seeing the light

AIM: To investigate the reflection of light and its transmission through a prism and lens

Materials:
ray box kit
power supply
several sheets of white paper
ruler and fine pencil

Method and results

- Connect the ray box to the power supply. Place a sheet of white paper in front of the ray box. Move the lens backwards and forwards until a beam of light with parallel edges is projected.
- Use one of the black plastic slides to produce a single thin beam of light that is clearly visible on the white paper.
1. Trace the path of this single beam of light as it meets the lens, prism or one of the mirrors shown in the diagram on the right.
   The path can be traced by using pairs of very small crosses along the centre of the beam before and after meeting each 'obstacle'. Trace and label the shape of each 'obstacle' before you trace the light paths.
- Change the slide in the ray box so that you can project several parallel beams towards each of the 'obstacles'.
2. Use a ruler to draw a small diagram showing the path followed by the parallel beams when they meet each of the 'obstacles'.
- Change the black plastic slide to produce a single thin beam of light from the ray box.
- Place the triangular prism in front of the beam and move it around until you can see a band of colours on the white paper.
3. Which colour is bent most by the triangular prism?
4. Which colour is bent least by the triangular prism?
5. Explain why droplets of water can cause white light from the sun to separate into the colours of the visible spectrum.

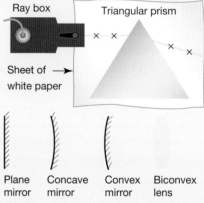

Tracing the path of a beam of light

Ray box — Triangular prism

Sheet of white paper →

Plane mirror   Concave mirror   Convex mirror   Biconvex lens

Discuss and explain

6. What happens to a beam of light when it meets a perspex surface:
   (a) 'head on'?
   (b) at an angle?
7. What happens to a beam of light when it meets a plane mirror surface:
   (a) 'head on'?
   (b) at an angle?
8. Describe your observations of the path followed by the three parallel beams when they meet each of the mirrors and the lens.

10.6.4 Where is the image?

Whenever light is reflected from a smooth, flat surface, it bounces away from the surface at the same angle from which it came. This observation is known as the Law of Reflection. This law can be used to find out where your image is when you look into a mirror.

The diagram on the right shows how the Law of Reflection can be used to find the image of the tip of your nose.

Almost all of the light coming from the tip of your nose and striking the mirror is reflected. (A very small amount of light is absorbed by the mirror.) All of the reflected light appears to be coming from the same point behind the mirror; and that is exactly where the image is. The image of the tip of your nose is the same distance behind the mirror as the real tip of your nose is in front of the mirror.

The reflected light appears to be coming from just one place. That's where the image is.

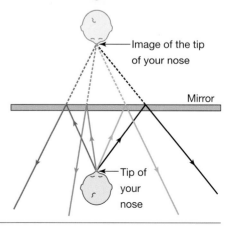

Image of the tip of your nose

Mirror

Tip of your nose

Reflection from curved mirrors

Flat mirrors are commonly found in the home. Curved mirrors have many applications too, including make-up mirrors, security mirrors in shops and safety mirrors at dangerous street intersections. Curved mirrors may be **concave** (curved inwards) or **convex** (curved outwards). Light reflecting from concave and convex mirrors also follows the law of reflection, such that the parallel rays of light are reflected to a **focal point** as shown far right.

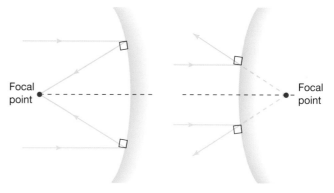

Reflected light rays converge from a concave mirror.

Reflected light rays diverge from a convex mirror.

Why can't you see your image in a wall?

When you look very closely at surfaces like walls, you can see that they are not as smooth as the surface of a mirror. The laws of reflection are still obeyed, but light is reflected from those surfaces in all directions. It doesn't all appear to be coming from a single point. There is no image.

Why is the word 'AMBULANCE' printed in reverse?

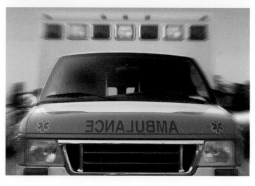

10.6.5 Lateral inversion

The sideways reversal of images that you see when you look at yourself in a mirror is called **lateral inversion**. The sign on the ambulance in the photograph at right is printed so that drivers in front of it can easily read the word 'AMBULANCE' in their rear-view mirrors.

10.6.6 A change of direction

When light is transmitted from one substance into another substance, it can slow down or speed up. This change in speed as light travels from one substance into another is called **refraction**. Refraction causes light to bend, unless it crosses at right angles to the boundary between the substances.

The best way to describe which way the light bends is to draw a line at right angles to the boundary. This line is called the **normal**. When light speeds up, as it does when it passes from water into air, it bends away from the normal. When light slows down, as it does when it passes from air into water, it bends towards the normal.

What happened to my legs?

Looks can be deceiving! The people in the photograph do not have unusually short legs. Everything you see is an **image**. An image of the scene you are looking at forms at the back of your eye. When light travels in straight lines, the image you see provides an accurate picture of what you are looking at. However, when light bends on its way to your eye, the image you see can be quite different.

The light coming from the swimmers' legs in the photograph bends away from the normal as it emerges from the water into the air. The light arrives at the eyes of an observer as if it were coming from a different direction.

Short legs? Not really.

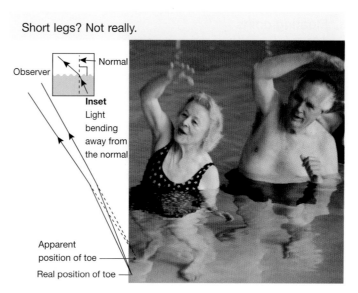

Observer

Normal

Inset
Light bending away from the normal

Apparent position of toe

Real position of toe

The diagram shows what happens to two rays of light coming from the swimmer's right toe. To the observer, the rays appear to be coming from a point higher than the real position of the toe. It can be seen by looking at the diagrams that the amount of bending depends on the angle at which the light crosses the boundary.

INVESTIGATION 10.6

Looking at images

AIM: To observe and compare the reflection of light from plane mirrors and curved mirrors

Materials:
plane mirror
shiny tablespoon or soup spoon

Method and results

- Look at your image in the back of a spoon. This surface is **convex**. Convex means curved outward. Move the spoon as close to your eyes as you can and then further away.
1. Record your observations in a table like the one below. Is the image small or large? Right-side up or upside down? Is there anything strange about the image?
   - Look at your image in the front of the spoon. This surface is **concave**. Concave means curved inward. Move the spoon closer to you and then further away.
2. Record your observations in the table.
   - Look at the image of your face in a plane mirror. Wink your right eye and take notice of which eye appears to wink in the image.
3. Write the word IMAGE on a piece of paper and place it in front of the mirror so that it faces the mirror. Write down the word as you see it in the image.
4. Write down how you think an image of the word REFLECTION would look in the mirror.

Discuss and explain

5. Which eye in the plane mirror image appears to wink?
6. Which letters in the image of the word IMAGE look different? Which look the same?
7. Test your hypothesis about the image of the word REFLECTION. Was your hypothesis correct?
8. List some places where you have seen curved mirrors. State whether the mirrors were convex or concave and explain why they are used.

	Observations of image		
	First observation	When you move closer	When you move further away
Convex side			
Concave side			

INVESTIGATION 10.7

Floating coins

AIM: To observe the refraction of light

Materials:
2 beakers
evaporating dish
coin

Method and results

- Place a coin in the centre of an evaporating dish and move back just far enough so you can no longer see the coin. Remain in this position while your partner slowly adds water to the dish.
1. Make a copy of the diagram at right. Use dotted lines to trace back the rays shown entering the observer's eye to see where they seem to be coming from. This enables you to locate the centre of the image of the coin.
2. Is the image of the coin above or below the actual coin?
3. What appears to happen to the coin while water is added to the evaporating dish?

The image of the coin is not in the same place as the actual coin.

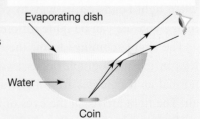

Evaporating dish

Water →

Coin

10.6 Exercises: Understanding and inquiring

To answer questions online and to receive **immediate feedback** and **sample responses** for every question, go to your learnON title at www.jacplus.com.au. *Note:* Question numbers may vary slightly.

Remember

1. You cannot usually see light as it travels through the air. What makes it possible to see a beam of light?
2. What does a mirror do to light in order to form an image?
3. In which type of mirror can your image be upside down?
4. How is your image in a plane mirror different from the real you?
5. What is refraction?
6. Which way does light bend when it slows down while passing from water into air: towards or away from the normal?
7. Explain how white light is separated into different colours by a triangular prism.
8. Why do dentists use concave mirrors to examine your teeth?
9. Which type of mirror is used to help you see around corners?
10. How would the word 'TOYOTA' on the front of a van look in the rear-view mirror of the driver in front of it?
11. The illustration at right shows a ray of light emerging from still water after it has been reflected from a fish. Should the spear be aimed in front of or behind the image of the fish? Use a diagram to explain why.

12. The Law of Reflection can be modelled with moving objects (as particles) in a number of sports. List as many of these sports as you can.
13. When you look down on a coin at the bottom of a glass of water it looks closer to you than it really is.
    (a) Draw a diagram to show why it looks closer.
    (b) In what other way is the image of the coin different from the real coin?
14. List six commonly recognised colours of the visible spectrum in order of the amount that they are bent (from least bent to most bent) as they pass through a triangular prism.

Imagine

15. Imagine that the world is plunged into darkness by a mysterious cloud of dust. What problems would be caused by the lack of visible light if the cloud lingered for:
    (a) one hour?
    (b) three days?
    (c) six weeks?
16. Imagine that you are the fish from question 11 in the illustration below.
    (a) Will the image of the girl's head be higher or lower than her real head?
    (b) Draw a sketch of how the girl might appear to you.

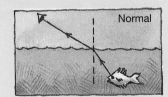

A periscope uses mirrors to enable you to see around corners or over objects.

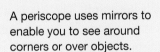

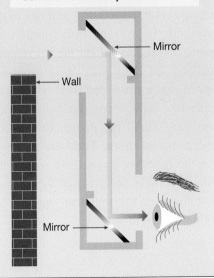

Design and create

17. Design and build a simple periscope like the one shown in the diagram on the right. You will need stiff card, scissors, two small mirrors, sticky tape or glue, a pencil and a ruler. Explain, with the aid of a diagram, how it works.

10.7 Seeing the light

Science as a human endeavour

10.7.1 The human eye

Everything that you see is an image, created when the energy of light waves entering your eyes is transmitted to a 'screen' at the back of each eye.

This screen, called the **retina**, is lined with millions of cells that are sensitive to light. These cells respond to light by sending electrical signals to your brain through the **optic nerve**.

The image formed on the retina is upside down.

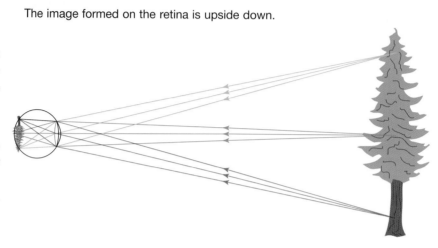

Some of the light reflected from your surroundings, along with light emitted from luminous objects such as the sun, enters your eye. It is refracted as it passes through the outer surface of your eye. This transparent outer surface, called the **cornea**, is curved so that the light converges towards the **lens**. Most of the bending of light done by the eye occurs at the cornea.

On its way to the lens, the light travels through a hole in the coloured **iris** called the **pupil**. The iris is a ring of muscle that controls the amount of light entering the lens. In a dark room the iris contracts to allow as much of the available light as possible through the pupil. In bright sunlight the iris relaxes, making the pupil small to prevent too much light from entering. The clear, jelly-like lens bends the light further, ensuring that the image formed on the retina is sharp. The image formed on the retina is inverted. However, the brain is able to process the signals coming from the retina so that you see things the right way up.

The human eye

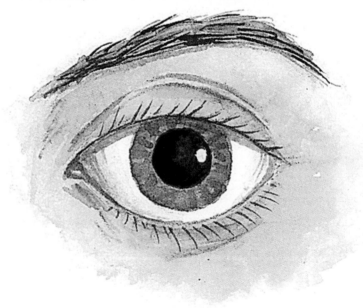

Side view of a human eye

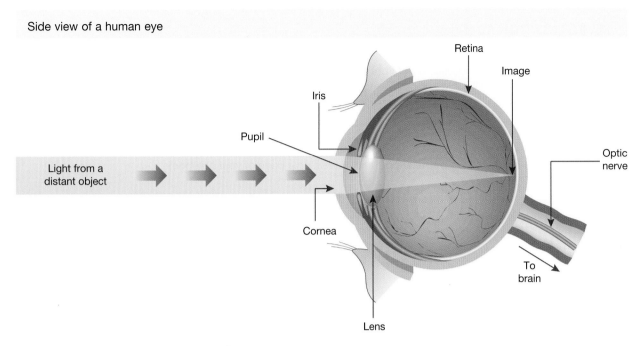

10.7.2 Getting things in focus

Although most of the bending of light energy done by the eye occurs at the cornea, it is the lens that ensures the image is sharp.

Two types of lenses

The lens in each of your eyes is a **converging lens**. Its shape is **biconvex** — that means it is curved outwards on both sides. A beam of parallel rays of light travelling through a biconvex lens 'closes in' (converges) towards a point called the focal point, or focus.

Another type of lens is a **diverging lens**, which spreads light outwards because of its biconcave shape. A biconcave lens does not have a real focal point. When the parallel light rays emerge from a biconcave lens, they do not converge to a focal point. However, if you trace the rays back to where they are coming from, you find that they do appear to be coming from a single point. That point is called the **virtual focal point**, or virtual focus.

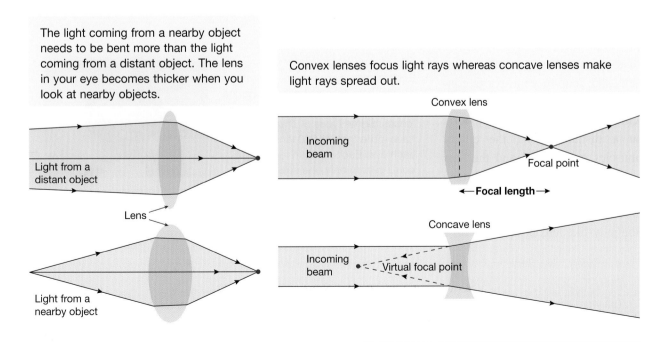

The light coming from a nearby object needs to be bent more than the light coming from a distant object. The lens in your eye becomes thicker when you look at nearby objects.

Convex lenses focus light rays whereas concave lenses make light rays spread out.

Focusing on light

AIM: To investigate the transmission of light through different lenses

Materials:

ray box kit

sheet of white paper

12 V DC power supply

ruler and fine pencil

Method and results

- Connect the ray box to the power supply and place it on a page of your notebook.

Part A: Biconvex lenses

- Place the thinner of the two biconvex lenses in the kit on the page and trace out its shape. Project three thin parallel beams of white light towards the lens.

1. Trace the paths of the light rays as they enter and emerge from the lens. Remove the lens from the paper so that you can draw the paths of the light rays through the lens.

   - Replace the thin biconvex lens with a thicker one and repeat the previous steps.

Part B: Biconcave lenses

- Place the thinner of the two biconcave lenses on your notebook page and trace out its shape.

2. Trace the path of each of the three thin light beams as they enter and emerge from the lens. Remove the lens from the page so that you can draw the paths of the light beams through the lens.

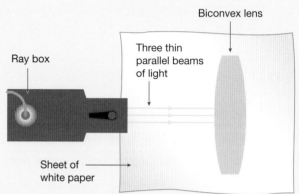

Discuss and explain

3. State the focal length (distance from the focal point to the centre of the lens) for each lens.
4. Which of the biconvex lenses bends light more, the thin one or the thicker one?
5. Explain why the middle light ray does not bend.
6. How many times do each of the other rays bend before arriving at the focal point?
7. Do the diverging rays come to a focus?
8. Do the diverging rays appear to be coming from the same direction? Use dotted lines on your diagram to check.
9. Predict where the diverging rays will appear to come from if you use a thicker biconcave lens. Check your prediction with the thicker biconcave lens in the ray box kit.

Accommodation

The exact shape of the clear jelly-like lens in your eye is controlled by muscles called the **ciliary muscles**. When you look at a distant object, the ciliary muscles are relaxed and the lens is thin, producing a sharp image on the retina. When you look at a nearby object, the light needs to be bent more to produce a sharp image. The ciliary muscles contract and the jelly-like lens is squashed up to become thicker. This process is called **accommodation**.

HOW ABOUT THAT!

Each human eye contains just one convex lens. Insects have compound eyes. Each eye contains many lenses. Some types of dragonfly have more than 10 000 lenses in each eye. Each eye can focus light coming from only one direction.

WHAT DOES IT MEAN?

The word *accommodation* comes from the Latin term *accommodatio*, meaning 'adjustment'.

10.7.3 Too close for comfort

As you get older, the tissues that make up the lens become less flexible. The lens does not change its shape as easily. Images of very close objects (like the words you are reading now) become blurred. The lens does not bulge as much as it should and the light from nearby objects converges to a point behind the retina instead of on the retina. You may have to hold what you are reading further away in order to obtain a clear image.

This change in accommodation with age is known as **presbyopia** and is a natural process. Some people are not inconvenienced at all while others need to wear reading glasses so that they can read more easily and comfortably. The table at right shows how the smallest distance at which a clear image can be obtained changes with age. The distances shown are averages and there is a lot of variation from person to person.

How the average smallest distance at which a clear image can be obtained changes with age

Age (years)	Distance (cm)
10	7.5
20	9
30	12
40	18
50	50
60	125

INVESTIGATION 10.9

Getting a clear image

AIM: To investigate accommodation

Materials:
ruler

Method and results

- Look closely at the **X** printed here from the smallest distance at which you can see it clearly and sharply with comfort. Quickly look away and focus on a distant object for a second or two and then focus on the 'X' again from the smaller distance.
- Try to feel the action of the muscles that allow you to see a sharp image of the 'X'.
- Use the following procedure to estimate the smallest distance at which you can obtain a clear image of a nearby object. (If you are wearing glasses, remove them during this part of the experiment.)
- Hold this book vertically at arm's length from your eyes and focus on it. Move the book to a position about three or four centimetres from your eyes and then gradually move the book further away until you can see the print clearly and sharply.
- Have a partner use the ruler to estimate the distance between the page and your eyes. The ruler should be placed carefully beside your head for this measurement.
1. Record the distance measured.
2. Collate the results for the whole class and determine the average smallest distance at which a clear image could be obtained.

Discuss and explain

3. How does your result compare with the average smallest distance for your class?
4. Write down the highest single result and lowest single result for your class. Comment on the range of results.

10.7.4 Improving the image

The eye is truly an amazing optical system. It is able to focus on distant objects many kilometres away as well as very close objects only centimetres away. However, the ability to obtain sharp images varies from person to person — as well as with age.

The most common conditions that reduce the ability to obtain sharp images are **short-sightedness** and **long-sightedness**.

Short-sightedness (myopia)

A person who is short-sighted is unable to obtain sharp images of distant objects. That happens when the cornea and lens bend the light too much. As a result, the light from a distant object focuses in front of the retina. The image formed on the retina is blurry.

Short-sightedness can be corrected by wearing glasses with convexo-concave lenses. These lenses diverge the light just a little before it enters the eye. As a result, the light from the object focuses on the retina instead of in front of it.

Long-sightedness (hyperopia)

A person who is long-sighted is unable to obtain sharp images of nearby objects. That happens when the cornea and lens don't bend the light enough. As a result, the light from a nearby object reaches the retina before it comes to a focus. The image formed on the retina is blurry.

Long-sightedness can be corrected by wearing glasses with concavo-convex lenses. These lenses converge the light just a little before it enters the eye. As a result, the light from the object focuses on the retina instead of behind it.

Multifocal lenses

As people get older they are more likely to have difficulties in forming clear images of distant as well as nearby objects. In days gone by the solution to this problem was to have two pairs of glasses – one with a converging lens for reading and the other with a diverging lens for distance vision.

Multifocal lenses remove the need for two separate pairs of glasses. They can be bifocal with two parts or trifocal with three parts that merge into each other. Bifocal lenses are shaped so that the top part of the lens converges light to assist reading. The bottom part is shaped to diverge light to assist distance vision. A clear boundary between the two parts of the lens is visible. Trifocal lenses gradually change in shape from bottom to top to assist with close-up vision, arms' length vision and distance vision. There are no visible boundaries between the different parts of the lenses.

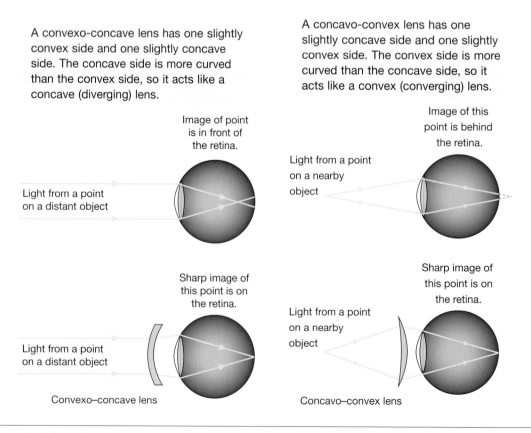

A convexo-concave lens has one slightly convex side and one slightly concave side. The concave side is more curved than the convex side, so it acts like a concave (diverging) lens.

A concavo-convex lens has one slightly concave side and one slightly convex side. The convex side is more curved than the concave side, so it acts like a convex (converging) lens.

Image of point is in front of the retina.

Light from a point on a distant object

Image of this point is behind the retina.

Light from a point on a nearby object

Sharp image of this point is on the retina.

Light from a point on a distant object

Sharp image of this point is on the retina.

Light from a point on a nearby object

Convexo–concave lens

Concavo–convex lens

10.7 Exercises: Understanding and inquiring

To answer questions online and to receive **immediate feedback** and **sample responses** for every question, go to your learnON title at www.jacplus.com.au. *Note:* Question numbers may vary slightly.

Analyse and evaluate

1. Use the data in the table on page 417 to draw a line graph to show how the ability to focus on nearby objects changes with age.
   (a) Use your graph to predict the smallest distance at which a clear image can be obtained by an average person of your age.
   (b) At what age does the decrease in focusing ability appear to be most rapid?

Remember

2. At which part of the human eye does most of the bending of light occur?
3. Describe the function that the iris and pupil work together to perform.
4. Name and sketch the shape of a lens that:
   (a) converges a beam of light to a single point
   (b) makes the rays in a beam of light diverge.
5. What is the focal length of a converging lens a measure of?
6. What is accommodation?
7. What is the name given to the shape of the lens in the human eye?
8. Sketch the shape of the lens in the eye when you are viewing:
   (a) a nearby object
   (b) a distant object.
9. How does the lens change its shape?
10. Why is it common to see older people holding a newspaper at arm's length while they are reading it?
11. How are messages sent from the eye to your brain?
12. Which eye condition occurs if the cornea and lens bend the light entering the eye too much?
13. Which type of lens would you expect to find in glasses worn by a long-sighted person — converging or diverging?
14. What are bifocals and why are they used?

Think

15. Does light slow down or speed up when it passes from the air into the cornea? How do you know this? (*Hint:* Refer to subtopic 10.5.)
16. List some commonly used inventions that contain lenses.
17. Explain why the focal point of a diverging lens is called a virtual focal point.
18. Why does the lens need to be thicker for viewing nearby objects?
19. Which condition of the eye is most likely to be responsible for each of these problems?
   (a) A student who can read the whiteboard from the back of the room but has to strain to read the print in a textbook and gets headaches while reading from a computer screen at home
   (b) A science teacher who has never had eye problems before begins to find it easier to read books and newspapers when they are held further away
   (c) A person who has no problem reading a newspaper can't read the numbers on the scoreboard at an AFL football match

Create and investigate

20. Use two or more lenses and lens holders to make a model microscope or telescope on a laboratory workbench. Investigate the effect of changing the distance between the lenses on the magnification and write a report on your findings.

Investigate

21. Research and report on astigmatism of the eye and how it is corrected.

22. Use the internet to research and report on the development of the bionic eye by Australian scientists. Include in your report information about:
    - macular degeneration
    - which patients it is designed to benefit
    - how it works
    - a comparison with the bionic ear.

 **learn on** RESOURCES — ONLINE ONLY

Try out this interactivity: Lenses (int-1017)

10.8 Sound energy

10.8.1 Good vibrations

Humans and other animals rely heavily on sound energy to communicate with each other. You can use your voice, whistle or tap something to make a sound. How else can you make a sound?

All sounds are caused by **vibrations**. When you speak or sing, the vocal cords in your throat vibrate. You can feel the vibrations if you put your hand over the front of your throat. When you strike a drum, the up and down movements of the drum skin cause the air around the drum to vibrate. When the drum skin moves down, the air particles near it are pulled back, spreading them out. A fraction of a second later the drum skin moves back up, squeezing the air particles together.

The energy of the air particles is transferred to nearby air particles, causing them to vibrate as well. This creates a moving series of **compressions** (air particles closer together than usual) and **rarefactions** (air particles further apart than usual) that move away from the source of the sound. These moving compressions and rarefactions are what we know as **sound waves**. If enough energy is transferred to the vibrating air, the sound waves reach your eardrum and you hear sound.

Sound waves consist of a series of vibrating air particles.

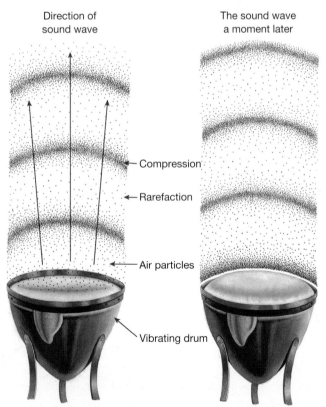

WHAT DOES IT MEAN?

The word *vibration* comes from the Latin word *vibrare*, meaning 'to shake'.

Modelling sound waves

AIM: To model sound waves on a slinky

Materials:
slinky spring

Method

- Pull the slinky spring from both ends to stretch it a couple of metres along the floor.
- Create vibrations at one end of the slinky by moving the coils in and out.
- Watch the series of compressions and rarefactions travel to the opposite end and reflect back.

Modelling sound waves using a slinky spring

Discuss and explain

1. Describe how your model is similar to real sound waves.
2. Describe how your model is different from real sound waves.

10.8.2 Measuring the speed of sound

The first known attempt to measure the speed of sound was made by French philosopher and scientist Pierre Gassendi in 1635. He measured the time taken between seeing the flash made by a cannon and hearing the sound it made from a long distance. He assumed (correctly) that the time taken for light from the flash to reach his eyes was very close to zero. In fact, we now know that light takes only 0.000 002 seconds to cover a distance of 500 metres. Gassendi measured the speed of sound to be 478 m/s, which is quite a bit more than the correct value of about 340 m/s at a temperature of 20 °C. But remember there were no accurate timing devices in 1635.

The flash of light from a cannon will reach you before the sound of the shot will.

The number of compressions (or rarefactions) that reach your ear per second is known as the **frequency** of the sound. The musical note middle C, for example, creates 256 compressions every second. Frequency is measured in **hertz** (Hz), a unit named after Heinrich Hertz, the German physicist who, in 1887, was the first person to detect radio waves. So, the frequency of the note middle C is 256 Hz.

The frequency of a sound determines its **pitch**. The faster a sound-producing object vibrates, the higher its frequency and the higher the pitch of the sound you hear. A short string vibrates faster than a long one and so has a higher frequency and pitch.

When you blow across the top of a straw, the air inside it vibrates. If the straw is shortened, the air inside vibrates faster, producing a higher frequency and a high pitch. A longer straw would produce slower vibrations, a lower frequency and a lower pitch.

The distance between two neighbouring compressions is known as the **wavelength**. The wavelength of the musical note middle C is about 1.3 metres. The wavelength of the sounds produced during normal speech varies between about 5 centimetres and about 2.5 metres. When the frequency of a sound increases, the compressions become closer together, decreasing the wavelength of the sound.

The particles of air that vibrate when a sound is made don't move from the source of the sound to your ear. They just move backwards and forwards, sideways or up and down. Each vibrating particle makes the next one vibrate — that's how the sound energy travels through the air. The distance that each vibrating particle moves from its usual resting place is known as the sound's **amplitude**. Higher amplitudes make louder sounds.

The wavelength of a sound is the distance between the centres of two neighbouring compressions. It is the same as the distance between two neighbouring rarefactions.

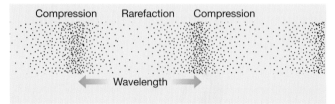

INVESTIGATION 10.11

Measuring the speed of sound

This investigation should be conducted on a school sporting oval or a safe open space on a calm day.

AIM: To measure the speed of sound in air

Materials:
stopwatches (Mobile phone stop watches are ideal.)
trundle wheel or long measuring tape
starting pistol that emits smoke, two large blocks of wood or pair of hand cymbals

---

CAUTION

Only a teacher wearing earplugs should fire a starting pistol.

Method

- Position as many students with stopwatches as possible at a distance of at least 150 metres from the location of the starting pistol or wooden blocks. A longer distance up to 300 metres could be used if possible.
- The sound is made by firing the starting pistol, banging the wooden blocks together or crashing the cymbals. Practise measuring the time taken between seeing the smoke, the wooden blocks being banged together or the hand cymbals crashed together. Repeat the practice several times to allow those with stopwatches to react as quickly as possible.
- Prepare for a final recording of the time taken between seeing the sound being made and hearing the sound.
- Conduct the timing and record the times measured by those with stopwatches.
- Calculate the average time for the sound to travel the distance.
- Use the formula speed = distance/average time to calculate the average speed of sound measured by the class.

Discuss and explain

1. How does your measurement compare with the known speed of sound of about 340 m/s?
2. What is the biggest cause of error in your measurement of the time taken for the sound to reach the observers?
3. Suggest how you could make your results more accurate.
4. What assumption about the speed of light has been made in your calculation of the speed of sound?

The need for air

When a mobile phone rings in a bell jar, the sound can be heard clearly. But if the air inside is sucked out by a vacuum pump, the sound can't be heard. Sound energy cannot travel through empty space — it can travel only by making particles vibrate. In empty space there are no particles to vibrate. Light energy, unlike sound, can travel through empty space. It doesn't need particles. So you can still see the ringing phone in the bell jar, even if you can't hear it.

10.8.3 Making it louder

If you pluck a stretched guitar string while it is not attached to a guitar, it vibrates but makes very little sound. If you strike a stretched drum skin while it is not attached to the drum, it makes very little noise. Even your own vocal cords make very little noise while they are vibrating. In each of these cases, a vibration is needed to create the sound but an enclosed region of air is needed to make the sound louder.

The air inside the body of an acoustic guitar is set vibrating by the strings. The air inside a drum vibrates when the drum skin is struck. The vibrating air inside your throat and mouth makes the sound created by your vocal cords loud enough to be heard.

Sound energy cannot travel through empty space. Sound waves require a medium to travel through; light does not.

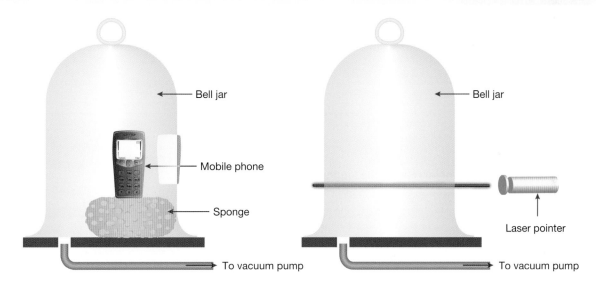

HOW ABOUT THAT!

During a thunderstorm, the flash of lightning and the crash of thunder occur only a tiny fraction of a second apart. So why do you always hear thunder one or more seconds after you see the lightning? The answer lies in one of the differences between sound and light. Sound energy travels through the air at a speed of about 340 m/s. Light energy travels through air at a speed of 300 000 km/s. The delay between seeing lightning and hearing thunder is about 3 seconds for each kilometre that you are away from the lightning.

INVESTIGATION 10.12

Vibrations and pitch

AIM: To investigate the relationship between the size of a vibrating object and pitch of the sound it produces

Materials:

ruler	2 straws
scissors	spatula
small beaker	large beaker

Method and results

- Hold a ruler over the edge of a table so that one end is firmly held down. Flick the overhanging end of the ruler.
- Move the ruler so that more of it is over the edge of the table and flick it again.
1. How does the sound change as the vibrating part of the ruler is made longer?
    - Cut one straw into two so that one part is twice as long as the other part. Place the top of the uncut straw lightly against your bottom lip and blow gently across the opening. Listen to the sound made.
    - Blow across the two shorter (cut) pieces of straw in the same way and listen to the sounds.
2. How does the sound change as the straws get shorter?
    - Tap the side of a small beaker gently with a spatula and listen to the sound. Do the same with a larger beaker.
3. How does the sound of the large beaker compare with the sound of the smaller one?

Discuss and explain

4. How would you change each of the following to make a higher pitched sound?
    (a) The length of a vibrating strip of wood
    (b) The length of a tube of air
    (c) The size of a cymbal

INVESTIGATION 10.13

Making it louder

AIM: To explore methods of increasing the loudness of sound

Materials:

guitar
guitar string
tuning fork

Method and results

- Pluck a stretched guitar string. Listen to the sound it makes.
- Pluck a similar string attached to a guitar.
1. How does the sound of a plucked string change when it is attached to a guitar?
    - Strike a tuning fork on the sole of your shoe and listen to the sound it makes. While it is still vibrating, place the base of the fork on a solid table surface.
2. How does the sound change when the tuning fork is placed on the table?

Discuss and explain

3. Explain why the sound changes in each case.

10.8.4 The sounds of music

How do musical instruments produce sound? The energy comes from the person playing the instrument — but what does the instrument do to convert that energy into sound?

With an acoustic guitar, the vibrations are made by plucking the strings. The air around the sound hole vibrates, causing the air inside the body of the guitar to vibrate. In an electric guitar, a microphone or pick-up detects the vibrating air and an amplifier is used to make the sound louder. The pitch of the sound made by a guitar is increased by shortening the strings using your fingers, tightening the strings or using lighter strings.

A saxophone's vibrations are first made when air is blown across a thin wooden reed. The air inside the saxophone then vibrates, making a loud sound. The pitch can be changed by using keys to open or close holes. When all the holes are closed, the saxophone contains a long column of air, producing a low-pitched sound. As holes are opened, the length of the air column becomes shorter, and the pitch increases.

On a stringed instrument, vibrations are made by plucking strings

The didgeridoo is a wind instrument that has no holes to change the length of the column of vibrating air. The player blows into the instrument using loosely vibrating lips to control how quickly the air inside vibrates.

10.8.5 Sounding great

Like light, sound energy can be transmitted, reflected or absorbed when it meets a new substance.
- All materials transmit some sound, some better than others. That's why you can sometimes hear conversations from the other side of a wall.
- Sound is reflected from hard substances like the tiles in your bathroom. Each note that you sing in the shower lasts longer because it is reflected over and over again from hard surfaces. This effect is called **reverberation**.
- Soft materials, like curtains and carpet, absorb much more sound than walls of plaster or tiles.

INVESTIGATION 10.14

Making music

AIM: To explore the ways in which musical instruments make sound

Materials:
a small selection of musical instruments

Method
- If musical instruments are available in your classroom, have someone demonstrate how they work.
- Look at the musical instruments illustrated at right.

Discuss and explain
1. For each instrument, write down:
   (a) what the player does to make the instrument work
   (b) what vibrates to make musical sounds.
2. What do all of the musical instruments have in common about the way they make sound? What differences are there in how they make sound?

Concert halls are designed to control the transmission, reflection and absorption of sound. For example, the timber in the panelling on the ceiling and walls of the concert hall in the Melbourne Recital Centre was selected because it minimises reflection and reverberation. In the Melbourne Concert Hall, Hamer Hall, heavy curtains behind the audience can be closed to increase the amount of sound absorbed.

The Elisabeth Murdoch Hall at the Melbourne Recital Centre

10.8 Exercises: Understanding and inquiring

To answer questions online and to receive **immediate feedback** and **sample responses** for every question, go to your learnON title at www.jacplus.com.au. *Note:* Question numbers may vary slightly.

Remember

1. What is the cause of all sounds?
2. When compared with air particles in a silent room, how are the particles in compressions and rarefactions different?
3. What is the unit of frequency and what does it measure?
4. Which quality of sound that you hear does the frequency determine?
5. What distance is the wavelength of a sound wave the measure of?
6. What happens to the wavelength of a sound wave when its frequency increases?
7. What characteristic of vibrating air does amplitude describe?
8. Explain why sound can't travel through empty space.
9. If you blow across the top of a straw, a sound is made. How could you increase the pitch of the sound? How do we know, without taking any measurements, that light travels through air faster than sound?
10. Which vibrates more quickly — a long string or a short string made of the same material?
11. Describe what can happen to sound energy travelling through the air when it meets a new substance.
12. Which types of surfaces cause reverberation?

Think

13. Is sound energy a form of kinetic energy? Explain your answer.
14. Explain why the mobile phone in the bell jar needed to be sitting on a sponge when the air was removed.
15. Imagine that light energy couldn't travel through empty space. What would you observe if the air was removed from a bell jar containing a ringing mobile phone?
16. How would you expect a carpeted classroom to sound compared with one with a hard vinyl floor? Give reasons for the differences.
17. How are different notes played on:
    (a) a single string of a guitar
    (b) a recorder
    (c) a xylophone?
18. Complete the gaps in the following table.

Musical instrument	What vibrates first?	What makes the sound louder?
Guitar	Plucked string	Air inside guitar
Trumpet	Player's lips	
Drum		Air inside drum
Saxophone		Air inside saxophone
	String hit by hammers	Air inside instrument

Create and explain

19. Make a string telephone. You will need about five metres of string and two open and empty cans. Punch a small hole in the bottom of each can. Thread the string through each hole and tie a knot to keep the string in place. Hold the cans far enough apart so that the string is tight. Talk into the can at one end while your partner listens at the other end.
    (a) How does the sound travel from one can to the other?
    (b) Does the sound change if you make the string tighter or looser?
    (c) Would a string telephone work without the cans? Why are the cans used?

Investigate

20. Find out more about the following careers that involve using and understanding sound energy.
    (a) Audiologist
    (b) Acoustic engineer
    (c) Audio engineer

10.9 Matrixes and Venn diagrams

10.9.1 Matrixes and Venn diagrams

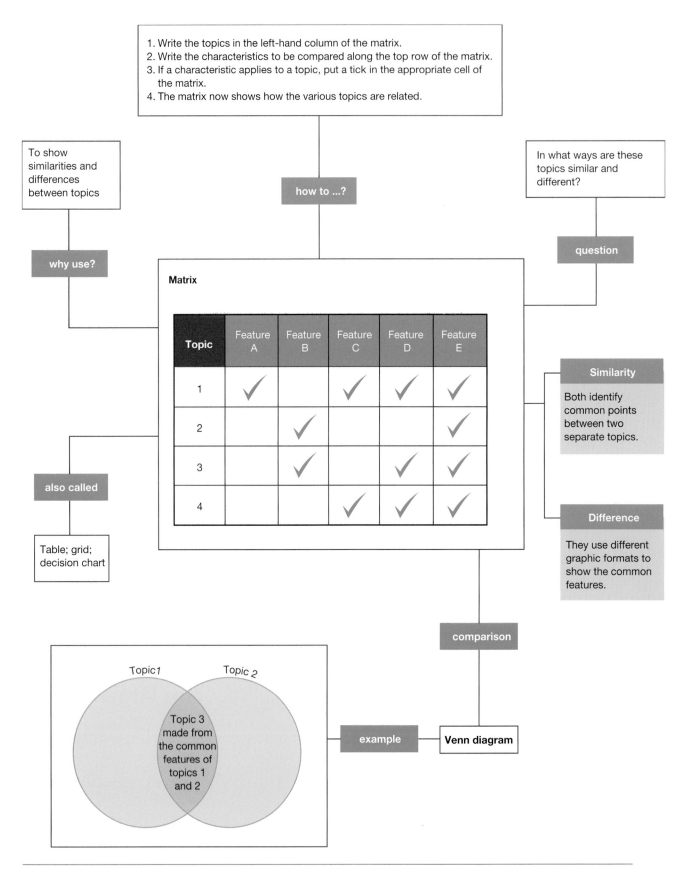

1. Write the topics in the left-hand column of the matrix.
2. Write the characteristics to be compared along the top row of the matrix.
3. If a characteristic applies to a topic, put a tick in the appropriate cell of the matrix.
4. The matrix now shows how the various topics are related.

To show similarities and differences between topics

why use?

In what ways are these topics similar and different?

question

how to ...?

Matrix

Topic	Feature A	Feature B	Feature C	Feature D	Feature E
1	✓		✓	✓	✓
2		✓			✓
3		✓		✓	✓
4			✓	✓	✓

Similarity

Both identify common points between two separate topics.

Difference

They use different graphic formats to show the common features.

also called

Table; grid; decision chart

comparison

Topic 1   Topic 2

Topic 3 made from the common features of topics 1 and 2

example

Venn diagram

10.9 Exercises: Understanding and inquiring

To answer questions online and to receive **immediate feedback** and **sample responses** for every question, go to your learnON title at www.jacplus.com.au. *Note:* Question numbers may vary slightly.

Think and create

1. Copy and complete the matrix below. Use ticks to indicate the forms of energy that electrical energy is transformed into by each of the electrical devices listed.

| Device | Electrical energy is converted into ... | | | | |
	Light energy	Sound energy	Thermal energy	Kinetic energy	Potential energy
Hairdryer					
Television					
Desk lamp					
Vacuum cleaner					
Home computer					
Incandescent light bulb					
Air-conditioner					
Elevator going up					

2. (a) Copy and complete the matrix below. Use ticks to show which statements refer to light and which refer to sound. Some of the statements refer to both light and sound.
   (b) The information in the matrix below can be represented in a Venn diagram. Convert the information in the matrix into a large Venn diagram based on the one at right.

The overlapping section of the Venn diagram contains the properties of both light and sound.

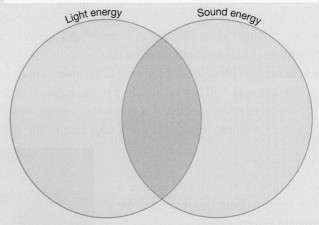

Statement	Light energy	Sound energy
Travels through air at 300 000 kilometres per second		
Travels faster than a speeding bicycle		
Can be reflected		
Can be absorbed		
Is always caused by vibrating objects or substances		
Is observed in an electrical storm		
Can travel through empty space		
Can be produced from another form of energy		
Is used by physiotherapists to treat some muscle problems		

10.10 Project: Going green

Scenario

As the supply of fossil fuels dwindles, cities become more crowded and human-caused global warming becomes an unavoidable reality, an increasing number of people are opting for a more self-sufficient life-style. To meet this need, an increasing number of architecture and building firms specialise in the design and construction of energy-efficient houses that are able to exist off the electricity grid indefinitely because they use electricity generation systems that meet all of the household's needs using renewable energy sources.

You and your team at Sustainable Housing Solutions have been approached by a potential client who wants to build a series of sustainable eco-tourist cottages in remote locations across the country. To see whether your company should be awarded the lucrative contract to oversee the work on the whole chain of cottages, the client has asked you to make a presentation detailing how you would make one of these cottages as energy efficient and self-sustaining as possible. You can place this trial cottage anywhere in the country for your presentation purposes, provided that it is at least 100 km away from any town with a population greater than 5000 people. Other criteria must also be met as follows:

- All of the cottages will have the same layout and will be constructed of mud bricks and have tiled roofs (you will be given a copy of the plan). While you can change the orientation and location of the cottage, you cannot change the design or these construction materials.
- Each cottage must have the following appliances: refrigerator, washing machine, stove, microwave, TV set, DVD player and stereo system. Smaller appliances such as toasters, shavers, hairdryers and computers may occasionally be used by guests as well.
- The cottages must be cool in summer and warm in winter; the client is not opposed to the idea of a reverse-cycle air-conditioner or fans.
- There must be sufficient lighting to be able to read in every room.
- The cottages will not be attached to the national electricity grid — all of the electricity needs of each cottage must be met using a renewable energy source in its area. (Water will be provided from rain-water tanks and septic tanks will take care of the sewage.)

Your task

Your team will prepare and deliver a report for the client that provides the following information:

- The best location to place the trial cottage (keeping in mind that it can be placed somewhere close to a source of renewable energy)
- Suggestions as to how the cottage can be made as energy efficient as possible
- A detailed estimate of how much electricity will need to be generated to power the cottage and run appliances

Estimated Crustal Temperature at 5km Depth

Map derived from the AUSTHERM05 database of Chopra & Holgate (2005)
Image is © 2007; Dr Prame Chopra, Earthinsite.com Pty Ltd

- A justified recommendation as to which renewable energy system should be used to generate that amount of electricity and how it would be supplied to the trial cottage
- An estimate of how much the energy system will cost, using costs for similar systems available on the internet as a guideline.

The report will take the form of an oral presentation with visuals (which may include PowerPoint slides and models). The presentation should be between six and eight minutes long.

10.11 Review

10.11.1 Study checklist

Energy transfers and transformations

- define the term 'energy'
- identify bodies that possess kinetic energy because of their motion
- define potential energy as stored energy
- differentiate between gravitational, chemical and elastic potential energy
- outline examples of transformations from potential and kinetic energy into other forms of energy
- recognise that heat energy is always produced as a by-product of energy transfers
- use flow diagrams to illustrate changes between different forms of energy

Heat, light and sound energy

- define heat as energy in transit from one object or substance to another object or substance with a lower temperature
- describe and compare the transfer of heat by conduction, convection and radiation
- differentiate between luminous and non-luminous objects
- relate the ability to see non-luminous objects to the scattering and reflection of light
- describe sound as a series of vibrating air particles
- recognise that heat, light and sound energy can be transmitted, reflected or absorbed
- describe the properties and uses of different forms of electromagnetic radiation
- measure the speed of sound
- describe the measured properties of wavelength, frequency and amplitude of sound waves
- describe how sounds are produced by a variety of musical instruments
- describe the spread and order of colours in the visible spectrum
- explain how images are formed in mirrors and lenses and how they can be changed
- explain how images are formed by the human eye and how corrective technologies can improve eyesight

Science as a human endeavour

- explain how the unwanted transfer of heat can be decreased to reduce the amount of energy needed for home heating, cooling and lighting
- describe the energy transformations involved in playing a variety of musical instruments
- investigate how energy efficiency can reduce energy consumption
- investigate the development of more energy-efficient motor vehicles

10.11 Review 1: Looking back

To answer questions online and to receive **immediate feedback** and **sample responses** for every question, go to your learnON title at www.jacplus.com.au. *Note:* Question numbers may vary slightly.

1. Replace each of the following descriptions with a single word.
   (a) Energy associated with all moving objects
   (b) Energy that is stored
   (c) The form of energy that causes an object to have a high temperature
   (d) The form of energy stored in a battery that is not connected to anything
   (e) The source of most of the Earth's light

2. Explain why the amount of energy in the universe never changes.

3. Describe an example of an object that has:
   (a) elastic potential energy
   (b) gravitational potential energy.

4. Draw a flowchart to illustrate the energy transformations that take place:
   (a) after you switch on a torch
   (b) when a firecracker is lit
   (c) when a ball rolls down a hill and then up another hill.

5. When a kettle of water is boiled on a gas cooktop, not all of the energy stored in the gas is used to heat the water. Where does the rest of the energy go?

6. Explain why it is not possible for an energy converter like a battery or car to have an efficiency of 100 per cent.

7. Explain how fibreglass batts are able to reduce the loss of heat through the ceiling by both conduction and convection.

8. Heat moves from regions of high temperature to regions of low temperature by conduction, convection or radiation. In which of these three ways is heat most likely to be transferred:
   (a) from a frying pan to an egg being fried
   (b) from the sun to the planets of the solar system
   (c) through water in a saucepan on a hotplate or gas burner
   (d) through a metal spoon being used to stir hot soup
   (e) from a very hot and bright light globe near the ceiling to your body directly beneath?

9. Each of the diagrams on the right shows radiated heat falling on a solid object. Which diagram shows heat being:
   (a) absorbed
   (b) reflected
   (c) transmitted?

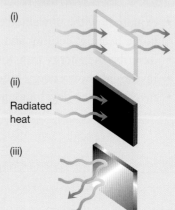

(i)

(ii)
Radiated heat

(iii)

10. Explain how your body keeps its core temperature at 37 °C even when the air temperature is greater than this.
11. Explain how you are able to see an object like a tree even though it doesn't produce its own light energy.
12. Make a list of as many luminous objects as you can.
13. Why is it incorrect to describe fluorescent lights as incandescent?
14. You can't normally see the beam of light coming from a car headlight. However you can see the beams if there is fog or smoke in the air. How does the fog or smoke make a difference?
15. Describe one example of evidence that white light is made up of many different colours.
16. To which form of electromagnetic radiation do microwaves belong?
17. Which form of electromagnetic radiation has:
    (a) the least energy?
    (b) the most energy?
18. Which part of the electromagnetic spectrum is used:
    (a) in remote control devices?
    (b) to help the human body produce vitamin D?
19. When a sound is made, what happens to the particles in the regions of the air nearby that are called:
    (a) compressions
    (b) rarefactions?
20. When an object vibrates faster what happens to the sound's:
    (a) frequency?
    (b) pitch?
    (c) amplitude?
21. How does amplitude of a sound wave affect the sound that you hear?
22. When you sing in the shower the sound of your voice reverberates.
    (a) What happens to sound energy to cause reverberation?
    (b) Why don't you observe reverberation when you sing in a room with carpet and soft curtains?
    (c) In some outdoor places, if you speak loudly you can hear an echo. For example, you might say 'hello' and a second or two later you hear the word 'hello' again. Explain how an echo is different from a reverberation.
23. Explain what is wrong with this cartoon.
24. Explain the difference in meaning of each of the following pairs of words.
    (a) Ray and beam
    (b) Scattering and reflection
    (c) Refraction and dispersion
25. Copy and complete the diagrams below to show the path of light after it meets the objects shown.

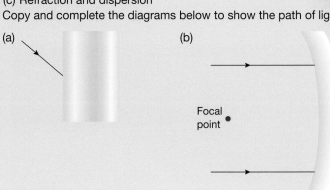

(a)

(b)

Focal point

concave mirror

Reflected light rays converge from a concave mirror.

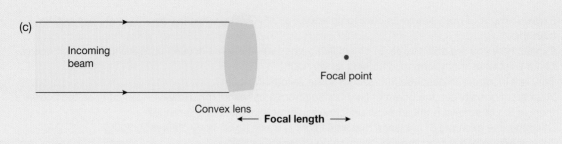

(c)

Incoming beam

Convex lens

Focal point

←—— **Focal length** ——→

26. Use a diagram to show why your legs appear to be shorter when you stand in clear, shallow water.
27. Describe the role of each of the following parts of the eye.
    (a) Cornea
    (b) Iris
    (c) Lens
    (d) Retina
    (e) Ciliary muscles
28. The diagram at right shows how light rays from a distant object arrive at the retina of a person with blurry distance vision.
    (a) What is the name of the condition illustrated at right?
    (b) What does the correcting lens need to do to the incoming light to correct the problem?

light from a distant object

Image on retina is blurry.

GLOSSARY

abrasive: a property of a material or substance that easily scratches another

absorbed: taken in

absorption: the taking in of a substance, for example from the intestine to the surrounding capillaries

accommodation: changing the lens shape to focus a sharp image on the retina

adult stem cells: undeveloped cells found in blood and bone marrow

alchemist: olden-day 'chemist' who mixed chemicals and tried to change ordinary metals into gold. Alchemists also tried to tell the future.

alimentary canal: passage from the mouth to the anus. Digestion of food occurs as it moves through the canal.

alloy: a mixture of a metal with a non-metal or another metal

alveoli: tiny air sacs in the lungs at the ends of the narrowest tubes. Oxygen moves from alveoli into the surrounding blood vessels, in exchange for carbon dioxide.

ammonia: a nitrogenous waste product of protein break down

amniocentesis: removal and testing of fluid from the amniotic sac surrounding the fetus

amplitude: maximum distance that a particle moves away from its undisturbed position

amygdala: emotional centre of the brain. It processes primal feelings, such as fear and rage. It may also be involved in emotional memories.

amylase: an enzyme in saliva that breaks starch down into sugar

Animalia: the kingdom of organisms that have cells with a membrane around the nucleus, but no cell wall, large vacuole or chloroplasts

anther: the male part of a flower that makes pollen

anti-fertility vaccines: injections to control sperm and testosterone levels

antibiotic: substance derived from a micro-organism and used to kill bacteria in the body

antiseptic: mild disinfectant used on body tissue to kill microbes

anus: the final part of the digestive system, through which faeces are passed as waste

aorta: a large artery through which oxygenated blood is pumped at high pressure from the left ventricle of the heart to the body

arteries: hollow tubes (vessels) with thick walls carrying blood pumped from the heart to other body parts

arterioles: vessels that transport oxygenated blood from the arteries to the capillaries

arthritis: a condition in which inflammation of the joints causes them to swell and become painful

asexual reproduction: reproduction that does not involve sex half-cells

asthma: narrowing of the air pipes that join the mouth and nose to the lungs

atom: a very small particle that makes up all things. Atoms have the same properties as the objects they make up.

atomic number: number of protons in the nucleus of an atom. The atomic number determines which element an atom is.

bactericidal: describes an antiseptic that kills bacteria

bacteriostatic: describes an antiseptic that stops bacteria from growing or dividing but doesn't kill them

ball and socket joints: joints where the rounded end of one bone fits into the hollow end of another

basalt: a dark, igneous rock with small crystals formed by fast cooling of hot lava. It sometimes has holes that once contained volcanic gases.

batholith: intrusive rock mass that measures more than 100 kilometres across

beaker: container for mixing or heating substances

benign: describes a tumour that does not spread to other parts of the body

bicuspid: a type of valve with two cusps (points). The valve between the heart's left atrium and left ventricle is a bicuspid valve.

bile: a substance produced by the liver that helps digest fats and oils

binary fission: reproduction by the division of an organism (usually a single cell) into two new organisms

binocular: a microscope with two eyepieces, so you use both eyes to look at the object

biodegradable: describes a substance that breaks down or decomposes easily in the environment

bioluminescent: describes living things that release light energy

bladder: sac that stores urine

blood pressure: measures how strongly the blood is pumped through the body's main arteries

blood vessels: the veins, arteries and capillaries through which the blood flows around the body

boiling point: the temperature at which a liquid changes to a gas

bolus: round, chewed-up ball of food made in the mouth that makes swallowing easier

bond: force that holds particles of matter, such as atoms, together

bone marrow: a substance inside bones in which blood cells are made

bones: the pieces of hard tissue that make up the skeleton of a vertebrate

Bowman's capsule: a cup-like structure at one end of a nephron within the kidney, surrounding the glomerulus. It serves as a filter to remove wastes and excess water.

breathing: movement of muscles in the chest causing air to enter the lungs and the altered air in the lungs to leave. The air entering the lungs contains more oxygen and less carbon dioxide than the air leaving the lungs.

breech: describes a birth in which the baby is born feet first

brittle: breaks easily into many pieces

bronchi: the narrow tubes through which air passes from the trachea to the smaller bronchioles and alveoli in the respiratory system. Singular = bronchus.

bronchioles: small branching tubes in the lungs leading from the two larger bronchi to the alveoli

budding: forming a new organism from an outgrowth (bud) of the parent

burning: combining a substance with oxygen in a flame

burping: release of swallowed gas through the mouth

caesarean: operation to remove a baby by cutting the mother's abdomen

calcium: an element occurring in limestone, chalk, also present in vertebrates and other animals as a component of bone, shell etc. It is necessary for nerve conduction, heartbeat, muscle contraction and many other physiological functions.

cancer: a disease resulting in the uncontrolled growth of body cells, forming tumours

capillaries: minute tubes carrying blood to body cells. Every cell of the body is supplied with blood through capillaries.

carbon dioxide: a gas in the air produced by respiration and used by plants as part of photosynthesis. The burning of fossil fuels releases carbon dioxide.

carcinogens: chemicals that cause cancer

cardiac muscle: special kind of muscle in the heart that never tires. It is involved in pumping blood through the heart.

carnivore: animal that eats other animals

cartilage: a waxy, whitish, flexible substance that lines or connects bone joints or, in some animals such as sharks, replaces bone as the supporting skeletal tissue. The ears and tips of noses of people are shaped by cartilage.

catalyst: chemical that helps to start or speed up a chemical reaction

cell: the smallest unit of life. Cells are the building blocks of living things. There are many different sized and shaped cells in animals and plants, as well as single-celled organisms.

cell membrane: structure that encloses the contents of a cell and allows the movement of some materials in and out

cell sap: the mixture inside a plant's vacuoles

cellular respiration: the chemical reaction involving oxygen that moves the energy in glucose into the compound ATP. The body is able to use the energy contained in ATP.

cellulose: a natural substance that keeps the cell wall of plants rigid

chemical change: change that results in a new substance being formed. During a chemical change, some chemical bonds break and others form.

chemical digestion: the chemical reactions changing food into simpler substances that are absorbed into the bloodstream for use in other parts of the body

chemical energy: potential energy derived from chemical reactions

chemical properties: properties that describe how a substance combines with other substances to form new chemicals or how a substance breaks up into two or more different substances

chemical reaction: a chemical change in which one or more new chemical substances are produced

chemical sterilisation: oral medication that results in male infertility

chemical symbol: the standard way that scientists write the names of the elements, using either a capital letter or a capital followed by a lowercase letter. For example, carbon is C and copper is Cu.

chlorophyll: the green-coloured chemical in plants that absorbs the light energy used in photosynthesis to make food from carbon dioxide and water

chloroplast: oval-shaped organelle found only in plant cells. Chloroplasts contain the pigment chlorophyll. They are the 'factories' in which carbon dioxide and water are changed by sunlight and water into food by the process of photosynthesis.

chorionic villus sampling: a procedure in which cells from the developing placenta are removed for testing at 10–12 weeks of pregnancy, allowing certain abnormalities to be detected

chromosome: tiny, thread-like structure inside the nucleus of a cell. Chromosomes contain the DNA that carries genetic information.

ciliary muscles: muscles that control the shape of the lens in the eye

circulatory system: the body system that circulates oxygen in blood to all the cells of the body. The circulatory system consists of the heart, the blood vessels and blood.

clone: identical copy

cloning: process used to produce genetically identical organisms

coal: a sedimentary rock formed from dead plants and animals that were buried before rotting completely

colon: the part of the large intestine where a food mass passes from the small intestine, and where water and other remaining essential nutrients are absorbed into the body

combustible: easily ignited

combustion: the process of combining with oxygen, most commonly burning with a flame

compound: substance made up of two or more different types of atoms that have been joined (bonded) together

compression: the process of pushing a material into itself

concave: refers to a lens that is curved inwards

conception: embedding of a fertilised egg in the uterus wall

conduction: transfer of heat through collisions between particles

conglomerate: sedimentary rock containing large particles of various sizes cemented together

contraception: prevention of fertilisation of an egg or prevention of a fertilised egg becoming embedded in the uterus wall

contraceptives: devices or substances that prevent fertilisation of an egg or prevent a fertilised egg becoming embedded in the uterus wall

contract: shorten or become smaller in size

convection: transfer of heat through the flow of particles

converging lens: lens that bends rays so that they move towards each other. Converging lenses are thicker in the middle than at the edges.

convex: refers to a lens that is curved outwards

copulation: the act of inserting sperm into the female; also called sexual intercourse or mating

cornea: clear, curved outer surface of the eye

corrosion: a chemical reaction that wears away a metal. Air, water or chemicals in air and water, as well as many household substances, can be corrosive. Many acids and other corrosive substances are dangerous.

corrosive: describes a chemical that wears away the surface of substances, especially metals

cotyledons: special leaves of the embryo plant inside a seed that provide food for the developing seedling

cross-pollination: transfer of pollen from stamens of one flower to the stigma of a flower of another plant of the same species

crystal: geometrically-shaped substance made up of atoms and molecules arranged in one of seven different shapes. The elements that make up a crystal and the conditions present during the crystal's growth determine the arrangement of atoms and molecules and the shape of the crystals.

cullet: used glass

cytokinesis: the process where the cytoplasm of a cell divides to form two daughter cells

cytoplasm: the jelly-like material inside a cell. It contains many organelles such as the nucleus and vacuoles.

cytosol: the fluid found inside cells

data: observations or measurements made and recorded during an investigation

deciduous: describes plants that lose their leaves during autumn and winter

decomposition: breaking up of a substance into smaller parts

denatured: describes the condition of proteins after they have been overheated

deoxygenated blood: describes blood from which some oxygen has been removed

deoxyribonucleic acid (DNA): the chemical substance found in all living things that encodes the genetic information of an organism

dependent variable: a variable that is expected to change when the independent variable is changed. The dependent variable is observed or measured during the experiment.

dermis: the medical name for the deeper part of the skin

diaphragm: flexible, dome-shaped, muscular layer separating the chest and the abdomen. It is involved in breathing.

diarrhoea: excessive discharge of watery faeces

diastolic pressure: the lower blood pressure reading during relaxation of the heart muscles

diffusion: the spreading of one substance through another due to the movement of their particles

digestion: breakdown of food into a form that can be used by an animal. It includes both mechanical digestion and chemical digestion.

digestive system: a complex series of organs and glands that processes food to supply the body with the nutrients it needs to function effectively

disinfectant: chemical used to kill bacteria on surfaces and non-living objects

disperse: scatter the seeds from plants

dispersion: separation of the colours that make up white light. Each colour is bent differently when it enters or leaves a glass prism.

diverging lens: lens that bends rays so that they spread out. Diverging lenses are thinner in the middle than at the edges.

donor eggs: ova from another woman

ductile: capable of being drawn into wires or threads; a property of most metals

durability: the quality of lasting, not easily being worn out

ectothermic: describes an animal whose body temperature is determined by its environment

efficiency: the fraction of energy supplied to a device as useful energy. It is usually expressed as a percentage.

ejaculation: release of sperm

elastic: describes a material that is able to return to its original size after being stretched

elasticity: the property that allows a material to return to its original size after being stretched

elastic potential energy: the potential energy stored in a stretched elastic material

electrical energy: the energy made available by the flow of electric charge through a conductor

electrocardiogram (ECG): graph made using the tiny electrical impulses generated in the heart muscle, giving information about the health of the heart

electromagnetic radiation: the radiant energy such as radio waves, infrared, visible light, X-rays and gamma rays released by magnetic or electric fields

electromagnetic spectrum: complete range of wavelengths of energy radiated as electric and magnetic fields

electron microscope: instrument for viewing very small objects. An electron microscope is much more powerful than a light microscope and can magnify things up to a million times.

electrons: very light, negatively charged particles inside an atom. Electrons move around the central nucleus of an atom.

element: pure substance made up of only one type of atom

embryo: group of cells formed from the zygote and developing into different body organs

embryonic stem cells: stem cells derived from the inner cell mass of a blastocyst. These cells are pluripotent and can give rise to most cell types.

emphysema: condition in which the air sacs in the lungs break open and join together, reducing the amount of oxygen taken in and carbon dioxide removed

emulsify: combine two liquids that do not normally mix easily

endocrine system: the body system of glands that produce and secrete hormones into the bloodstream to regulate processes in various organs

endoskeleton: skeleton that lies inside the body

endosperm: food supply for the embryo plant in a seed

endothermic: describes an animal that requires heat input to maintain its body temperature

environmental impact statement (EIS): study of the possible effects of a planned project on the environment

enzymes: special chemicals that speed up reactions but are themselves not used up in the reaction

epidermis: outermost layer of the skin

epigenetics: a new branch of science that involves studying the effect of our experiences on the expression of our genetic information

epiglottis: leaf-like flap of cartilage behind the tongue that closes the air passage during swallowing

erosion: the wearing away and removal of soil and rock by natural elements, such as wind, waves, rivers and ice, and by human activity

erythrocytes: red blood cells

eukaryote: member of the group of organisms that has a membrane around the nucleus in each of their cells

evaporate: change state from a liquid to a gas. Evaporation occurs only from the surface of a liquid.

excretion: removal of wastes from the body

excretory system: the body system that removes waste substances from the body

exoskeleton: skeleton or shell that lies outside the body

expand: increase in size due to the movement of particles in a substance

external fertilisation: reproduction where the eggs are released by the female into water and fertilised by sperm released nearby

extinct: describes volcanoes that are no longer active. Extinct volcanoes have not erupted for thousands of years and show no sign of future eruption.

extrusive rock: igneous rock that forms when lava cools above the Earth's surface

fair test: a method for determining an answer to a problem without favouring any particular outcome — another name for a controlled experiment

fallopian tubes: tiny tubes joining the ovaries to the uterus. Fertilisation occurs in one of these tubes.

fertilisation: penetration of the ovum by a sperm

fetus: the unborn young of an animal that has developed a distinct head, arms and legs

Filshie clip: type of fallopian tube clamp

filter funnel: used with filter paper to separate solids from liquids

first-floor thinkers: thinkers who gather information

flaccid: describes cells that are not firm due to loss of water

flammability: an indicator of how easily a substance catches fire

flammable: describes substances such as methylated spirits that burn easily

flatulence: release of gas through the anus. This gas is produced by bacteria in the large intestine.

flint: a fine-grained sedimentary rock which leaves a very sharp edge when broken

floodplain: flat, open land beside a river where sediments are deposited during floods

flower: the reproductive part of angiosperms containing petals, stamens and carpels. They are often colourful to attract pollinating insects.

fluid: a substance that flows and has no fixed shape. Gases and liquids are fluids.

folding: the buckling of rocks. It is caused when rocks are under pressure from both sides.

follicles: sacs containing egg cells

fossil: any remains, impression, or trace of an animal or plant of a former geological age; evidence of life in the past

fossil fuel: substance, such as coal, oil and natural gas, that has formed from the remains of ancient organisms. Coal, oil and natural gas are often used as fuels; that is, they are burnt in order to produce heat.

fracture: a break in a bone

fraternal twins: twins developed from different fertilised eggs

frequency: number of vibrations in one second, or the number of wavelengths passing in one second

fruit: ripened ovary of a flower, enclosing seeds

fuels: substances that are burned in order to release energy, usually in the form of heat

Fungi: the kingdom of organisms, such as mushrooms and moulds, that help to decompose dead or decaying matter

gall bladder: a small organ that stores and concentrates bile within the body

galvanising: protecting a metal by covering it with a more reactive metal that will corrode first

gametes: reproductive cells (sperm or ova) containing half the genetic information of normal cells

gamma ray: high energy electromagnetic radiation produced during nuclear reaction

gas: state of matter with no fixed shape or volume

germination: first sign of growth from the seed of a plant

gestation period: time spent by offspring developing in the uterus

glaciers: large bodies of ice that move down slopes and push boulders, rocks and gravel in front of them

glomerulus: a cluster of capillaries in the kidney that acts as a filter to remove wastes and excess water

gneiss: a coarse-grained metamorphic rock formed mainly as a result of great pressure on granite

gossypol: a non-hormonal substance that occurs naturally in unrefined cottonseed oil. When taken orally, it reduces sperm count.

granite: a hard, igneous rock with different-coloured crystals large enough to see. It forms slowly below the Earth's surface.

gravitational potential energy: energy stored due to the height of an object above a base level

greenstick fracture: a break that is not completely through the bone, often seen in children

group: in the periodic table of elements, a single vertical column of elements with a similar nature

guard cells: cells on either side of a stoma that work together to control the opening and closing of the stoma

habits of mind: reactions to gaining knowledge, studied by Bena Kallick and Arthur Costa

haemodialysis: the process of passing blood through a machine to remove wastes

haemoglobin: the red pigment in red blood cells that carries oxygen

hardness: a measure of how difficult it is to scratch the surface of a solid material. The hardness rating of a mineral is determined by comparison with ten standard minerals. Diamond has a hardness rating of 10 and can scratch other minerals with a lower hardness rating.

heart: a muscular organ that pumps blood through the circulatory system so that oxygen and nutrients can be transported to the body's cells and wastes can be transported away

heartbeat: contraction of the heart muscle occurring about 60–100 times per minute

heartburn: burning sensation caused by stomach acid rising into the oesophagus

herbivore: animal that eats only plants

hermaphrodite: organism that has both male and female reproductive organs

hertz: unit of frequency; its abbreviation is Hz. One hertz is equal to one vibration every second.

hinge joints: joints in which two bones are connected so that movement occurs in one plane only

hippocampus: area of the brain stem able to transfer information between short- and long-term memory

hormones: chemical substances produced by glands and circulated in the blood. Hormones have specific effects in the body.

hydrogen: the element with the smallest atom. By itself, it is a colourless gas and combines with other elements to form a large number of substances, including water. It is the most common element in living things.

hypothesis: a suggested explanation for past observations that is tested in an experiment

identical twins: twins developed from the same fertilised egg

igneous rocks: rocks that form from the cooling of lava or magma as it is thrown through the air from a volcanic eruption

image: picture of an object

immovable joints: joints that allow no movement except when absorbing a hard blow

implantation: the process whereby the embryo becomes embedded in the wall of the uterus

incandescent: describes objects that emit light when they are hot

independent variable: the variable that the scientist changes to observe its effect on another variable

inert: not reactive

infectious disease: disease that can be transferred from one organism to another

infertility: the inability to have children

infra-red radiation: low energy electromagnetic waves with a much lower frequency and longer wavelength than visible light

insect-pollinated flowers: flowers that receive pollen carried on the body parts of insects from other flowers

insulator: a material that is a poor conductor of heat

internal fertilisation: reproduction where the egg remains in the female and is fertilised by sperm inserted into the female

interpersonal communication: communication with others

interpersonal intelligence: one of the multiple intelligences that involves 'we' learning, which can be developed through team projects, social interactions, debates and games

intrapersonal intelligence: one of the multiple intelligences that involves 'I' learning, which can be developed using blogs or journals, reflecting, self-assessment and creative expression

intrusive rock: igneous rock that forms when magma cools below the Earth's surface

investigation: activity aimed at finding information

involuntary muscles: muscles not under the control of the will; they contract slowly and rhythmically. These muscles are at work in the heart, intestines and lungs.

iris: coloured part of the eye that opens and closes the pupil to control the amount of light that enters the eye

joint: region where two bones meet

kidneys: body organs that filter the blood, removing urea and other wastes

kinesics: the use of bodily movements or actions to convey a specific meaning or idea

kinetic energy: energy due to the motion of an object

labour: childbirth

large intestine: the penultimate part of the digestive system, where water is absorbed from the waste before it is transported out of the body

lateral inversion: sideways reversal of images in a mirror

Law of Conservation of Energy: a law that states that energy cannot be made or lost. However, energy can be transformed from one type to another or transferred from one object to another.

left atrium: upper left section of the heart where oxygenated blood from the lungs enters the heart

left ventricle: lower left section of the heart, which pumps oxygenated blood to all parts of the body

leucocytes: white blood cells

ligament: band of tough tissue that connects the ends of bones or keeps an organ in place

light microscope: instrument for viewing very small objects. A light microscope can magnify things up to 1500 times.

lignin: a hard substance in the walls of dead xylem cells that make up the tubes carrying water up plant stems. Lignin forms up to 30 per cent of the wood of trees.

limestone: a sedimentary rock formed from the remains of sea organisms. It consists mainly of calcium carbonate.

lipases: enzymes that break fats and oils down into fatty acids and glycerol

lipids: type of nutrients that include fats and oils

liquid: state of matter that has a fixed volume, but no fixed shape

lithosphere: the outermost layer of the Earth, includes the crust and uppermost part of the mantle

liver: largest gland in the body. The liver secretes bile for digestion of fats, builds proteins from amino acids, breaks down many substances harmful to the body and has many other essential functions.

long-sightedness: A person who is long-sighted is unable to obtain sharp images of nearby objects.

long-term storage: a function of the brain that enables memories to be stored, sorted and retrieved

'lub dub': the sound made by the heart valves as they close

luminous: releasing its own light

lungs: the organ for breathing air. Gas exchange occurs in the lungs.

lustre: appearance of a mineral caused by the way it reflects light. A mineral can appear glassy, waxy, metallic, dull, pearly, silky or brilliant.

magma: a very hot mixture of molten rock and gases, just below the Earth's surface, that has come from the mantle

magnification: the number of times the image of an object has been enlarged using a lens or lens system. For example, a magnification of two means the object has been enlarged to twice its actual size.

malignant: describes a type of tumour that damages cells and can spread to other parts of the body

malleable: able to be beaten, bent or flattened into shape

marble: a metamorphic rock formed as a result of great pressure on limestone

mass: the quantity of matter in an object (usually measured in grams or kilograms)

measuring cylinder: used to measure volumes of liquids accurately

mechanical digestion: digestion that uses physical factors such as chewing with the teeth

meiosis: cell division process that results in new cells with half the number of chromosomes of the original cell

melting point: the temperature at which a solid substance turns into a liquid (melts) or a liquid turns into a solid (freezes)

menarche: the first menstrual period

menstrual cycle: beginning of one period to the beginning of the next period

menstruation: monthly bleeding, sometimes called 'periods'. It occurs as the lining of the uterus is released.

metabolism: the chemical reactions occurring within an organism that enable the organism to use energy and grow and repair cells

metalloids: elements that have the appearance of metals but not all the other properties of metals

metals: elements that conduct heat and electricity; shiny solids that can be made into thin wires and sheets that bend easily. Mercury is the only liquid metal.

metamorphic rock: rock formed from another rock that has been under great heat or pressure (or both)

metamorphism: the process that changes rocks by extreme pressure or heat (or both)

meteorologist: scientist who uses observation of the atmosphere to predict or explain the weather

micrometre: a length of one millionth of a metre

microscope: an instrument for viewing small objects

mineral: any of the inorganic elements that are essential to the functioning of the human body and are obtained from foods

mineral ores: rocks mined to obtain a metal or other chemical within them

minerals: substances that make up rocks. Each mineral has its own chemical make-up.

mining: extraction of natural resources from the Earth

mitochondria: small rod-shaped organelles that supply energy to other parts of the cell. They are usually too small to be seen with light microscopes. Singular = mitochondrion.

mitochondrial DNA (mtDNA): genetic material from the mitochondria, passed on only from the mother

mitosis: cell division process that results in new cells with the same number of chromosomes as the original cell

mixture: a combination of substances in which each keeps its own properties

mnemonic: a strategy to help you to remember things

molecule: two or more atoms joined (bonded) together, forming a small particle

monocular: describes a microscope through which the specimen is seen using one eye only

moraine: deposit left by movement of a glacier

mould: cavity in a rock that shows the shape of the hard parts of an organism; types of fungi found growing on the surface of foods

mudstone: a fine-grained, sedimentary rock without layering

multicellular: having many cells. Most plants and animals are multicellular.

multifocal lenses: lenses that are composed of different sectors of lens with differing focal length, which allow people with both short-sightedness and long-sightedness to see clearly

multiple fission: reproduction method where a single-celled organism divides into more than two cells

multipotent: describes stem cells that can give rise to only certain cell types, e.g. various types of blood or skin cells

muscle: tissue consisting of cells that can shorten

musculoskeletal system: consists of the skeletal system (bones and joints) and the skeletal muscle system (voluntary or striated muscle). Working together, these two systems protect the internal organs, maintain posture, produce blood cells, store minerals and enable the body to move.

nanometre: a unit of measurement equal to one billionth of a metre

nanotechnology: a rapidly developing field that includes studying and investigating cells and other objects of the smallest dimensions

native elements: elements found uncombined in the Earth's crust

natural fibres: fibres that form naturally; that is, they have not been made by humans. Natural fibres include wool and silk from animals and cotton from plants.

nectaries: parts of a flower, at the base of the petals, that secrete nectar

nephrons: the filtration and excretory units of the kidney

nervous system: the system of nerves and nerve centres in an animal in which messages are sent

neutron: tiny, but heavy, particle found in the nucleus of an atom. Neutrons have no electrical charge.

nitrogenous wastes: waste products from protein break down, including ammonia, urea and uric acid

noble gases: elements in the last column of the periodic table. They are extremely inert gases.

non-infectious disease: disease that cannot be transferred from one organism to another

non-luminous: describes objects that do not emit their own light, but can be seen by reflected light

non-metals: elements that do not conduct electricity or heat. They melt and turn into gases easily and are brittle and often coloured.

normal (the): line drawn perpendicular to a surface at the point where a light ray meets it

nuclear DNA: genetic material passed on from both parents, from the nucleus of both the sperm and the ovum

nuclear energy: the energy stored at the centre of atoms, the tiny particles that make up all substances. Nuclear energy can be released from the radioactive metals uranium or plutonium, and transformed into electrical energy in a nuclear power station.

nucleus: central part of an atom, made up of protons and neutrons. Also roundish structure inside a cell that acts as the control centre for the cell. Plural = nuclei.

nutrients: substances that provide energy and chemicals that living things need to stay alive, grow and reproduce

nylon: synthetic fibre. The monomers are joined by the elimination of water molecules at the joins.

observations: information obtained by the use of our senses or measuring instruments

obsidian: a black, glassy rock that breaks into pieces with smooth shell-like surfaces

oesophagus: part of the digestive system composed of a tube connecting the mouth with the stomach

omnivore: animal that eats plants and other animals

open-cut mining: mining that scours out soil and rocks on the surface of the land

optic nerve: large nerve that sends signals to the brain from the sight receptors in the retina

organelle: small structure in a cell with a special function

organisms: living things

oscillation: one complete swing of a pendulum

ossification: hardening of bones

osteoporosis: loss of bone mass that causes bones to become lighter, more fragile and easily broken

ova: female gametes or sex cells. Singular = ovum.

ovary: in plants, the hollow, lower end of the carpel containing the ovules (the female egg cells); in animals, the female organ that produces ova and reproductive hormones

overburden: waste rock removed from below the topsoil. This rock is replaced when the area is restored.

ovule: receptacle within an ovary that contains egg cells

oxidation: chemical reaction involving the loss of electrons by a substance

oxygen: a gas in the air (and water) that animals need to breathe in; made up of particles with two oxygen atoms. Plants produce oxygen as part of photosynthesis.

oxygenated blood: describes the bright red blood that has been supplied with oxygen in the lungs

oxyhaemoglobin: haemoglobin with oxygen molecules attached

oxytocin: a hormone

pacemaker: electronic device inserted in the chest to keep the heart beating regularly at the correct rate. It works by stimulating the heart with tiny electrical impulses.

palaeontologist: a scientist who studies fossils

pancreas: a large gland in the body that produces and secretes the hormone insulin and an important digestive fluid containing enzymes

paralinguistics: involves how something is said

parthenogenesis: development of new individuals from unfertilised eggs

particle model: a description of the moving particles that make up all matter and how they behave. The model explains the properties of solids, liquids and gases.

pendulum: an object swinging on the end of a string, chain or rod

percussion flaking: a process in which tool stones such as flint or obsidian were struck with harder stones, such as quartzite, to shear large flakes that could be used to make small tools

period: the time taken for one oscillation of a pendulum

periodic table: a table listing all known elements. The elements are grouped according to their properties and in order of the number of protons in their nucleus.

peristalsis: the process of pushing food along the oesophagus or small intestine by the action of muscles

petals: the coloured parts of a flower that attract insects

pheromones: chemicals that are important in communication between members of the opposite sex

phloem: type of tissue that transports sugars made in the leaves to other parts of a plant

phosphorus: a substance that plays an important role in almost every chemical reaction in the body. Together with calcium, it is required by the body to maintain healthy bones and teeth.

photosynthesis: the food-making process in plants that takes place in chloroplasts within cells. The process uses carbon dioxide, water and energy from the sun.

physical change: change in which no new chemical substances are formed. A physical change may be a change in shape, size or state. Many physical changes are easy to reverse.

physical properties: properties that you can either observe using your five senses — seeing, hearing, touching, smelling and tasting — or measure directly

pitch: the highness or lowness of a sound. The pitch that you hear depends on the frequency of the vibrating air.

pivot joint: joint that allows a twisting movement

placenta: an organ formed in the mother's womb through which the baby receives food and oxygen from the mother's blood and the baby's wastes are removed

Plantae: the kingdom of organisms that have cells with a membrane around the nucleus, cell wall, large vacuole and chloroplasts, commonly called plants

plasma: the yellowish, liquid part of blood that contains water, minerals, food and wastes from cells

plastic: synthetic substance capable of being moulded

platelets: small bodies involved in blood clotting. They are responsible for healing by clumping together around a wound.

plumule: small bud at the tip of the embryo plant in a seed

pluripotent: describes stem cells that can give rise to most cell types, e.g. blood cells, skin cells and liver cells

pollen: fine powder containing the pollen grains (the male sex cells of a plant)

pollen grains: the male gametes of a flower

pollen tube: long tube growing from a pollen grain through the style to the ovule

pollination: transfer of pollen from the stamen (the male part) of a flower to the stigma (the female part) of a flower

polyester: synthetic fibre. The monomers are joined together by the elimination of water molecules at the joins.

polymer: substance made by joining smaller identical units. All plastics are polymers.

pore: small opening in the skin. Perspiration reaches the surface of the skin through pores.

potential energy: energy that has the potential to do work and so the energy is 'stored', such as gravitational energy, elastic energy and chemical energy

pre-implantation genetic diagnosis: a procedure to diagnose and exclude genetic abnormalities in embryos before implantation into the mother's body

precipitate: solid product of a chemical reaction that does not dissolve in water

precipitation: falling water in solid or liquid form. The type of precipitation depends mostly on the temperature in the clouds and the air around them.

prefix: part of a word that is put before another word, stem, or word element to add to or qualify its meaning

premature: describes a baby born less than 37 weeks after conception

presbyopia: blurring of images of very close objects caused by loss of flexibility of the lens in the eyes

product: new chemical substance that results from a chemical reaction

Prokaryotae: the kingdom of organisms that consist of single cells with a nucleus not surrounded by a membrane or a cell wall, commonly called bacteria

prokaryote: organism classified as belonging to the Prokaryotae kingdom

properties: characteristics or features of an object or substance

proteases: enzymes that break proteins down into amino acids

protein: chemical made up of amino acids needed for growth and repair of cells in living things

Protoctista: the kingdom of organisms, including algae and protozoans, that do not fit into other groups

proton: tiny, but heavy, particle found in the nucleus of an atom. Protons have a positive electrical charge.

puberty: life stage when the sex glands become active and bodily changes occur to enable reproduction

pulmonary artery: the vessel through which deoxygenated blood, carrying wastes from respiration, travels from the heart to the lungs

pulmonary vein: the vessel through which oxygenated blood travels from your lungs to the heart

pulse: alternating contraction and expansion of arteries due to the pumping of blood by the heart

pumice: a pale rock that forms when frothy lava cools in the air. Pumice often floats on water as it is very light and full of holes that once contained gas.

pupil: hole through which light enters the eye

radiant heat: heat transferred by radiation, as from the sun to the Earth

radiation: a method of heat transfer that does not require particles to transfer heat from one place to another

radicle: the beginnings of a root making up part of a plant embryo inside a seed

rapport: a relationship of mutual trust and understanding

rarefactions: in sound waves, the layers of air particles that are spread apart (between compressions)

ray: narrow beam of light

reactant: chemical substance used up in a chemical reaction. Some chemical bonds in a reactant are broken during the reaction.

reaction rate: speed at which a reaction takes place

reactivity: a measure of how likely a particular metal is to take part in a displacement reaction

receptors: special cells that detect energy and convert it to electrical energy that is sent to the brain

record: preserve information

rectum: the final section of the digestive system, where waste food matter is stored as faeces before being excreted through the anus

recycled: reused an unwanted substance or object for another purpose

red blood cells: living cells in the blood that transport oxygen to all other living cells in the body

reflected: bounced off

refraction: change in the speed of light as it passes from one substance into another. It usually involves a change in direction.

rehabilitated: restored to its previous condition

relative age: the age of a rock compared with the age of another rock

reproductive system: the body system involving the reproductive organs, which differ between males and females

respiration: the chemical process that takes place in every cell to release energy. Glucose reacts with oxygen to produce carbon dioxide and water.

respiratory system: the body system involving the lungs and associated structures, which take in air and supply the blood with oxygen to deliver to the body's cells so they can carry out their essential functions; it also performs gas exchange to remove the waste gas carbon dioxide

reticular activation system: a function of the brain that sorts information according to its importance to your survival

retina: curved surface at the back of the eye. It is lined with sight receptors.

retrieve: bring back

reverberation: longer-lasting sound caused by repeated reflection from hard surfaces

ribosomes: small structures within a cell in which proteins such as enzymes are made

right atrium: upper right section of the heart where deoxygenated blood from the body enters

right ventricle: lower right section of the heart, which pumps deoxygenated blood to the lungs

rock salt: a sedimentary deposit formed when a salt lake or seabed dried up. Its main chemical is sodium chloride.

rust: a brown substance formed when iron reacts with oxygen and water

rusting: the corrosion of iron

safety glasses: plastic glasses used to protect the eyes during experiments

saliva: watery substance in the mouth that moistens food before swallowing

salivary glands: glands in the mouth that produce saliva

sandstone: a sedimentary rock with medium-sized grains. The sand grains are cemented together by silica, lime or other salts.

scattered: describes light sent in many directions by small particles within a substance

scavengers: animals that eat dead plant and animal material

scientists: people skilled in or working in the fields of science; scientists use experiments to find out about the material world around them

scoria: a dark, igneous rock formed from gassy lava that cools quickly

scrotum: pouch of skin containing testes. It hangs outside the body to keep the sperm cooler than normal body temperature.

second-floor thinkers: thinkers who process information

sediment: material broken down by weathering and erosion that is moved by wind or water and collects in layers

sedimentary rocks: rocks formed from sediments deposited by water, wind or ice. The sediments are cemented together in layers, under pressure.

seed: product of a fertilised ovule

seed coat: the protective layer around a seed

seedling: young plant produced from the embryo in the seed after germination

self-pollination: transfer of pollen from the flower's own stamen to its stigma

semen: combination of fluid and sperm

semilunar: a type of valve that is half-moon shaped. The aortic valve is semilunar and is located between the heart's left ventricle and aorta.

sensor: device connected to an instrument such as a data logger that measures and sends information

sensory register: a function of the brain that filters incoming information on the basis of importance

sexual intercourse: act of inserting sperm into the female; also called copulation or mating

sexual reproduction: joining together of male and female reproductive cells

shale: a fine-grained sedimentary rock formed in layers by the consolidation of clay

short-sightedness: A person who is short-sighted is unable to obtain sharp images of distant objects.

siltstone: a sedimentary rock with a particle size between that of sandstone and mudstone

skeletal muscle system: voluntary or striated muscle

skeletal system: consists of the bones and joints

skeleton: the bones or shell of an animal that support and protect it as well as allowing movement

skin: external covering of an animal body

slate: a fine-grained metamorphic rock formed as a result of moderate heat and pressure on shale

small intestine: the part of the digestive system between the stomach and large intestine, where much of the digestion of food and absorption of nutrients takes place

solid: state of matter that has a fixed shape and volume

somatic stem cells: stem cells derived from bone marrow, skin and umbilical cord blood. These cells are multipotent and can give rise to only certain types of cells.

sound waves: vibrations of particles in the air

sperm: male reproductive cell. It consists of a head, a middle section and a tail used to swim towards the egg.

sperm duct plug: injection of a liquid plastic to block the sperm duct

spinneret: device like a nozzle with small holes through which a plastic material passes, forming threads

sprain: injury caused by tearing a ligament

state: condition or phase of a substance. The three main states of matter are solid, liquid and gas.

stem cells: undeveloped cells found in blood and bone marrow

stereo: describes a microscope through which the specimen is viewed using both eyes

sterile: describes a person unable to produce reproductive cells

stigma: the female part of a flower, at the top of the carpel, that catches the pollen during pollination

stomach: a large muscular organ that churns and mixes food with gastric juice to start to break down protein

stomata: small openings mainly on the lower surface of leaves. These pores are opened and closed by guard cells. Singular = stoma.

Stone Age: the time beginning about two million years ago during which early humans made implements of stone

store: (verb) put away

streak: colour of a mineral as a fine powder, found by rubbing it onto an unglazed white ceramic tile

style: the supporting part of a flower that holds the stigma

sublimation: the change in state from a solid into a gas (or from a gas into a solid) without first becoming a liquid

substrate: substance acted upon by an enzyme

suffix: part of a word that follows another word, stem, or word element to add to or qualify its meaning

surface protection: coating over a metal surface to prevent corrosion

surrogacy: pregnancy that occurs when the fertilised egg is placed in the uterus of another woman who did not produce the egg

sustainable: describes the concept of using the Earth's resources so that the needs of the world's present population can be met, without damaging the ability of future populations to meet their needs

sweat gland: tiny, coiled tube in the skin through which water and salt are removed from the body, helping to control body temperature

synovial fluid: the liquid inside the cavity surrounding a joint that helps bones to slide freely over each other

system: group of organs working together

systolic pressure: the higher blood pressure reading during contraction of the heart muscles

tendons: tough rope-like tissue connecting a muscle to a bone

tennis elbow: an injury due to strain or overuse that causes the elbow's lining to become inflamed and painful

test tube: thin glass container for holding, heating or mixing small amounts of substances

test-tube babies: babies resulting from fertilisation of eggs by sperm in the laboratory

testes: organs that produce sperm and sex hormones

testosterone injections: hormone injections that reduce sperm production

thalamus: part of the brain at the top of the brain stem. It passes sensory information to key areas in the brain for processing.

third-floor thinkers: thinkers who apply information

tissue: groups of cells performing the same function

torn hamstring: a common sporting injury caused by overstretching the hamstring muscle, which joins the pelvis to the knee joint

totipotent: describes the most powerful stem cells that can give rise to all cell types

toxic: describes chemicals that are dangerous to touch, inhale or swallow

toxicity: the danger to your health caused when poisonous substances combine with chemicals in your body to produce new substances and damaging effects

trace fossils: fossils that provide evidence, such as footprints, that an organism was present when the rock was formed

trachea: narrow tube from the mouth to the lungs through which air moves

transdermal patches: skin applications containing hormones that are absorbed through the skin

translocation: transport of materials, such as water and glucose, in plants

transmitted: passed through something, such as light or sound passing through air

transpiration: loss of water from plant leaves through their stomata

transpiration stream: movement of water through a plant as a result of loss of water from the leaves

tricuspid: a type of valve with three cusps (points). The valve between the heart's right atrium and right ventricle is a tricuspid valve.

tumour: an abnormal growth

turgid: describes something that is firm

ultrasound: sound with frequencies too high for humans to hear

ultraviolet radiation: invisible radiation similar to light but with a slightly higher frequency and more energy

umbilical cord stem cells: undeveloped cells found in the umbilical cord attaching an unborn baby to the placenta

underground mining: mining that uses shafts and tunnels to remove rock from deep below the surface

unicellular: describes an organism having only one cell

urea: a nitrogen-containing substance produced by the breakdown of proteins and removed from the blood by the kidneys

ureters: tubes from each kidney that carry urine to the bladder

urethra: tube through which urine is emptied from the bladder to the outside of the body

uric acid: a nitrogenous waste product of protein break down

urination: passing of urine from the bladder to the outside of the body

urine: yellowish liquid, produced in the kidneys. It is mostly water and contains waste products from the blood such as urea, ammonia and uric acid.

vacuoles: sacs within a cell used to store food and wastes. Plant cells usually have one large vacuole. Animal cells have several small vacuoles or none at all.

vaginal pills: soluble capsules containing spermicides that are placed in the vagina to prevent conception

valves: flap-like folds in the lining of a blood vessel or other hollow organ that allow a liquid, such as blood, to flow in one direction only

variables: quantities or conditions in an experiment that can change

varicose veins: expanded or knotted blood vessels close to the skin, usually in the legs. They are caused by weak valves that do not prevent blood from flowing backwards.

vascular bundles: groups of xylem and phloem vessels within plant stems

vegetative propagation: reproduction of plants using parts other than sex cells

veins: blood vessels that carry blood back to the heart. They have valves and thinner walls than arteries.

vena cava: large vein leading into the top right chamber of the heart

venules: small veins

vibrations: repeated fast, back-and-forth movements

villi: tiny finger-like projections from the wall of the intestine that maximise the surface area of the structure to increase the efficiency of nutrient absorption. Singular = villus.

virtual focal point: a single point from which light rays seem to be coming after passing through a concave lens

virgin births: births that do not involve the joining of eggs and sperm

visible spectrum: different colours that combine to make up white light; they are separated in rainbows

vital capacity: the largest volume of air that can be breathed in or out at one time

vitamin D: a nutrient that regulates the concentration of calcium and phosphate in the bloodstream and promotes the healthy growth and remodelling of bone

volume: the amount of space taken up by an object or substance

voluntary muscle: muscle attached to bones; it moves the bones by contracting and is controlled by an animal's thoughts

vomiting: the forceful ejection of matter from the stomach through the mouth

wavelength: distance between two neighbouring crests or troughs of a wave. This is the distance between two particles vibrating in step.

white blood cells: living cells that fight bacteria and viruses as part of the human body's immune system

wilt: droop. Plant stems and leaves wilt when there is insufficient water in their cells.

wind-pollinated flower: flower that receives pollen carried by the wind from another flower

xylem vessels: pipelines for the flow of water up plants. They are made up of the remains of dead xylem cells fitted end to end with the joining walls broken down. Lignin in the cell walls gives them strength.

X-ray: high energy electromagnetic waves that can be transmitted through solids and provide information about their structure

zygote: cell formed by the fusion of two gametes; fertilised ovum before it starts to divide into more cells

INDEX